SILICON-ON-INSULATOR TECHNOLOGY:
Materials To VLSI

2nd Edition

SILICON-ON-INSULATOR TECHNOLOGY:
Materials to VLSI

2nd Edition

by

Jean-Pierre Colinge

Université catholique de Louvain, Belgium

KLUWER ACADEMIC PUBLISHERS
Boston / Dordrecht / London

Distributors for North America:
Kluwer Academic Publishers
101 Philip Drive
Assinippi Park
Norwell, Massachusetts 02061 USA

Distributors for all other countries:
Kluwer Academic Publishers Group
Distribution Centre
Post Office Box 322
3300 AH Dordrecht, THE NETHERLANDS

Library of Congress Cataloging-in-Publication Data

Colinge, Jean-Pierre
    Silicon-on-insulator technology : materials to VLSI / by Jean-Pierre
Colinge. -- 2nd ed.
    P.   Cm. --
    Includes bibliographical references and index
    ISBN 0-7923-8007-X (alk. paper)
    1. Semiconductors. 2. Silicon-on-insulator technology.
    3. Integrated circuits--Very large scale integration--Materials.
    I. Title.
TK7871.85.C758 1997
651.3815'2--dc21   97-35777
                                                                  CIP

Printed on acid-free paper.

Printed in the United States of America

This printing is a digital duplication of the original edition.

CONTENTS

PREFACE

$\mathcal{D}$ear SOI addict, dear enslaved Ph.D. student, dear SOI skeptic, dear friend met at some conference, dear tortured manager, welcome or welcome back! The news is: Kluwer allowed me to commit a second edition of this Silicon-on-Insulator (SOI) Book. The reason for this is quite simple: SOI technology has boomed since the first edition... SOI chips are commercially available and SOI wafer manufacturers have gone public. SOI has finally made it out of the academic world and is now a big concern for every major semiconductor company. SOI technology indeed deserves serious recognition: high-temperature (400°C), extremely rad-hard (500 Mrad(Si)), high-density (16 Mb, 0.9-volt DRAM, and even a 2V, 1 Gb DRAM!), high-speed (several GHz) and high-performance, low-voltage (260 MHz @ 0.5 V) SOI circuits have been demonstrated. Not bad for a former lab curiosity!

The SOI community has been extremely productive in recent years, and a large amount of papers have been published with SOI materials, devices and circuits. It is, unfortunately, not practical to take all these contributions into account, and a selection had to be made in order to present the most significant results of this research effort in a clear and concise manner. Papers published after 1996 are not included in this edition because of the time constraint, with a few exceptions.

The material covered by this Book is divided in eight Chapters, which are summarized below:

CHAPTER 1: INTRODUCTION briefly describes some straightforward advantages of SOI technology, such as the absence of latchup in CMOS structures and the reduction of parasitic source and drain capacitances.

CHAPTER 2: SOI MATERIALS lists the different approaches for producing SOI materials. The basic mechanisms behind the fabrication of thin silicon films on an insulating substrate using epitaxy, melting and recrystallization, implantation or bonding are described. The issues of material quality are addressed as well.

CHAPTER 3: SOI MATERIALS CHARACTERIZATION describes different techniques which have been developed to characterize the physical and electrical properties of SOI materials. While some "universal" characterization techniques such as SIMS and TEM can, of course, be used to assess the quality SOI materials, some techniques have been developed especially to evaluate the SOI crystal quality and its interface properties, and to measure the film thickness and the carrier lifetime in SOI materials.

CHAPTER 4: SOI CMOS TECHNOLOGY deals with the basics of SOI CMOS processing. Thin-film and thicker-film SOI CMOS processes are compared with bulk CMOS processing. Some process steps which are particular to SOI as well as different MOSFET structures are described.

CHAPTER 5: THE SOI MOSFET describes with the physics of the SOI MOSFET. The electrical characteristics (capacitances, threshold voltage, body effect and output characteristics) of thick- and thin-film MOSFETs are derived and compared to those of bulk devices. The subthreshold slope and the transconductance of SOI transistors are analyzed in detail. The effects caused by the parasitical bipolar transistor are reviewed.

CHAPTER 6: OTHER SOI DEVICES describes other types of devices fabricated on SOI substrates (double-gate transistors, bipolar transistors, high-voltage devices, JFETs, optical modulators, quantum devices,...).

CHAPTER 7: THE SOI MOSFET OPERATING IN A HARSH ENVIRONMENT describes the performances of SOI devices operating in a harsh environment (high temperature or radiations).

And, finally, **CHAPTER 8: SOI CIRCUITS** reviews the performance of modern SOI circuits, such as low-voltage, low-power CMOS, rad-hard, high-temperature, RAM and smart-power circuits.

Acknowledgements

I would like to express my appreciation to many colleagues who have supported SOI activity over the years. My gratitude especially goes to the organizers of SOI Workshops/Conferences and to those who, in Government or private institutions, believed in and gave support to SOI research.

I would like to thank the individuals who helped me collect information, published or unpublished, about SOI technology and/or reviewed the manuscript: Drs. S. Cristoloveanu (basically/LPCS/ENSERG), G.E. Davis (DNA/RAEE/Marge), C.A.C. Desmond (CSUS/IMY), D. Ioannou (most likely/GMU), B.Y. Nguyen (Motorola/APRDL), and A. Thanailakis (GMU). I would also like to thank my colleagues and Ph.D. students for their support and dedication: X. Baie, J. Chen, L. Demeûs, V. Dessard, J.P. Eggermont, L. Ferreira, D. Flandre, P. Francis, B. Gentinne, R. Gillon, I. Huynen, B. Iñiguez, P.G.A. Jespers, P. Smeys, J.P. Raskin, X. Tang, A. Vandooren, A. Viviani, F. Van de Wiele, A. Vander Vorst, and D. Vanhoenacker. This book is dedicated to all the hard-working SOI enthusiasts who will find their names referenced in the different chapters (those who are not referenced are not forgotten). Their tenacity has, over the years, turned an exotic laboratory curiosity into successful technology which *boldly goes where no no-one has gone before* and which is becoming a mainstream technology for low-voltage, low-power applications.

To Cindy, Colin-Pierre and Eliott

*"HELL WAS FULL, **SO I** CAME BACK" (bumper sticker purchased during the 1996 IEEE International SOI Conference in Florida and posted in Sorin Cristoloveanu's office...).*

CHAPTER 1 - Introduction

The idea of realizing semiconductor devices in a thin silicon film is mechanically supported by an insulating substrate has been around for several decades. The first description of the insulated-gate field-effect transistor (IGFET), which evolved into the modern silicon metal-oxide-semiconductor field-effect transistor (MOSFET), is found in the historical patent of Lilienfield dating from 1926 [1]. This patent depicts a three-terminal device where the source-to-drain current is controlled by a field effect from a gate, dielectrically insulated from the rest of the device. The piece of semiconductor which constituted the active part of the device was a thin semiconductor film deposited on an insulator. In a sense, it can thus be said that the first MOSFET was a Semiconductor-on-Insulator (SOI) device. The technology of that time was unfortunately unable to produce a successfully operating Lilienfield device. IGFET technology was then forgotten for a while, completely overshadowed by the enormous success of the bipolar transistor discovered in 1947 [2].

It was only years later, in 1960, that Kahng and Atalla realized the first working MOSFET [3], when technology had reached a level of advancement sufficient for the fabrication of good quality gate oxides. The advent of the monolithic integrated circuits gave MOSFET technology an increasingly important role in the world of microelectronics. CMOS technology is currently the driving technology of the whole microelectronics industry.

CMOS integrated circuits are almost exclusively fabricated on bulk silicon substrates, for two well-known reasons: the availability of electronic-grade material produced either by the Czochralski of by the floating-zone technique, and because a good-quality oxide can be readily grown on silicon, a thing which is not possible on germanium or on compound semiconductors. Yet, modern MOSFETs made in bulk silicon are far from the ideal structure described by Lilienfield. Bulk MOSFETs are made in silicon wafers having a thickness of approximately 800 micrometers, but only the first micrometer at the top of the wafer is used for transistor fabrication. Interactions between the devices and the substrate give rise to a range of unwanted parasitic effects.

One of these is the parasitic capacitance between diffused source and drain and the substrate. This capacitance increases with substrate doping, and becomes larger in modern submicron devices where doping concentration in the substrate is higher than in previous MOS technologies. Source and drain capacitance consists not only in the obvious capacitance of the depletion regions associated with the junctions, but also in the capacitance between the junction and the heavily-doped channel stop located underneath the field oxide. In addition, latchup, which consists of the unwanted triggering of a PNPN thyristor structure inherently present in all bulk CMOS structures, becomes a severe problem in devices with small dimensions, where the gain of the parasitic bipolar devices involved in the parasitic thyristor becomeslarger.

Of course, some "tricks" have been found to reduce these parasitic components. The area of the source and drain junctions can be minimized by creating local interconnections and placing contacts over the field area, and occurrence of latchup can be reduced by using epitaxial substrates or deep trench isolation. These different techniques, however, necessitate sophisticated processing, which impacts both the cost and the yield of manufacturing.

If an SOI substrate is used, quasi-ideal MOS devices can be fabricated. The SOI MOSFET contains the traditional three terminals (a source, a drain, and a gate which controls a channel in which current flows from source to drain). However, the full dielectric isolation of the devices prevents the occurrence of most of the parasitic effects experienced in bulk silicon devices. To illustrate this, Figure 1.1 schematically represents the cross-sections of both a bulk CMOS inverter and a CMOS SOI inverter. Most parasitic effects in bulk MOS devices find their origin in the interactions between the device and the substrate. Latchup in bulk devices finds its origin in the parasitic PNPN structure of the CMOS inverter represented in Figure 1.1. The latchup path can be symbolized by two bipolar transistors formed by the substrate, the well and the source and drain junctions. Latchup can be triggered by different mechanisms, such as node voltage overshoots, displacement current, junction avalanching and photocurrents. A necessary condition for latchup occurrence is that the current gain of the loop formed by the two bipolar transistors be larger than unity ($\beta > 1$)[4]. In an SOI CMOS inverter the silicon film containing the active devices is thin enough for the junctions to reach through to the buried insulator. A latchup path is ruled out because there is no current path to the substrate. In addition, the lateral PNPN structures contain heavily doped bases (the N^+ and P^+ drains), which reduces the gain of the bipolar devices to virtually zero.

Bulk circuits utilize reverse-biased junctions to isolate devices from one another. Consider the drain of the n-channel transistor of Figure 1.1. The drain is always positively or zero biased with respect to the substrate (the drain voltage can range between GND and $+V_{DD}$). A depletion capacitance is associated with the drain junction. Its capacitance maximum value is reached when the drain voltage is zero and is a function of the substrate doping concentration. The higher this doping concentration, the higher the capacitance. Modern submicron circuits use higher doping concentrations, which contributes to increase the junction capacitance. In addition, a parasitic capacitance between the junctions and the channel stop implant exists underneath the field oxide to prevent surface leakage between bulk devices. In SOI circuits, on the other hand, the maximum capacitance between the junctions and the substrate is the capacitance of the buried insulator (which tends towards zero if thick insulators are used, such in SOS technology). This capacitance is proportional to the dielectric constant of the capacitance material. Silicon dioxide, used as buried insulator, has a dielectric constant ($\varepsilon_{ox} = 3.9 \times \varepsilon_0$) which is three times smaller than that of silicon ($\varepsilon_{si} = 11.7 \times \varepsilon_0$), therefore, a junction located on a buried oxide layer gives rise to a parasitic capacitance approximately three times smaller than that of a bulk junction. Since SOI buried insulator thickness does scale with device scaling, parasitic capacitances do not necessarily increase as technology progresses. In addition, a lightly-doped, p-type silicon wafer can be utilized as mechanical support. In that case, a depletion layer can be created beneath the insulator, which further contributes to reduce the junction-to-substrate capacitances.

2

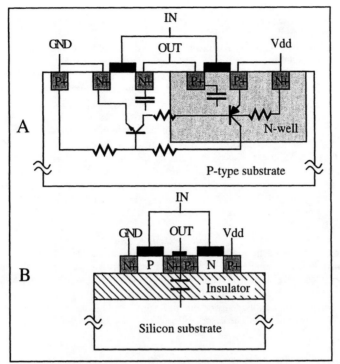

Figure 1.1: Cross section of a bulk CMOS inverter (A) showing a latchup path and an SOI CMOS inverter (B). The drain parasitic capacitances are also presented.

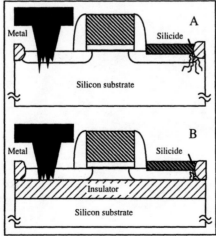

Figure 1.2: Formation contact or silicide on shallow junctions in the case of bulk silicon (A) and thin-film SOI (B).

3

Silicon-on-Insulator CMOS technology is also attractive because it is involves less processing steps than bulk CMOS technology and because it suppresses some yield hazard factors present in bulk CMOS. To illustrate this we can take the example of realizing a shallow junction and making contact to it (Figure 1.2). Making a shallow junction is not an obvious task in bulk CMOS. If a thin (100 nm) SOI substrate is used, on the other hand, the depth of the junction will automatically be equal to the thickness of the silicon film. Electrical contact to a shallow junction can be made using a metal (*e.g.*. tungsten), an alloy (*e.g.* Al:Si or Ti:W) or a metal silicide (*e.g.* $TiSi_2$). In bulk silicon devices, unwanted reactions can take place between the silicon and the metal or the silicide, such that the metal "punches through" the junction (Figure 1.2.A). This effect is well known in the case of aluminum (aluminum spiking), but can also occur with other metal or silicide systems. Such a junction punch-through gives rise to uncontrolled leakage currents. If the devices are realized in thin SOI material, the N^+ or P^+ source and drain diffusions extend to the buried insulator (reach-through junctions). In that case, there is no metallurgical junction underneath the metal-silicon contact area, and no leakage will be produced if some uncontrolled metal-silicon reaction occurs (Figure 1.2.B).

The absence of latch-up, the reduced parasitic source and drain capacitances, and the ease of making shallow junctions are merely three obvious examples of the advantages presented by SOI technology over bulk. There are many other properties which allow SOI devices and circuits to exhibit performances superior to those of their bulk counterparts (radiation hardness, high-temperature operation, improved transconductance and subthreshold slope, ...). These properties will be described in detail in Chapters 5, 7 and 8.

References

1   J.E. Lilienfield, U.S. patents 1,745,175 (filed 1926, issued 1930), 1,877,140 (filed 1928, issued 1932), and 1,900,018 (filed 1928, issued 1933)

2   see for example: W. Shockley, "The path to the conception of the junction transistor", IEEE Trans. on Electron Devices, Vol. 23, No. 7, p. 597, July 1976

3   see for example: D. Kahng, "A historical perspective on the development of MOS transistors and related devices", IEEE Trans. on Electron Devices, Vol. 23, No. 7, p. 655, July 1976

4   R.R. Troutman, Latchup in CMOS Technology, Kluwer Academic Publishers, 1986

CHAPTER 2 - SOI Materials

Many techniques have been developed for producing a film of single-crystal silicon on top of an insulator. Some of them are based on the epitaxial growth of silicon on either a silicon wafer covered with an insulator (homoepitaxial techniques) or on a crystalline insulator (heteroepitaxial techniques). Other techniques are based on the crystallization of a thin silicon layer from the melt (laser recrystallization, e-beam recrystallization and zone-melting recrystallization). Silicon-on-insulator material can also be produced from a bulk silicon wafer by isolating a thin silicon layer from the substrate through the formation and oxidation of porous silicon (FIPOS) or through the ion beam synthesis of a buried insulator layer (SIMOX, SIMNI and SIMON). Finally, SOI material can be obtained by thinning a silicon wafer bonded to an insulator and a mechanical substrate (wafer bonding, BESOI). Every approach has its advantages and its pitfalls, and the type of application to which the SOI material is destined dictates the material to be used in each particular case. SIMOX, and UNIBOND® materials for instance, seem to be ideal candidates for VLSI CMOS applications, while wafer bonding is more adapted to bipolar and power applications. This Chapter will review the different techniques used for producing SOI material.

2.1. Heteroepitaxial techniques

Heteroepitaxial Silicon-on-Insulator films are obtained by epitaxially growing a silicon layer on a single-crystal insulator. Reasonably good epitaxial growth is possible on insulating materials which have lattice parameters sufficiently close to those of single-crystal silicon. Substrates can either be single-crystal bulk material, such as $(01\overline{1}2)$ Al_2O_3 (sapphire) or thin insulating films grown on a silicon substrate (epitaxial CaF_2). Heteroepitaxial growth of a silicon film can never produce a defect-free material by itself if the lattice parameters of the insulator do not perfectly match those of silicon. Table 2.1.1 illustrates the differing materials properties. Furthermore, the silicon film will never be stress-free if the thermal expansion coefficients of the silicon and the insulating substrate are not equal.

Heteroepitaxial silicon-on-insulator films are grown using silane or dichlorosilane at temperatures around 1000°C. All the insulating substrates have thermal expansion coefficients which are 2 to 3 times higher than that of silicon. Therefore, thermal mismatch is the single most important factor determining the physical and electrical properties of heteroepitaxial silicon films grown on bulk insulators. Indeed, the silicon films have a thickness which is typically 1000 times smaller than that of the insulating substrate. While the films are basically stress-free at growth temperature, the important thermal coefficient mismatch results in a compressive stress in the silicon film which reaches $\cong -7 \times 10^9$ dynes/cm^2

at the surface of a 0.5 μm silicon-on-sapphire (SOS) film, for instance, while a still higher value is reached at the Si-sapphire interface. Such stresses may equal or exceed the yield stress of silicon, resulting in relaxation in the silicon film via generation of crystallographic defects such as microtwins, stacking faults and dislocations.

Material	Crystal Structure	Dielectric constant	Lattice parameter (nm)	Mean thermal expansion coefficient 20-1000°C, (/°C)
Si	Cubic	11.7	0.54301	3.8 E-6
Sapphire (0 1 -1 2)	Rhombo-hedral	9.3	0.4759	9.2 E-6
Cubic Zirconia	Cubic	38	0.5206	11.4 E-6
Spinel	Cubic	8.4	0.808	8.1 E-6
CaF2	Cubic	6.8	0.5464	26.5 E-6

Table 2.1.1: Parameters of the main materials involved in heteroepitaxial SOI technologies.[1]

The most important heteroepitaxial SOI techniques will now be briefly reviewed, and their advantages and pitfalls will be outlined.

2.1.1. Silicon-on-sapphire

There is no doubt that silicon-on-sapphire (SOS) is the single most mature of all heteroepitaxial SOI materials. Until the eighties, it was the only SOI material used to produce LSI-VLSI circuits. Several milestones of SOS technology are listed in Table 2.1.2.[2]

1963	Idea of SOS (Manasevit and Simpson) [3]
1971	Wafers commercially available
1975	1k SRAM (RCA) [4]
1976	16-bit microprocessor (HP) [5]
1977	4k SRAM (RCA) [6]
1978	16-bit, 7K gate microprocessor (Toshiba)
1978	16k CMOS SRAM (RCA) [7]
1980	16-bit high-speed microprocessor (Toshiba)
1982	5-inch SOS wafers (Kyocera)
1984	Subnanosecond CMOS gate array (Toshiba)
1987	64k CMOS SRAM (Westinghouse) [8]
1988	4-bit, 1 GHz flash ADC (Hughes) [9]
1988	Thin-film SOS devices [10]
1995	Thin-film, 60 GHz f_{max} MOSFETs [11]

Table 2.1.2: Some milestones of SOS technology

The sapphire (α-Al_2O_3) crystals are produced using either the flame-fusion growth technique, Czochralski growth, or edge-defined film-fed growth.[12] The first two techniques provide sapphire boules which must be sliced before polishing, while the third technique produces thin rectangular sapphire ribbons which are susequently be cut into circular wafers. After mechanical and chemical polishing, the sapphire wafers receive a final hydrogen etching at 1150°C in an epitaxial reactor, and a silicon film is deposited using the pyrolysis of silane at temperatures between 900 and 1000°C. Due to lattice and thermal mismatch, the defect density in the films is quite high, especially in very thin films. As the film thickness increases, however, the defect density appears to decrease as a simple power law function of the distance from the Si-Sapphire interface. The main defects present in the as-grown SOS films are: aluminum autodoping from the Al_2O_3 substrate, stacking faults and microtwins. Typical defect densities near the Si-Sapphire interface reach values as high as 10^6 planar faults/cm and 10^9 line defects/cm^2. These account for the low values of resistivity, mobility, and lifetime near the interface. The lowest aluminum autodoping is obtained at deposition rates of 2-2.5 μm/min [13], such that aluminum doping concentrations below the SIMS detection limit ($\cong 10^{15}$ cm^{-3}) can be obtained.

The electron mobility observed in SOS devices is lower than bulk mobility. This is a result of both the high defect density found in as-grown SOS films and the compressive stress measured in the silicon film. Indeed, in the case of (100) SOS, the compressive stress causes the kx and ky ellipsoids to become more populated with electrons than the kz ellipsoid (which is normal to the silicon surface). As a result, the effective mass of electrons becomes larger than in bulk silicon [14] and a relatively low channel electron mobility is observed in SOS n-channel MOSFETs ($\cong 250$-350 cm^2/V.s). On the other hand, the effective mass of holes is smaller than in bulk silicon, due to the same compressive stress. The hole surface mobility, which could in principle be higher than in bulk because of the compressive stress, is, however, affected by the presence of defects, such that the value of surface mobility for holes in SOS p-channel MOSFETs is comparable to that in bulk silicon. A beneficial consequence of the very low electron mobility at the Si-Sapphire interface is the reduction of the back-channel leakage current in n-channel devices. The minority carrier lifetime found in as-grown SOS films is a fraction of a nanosecond. As a result, relatively high junction leakage currents ($\cong 1$ pA/μm) are observed.[15]

Several techniques have been developed to reduce both the defect density and the stress in the SOS films. The first of these is melting and recrystallization of the silicon layer by means of a pulsed laser.[16] This technique brings about a 33% increase of the electron field-effect mobility, but it can have detrimental effects on both the leakage current and the threshold voltage uniformity, due to generation of interface states at the Si-sapphire boundary.

The Solid-Phase Epitaxy and Regrowth (SPEAR) and the Double Solid-Phase Epitaxy (DSPE) techniques are other more successful methods for improving the crystal quality of SOS films.[17,18,19] These techniques employ the following steps. First, silicon implantation is used to amorphize the silicon film, with the exception of a thin superficial layer, where the original defect density is lowest. Then a thermal annealing step is used to induce solid-phase regrowth of the amorphized silicon, the top silicon layer acting as a seed. A second silicon implant is then used to amorphize the top of the silicon layer, which is subsequently recrystallized in a solid-phase regrowth step using the bottom of the film as a seed. In the SPEAR process, an additional epitaxy step is performed after solid-phase regrowth. Using such techniques, substantial improvement of the defect density is obtained. Noise in MOS devices is reduced, and the minority carrier lifetime is increased by two to three orders of magnitude, up to 50 ns.[20] Typical improvements brought about by the DSPE process are: an increase of the electron mobility from 300 to 450 $cm^2/V.s$, an increase of the hole mobility from 185 to 250 $cm^2/V.s$, a decrease of the inverse subthreshold slope from 110 to 92 mV/dec, a decrease of the n-channel leakage (at V_{DS}=3 V) from 1 to 0.4 pA/μm, and an increase of the drive current from 47 to 63 μA/μm in nMOS devices (L_{eff}=1.4 μm, V_G=V_{DS}=3V).[21] More recently, thin (0.1-0.2 μm), high-quality SOS films have been produced, and MOSFETs with excellent performances have been fabricated in these films.[22] Field-effect mobilities of 800 and 250 $cm^2/V.s$ have been reported for electrons and holes, respectively, in devices made in these thin SOS films.

Silicon-on-sapphire is not only a subset of SOI technology but it is also a mature technology in itself. However, in this Book attention will be focused on new emerging SOI technologies. The reader who is interested in more information about SOS technology can always refer to the excellent review article written by A.C. Ipri in Reference. [23]

2.1.2. Cubic zirconia

Yttrium-stabilized cubic zirconia [$(Y_2O_3)_m\cdot(ZrO_2)_{1-m}$] can also be used as an alternative dielectric substrate for silicon epitaxy.[24] Indeed, zirconia is a superionic oxygen conductor at high temperature. This means that, while being an excellent insulator at room temperature ($\rho > 10^{13}$ Ω.cm), cubic zirconia allows very rapid transport of oxygen at high temperatures. This unique property has been used to grow an SiO_2 layer at the silicon-zirconia interface (*i.e.* to oxidize the most defective part of the silicon film) by the transport of oxygen through a 500 μm-thick zirconia substrate.[25,26] The growth of a 160 nm-thick film at the interface necessitates only 100 min at 925°C in pyrogenic steam. As-deposited, (100) oriented silicon films grown on cubic zirconia have substantially higher crystal quality than state-of-the art SOS, as indicated by the following RBS surface channeling yields: χ_o=0.048 in Si on cubic zirconia (SOZ), 0.12 in SOS, and 0.034 in bulk silicon.

2.1.3. Silicon-on-spinel

Spinel [$(MgO)_m\cdot(Al_2O_3)_{1-m}$] can be used as a bulk insulator material or can be grown epitaxially on a silicon substrate at a temperature between 900 and 1000°C.[27] Stress-free 0.6 μm silicon-on-spinel films have been grown, but the properties of MOSFETs made in this material are inferior to those of devices made in SOS films due to higher defect density.[28] Better device performance is obtained when much thicker 3-40 μm silicon-on-spinel films are used. Again as with SOZ, oxygen can diffuse at high temperature through thin spinel films. Using this property, a Si/spinel/SiO_2/Si structure has been produced.[29]

2.1.4. Epitaxial calcium fluoride

Like spinel, calcium fluoride (CaF_2) can be grown epitaxially on silicon. Fluoride mixtures can also be formed and their lattice parameters can be matched to those of most semiconductors. For example, $(CaF_2)_{0.55}\cdot(CdF_2)_{0.45}$ has the same lattice parameters as silicon at room temperature, and $(CaF_2)_{0.42} \cdot (SrF_2)_{0.58}$ is matched to germanium. Unfortunately, the thermal expansion coefficient of these fluorides is quite different from that of silicon, and lattice match cannot be maintained over any appreciable temperature range. Silicon can, in turn, be grown on CaF_2 using MBE or e-gun evaporation at a temperature of approximately 800°C.[30,31] As in the case of Si films on epitaxial spinel, Si/CaF_2/Si films are essentially stress-free, which can be readily understood by noticing that the mechanical support is a silicon wafer, which, of course, has the same thermal expansion coefficient as the top silicon film. MOSFETs have been made in Si/CaF_2/Si material and exhibit surface electron and hole mobilities of 570 and 240 $cm^2/V.s$, respectively.[32]

2.1.5. Other heteroepitaxial SOI materials

Cubic boron phosphide (BP) and rhombohedral $B_{13}P_2$ have also been epitaxially grown on silicon, and silicon has subsequently been grown on these materials.[33,34] BP has a lattice parameter of 0.453 nm and a thermal expansion coefficient of $(4-6.2)x10^{-6}/C$ in the temperature range 127-527°C. BP has a bandgap of only 2 eV, but it can be obtained in a semi-insulating form with a resistivity of 10^{12} $\Omega.cm$ upon a thermal annealing step at 1050°C, which also happens to be the temperature used for epitaxially growing silicon on BP. P-channel MOSFETs with near-bulk mobilities have been fabricated in Si/BP/Si films. A major problem in using BP as an insulator is the narrow window of temperatures which can be used during device processing, due to outdiffusion of either boron or phosphorus into the silicon film. For completeness, it can also be mentioned that silicon has been grown epitaxially on a wide variety of insulators, such as AlN [35], cubic β-SiC [36], BeO [37], NaCl [38], ...

2.1.6. Problems of heteroepitaxial SOI

In addition to the stress in the films and the high crystalline defect density inherent to heteroepitaxy, these SOI materials suffer from a basic problem: they cannot be processed in standard silicon IC fabrication lines because of contamination problems. It is, indeed, highly undesirable to introduce materials such as sapphire, zirconia, calcium fluoride or boron phosphide in the ovens of a bulk silicon clean room. Furthermore, these substrates, in particular SOS, are extremely brittle, and wafer breakage is an important factor limiting the use of this technology. In addition, SOS, zirconia and bulk spinel substrates are transparent. This poses problems for using commercial steppers for lithography, particle counters, and equipment for measuring the thickness of layers grown or deposited on SOS.

2.2. Laser recrystallization

MOS transistors can be fabricated in a layer of polysilicon deposited on an oxidized silicon wafer [39,40], but the presence of grain boundaries brings about low surface mobility values ($\cong 10$ $cm^2/V.s$) and high threshold voltages (several volts). Grain boundaries contain silicon dangling bonds giving rise to a high density of surface states (several 10^{12} cm^{-2}) which must be filled with channel carriers before threshold voltage is reached. Above threshold, once the traps are filled, the grain boundaries generate potential barriers which have to be overcome by the channel carriers flowing from source to drain. This gives rise to the low values of mobility observed in polysilicon devices.[41] Mobility can be improved and more practical threshold voltage values can be reached if the silicon dangling bonds in the grain boundaries are

11

passivated. This can be performed by exposing the wafers to a hydrogen plasma, during which the fast diffusing atomic hydrogen can penetrate the grain boundaries and saturate (passivate) the dangling bonds.[42] This treatment can improve the drive current of polysilicon devices by a factor of 10, due to both an increase of mobility and a reduction of threshold voltage. Hydrogen passivation also significantly improves the leakage current of the devices.[43] Such devices can find application as active polysilicon loads, and 64k SRAMs with p-channel polysilicon loads have been demonstrated.[44] High-performance IC applications, however, require much better device properties, and grain boundaries must be eliminated from the silicon film. This is the goal of all the recrystallization techniques which will be described next.

2.2.1. Types of lasers used

Experiments of laser recrystallization of polycrystalline silicon have been carried out with pulsed lasers (ruby lasers and Nd:YAG lasers) [45] at a time where pulsed laser annealing was popular for doping impurity activation.[46] The technique, however, was rapidly abandoned for the fabrication of device-worthy SOI layers because of its lack of controllability and because of the extreme difficulty of growing large silicon grains. Continuous-wave (cw) lasers such as CO_2, Ar and YAG:Nd lasers have proven to be much more effective for producing SOI films. Silicon is transparent at the 10.6 μm wavelength produced by CO_2 lasers. Therefore, silicon films cannot be directly heated by such lasers. SiO_2, on the other hand, absorbs the 10.6 μm wavelength with a penetration depth of $\cong 10$ μm. Polysilicon films deposited on SiO_2 (quartz or an oxidized silicon wafer) or covered by an SiO_2 cap can thus be melted through "indirect" heating produced by CO_2 laser irradiation.[47] CO_2 lasers have the advantage of a high output power (> 100 W), such that wide elliptical beams can be produced, but manipulation of the beam is difficult because the CO_2 infrared radiation is invisible to the human eye. This coupled with the fact that the polysilicon layer has to be surrounded by relatively thick SiO_2 layers (to efficiently absorb the 10.6 μm wavelength) have contributed to the limited use of CO_2 lasers in SOI materials fabrication. CO_2 lasers are also unpractical for 3D applications, where heating of only the top silicon layer is needed. Continuous-wave (CW) YAG:Nd lasers can output high power beams (300 W) at a wavelength of 1.06 μm. Silicon is transparent at this wavelength, but if the wafer is preheated to a temperature of 1200-1300°C, free carriers are generated in the silicon and the 1.06 μm wavelength can be absorbed.[48] CW Ar lasers, on the other hand, emit two main spectral lines at 488.0 and 514.5 nm (blue and green), and can reach an output power of 25 W when operated in the multiline mode. These wavelengths are well absorbed by silicon. In addition to this, the reflectivity of silicon increases abruptly once melting is reached. This effect is very convenient since it acts as negative feedback on the power absorption and prevents the silicon from overheating above melting point. This "self-limiting" absorption increases process flexibility during recrystallization with an Ar^+ laser (the effect is opposite when a CO_2 laser is used, since solid silicon is transparent at the 10.6 μm wavelength, while molten silicon absorbs CO_2 laser energy through its free carriers). As a consequence of these advantages, the vast majority of SOI-producing experiments based on laser recrystallization have been carried out using Ar^+ lasers. The laser beam is focused on the sample by means of an achromatic lens (or a combination of lenses) into a circular or, more often, an elliptical spot. Scanning of the beam is achieved through the motion of galvanometer-driven mirrors. The size of the molten zone and the texture of the recrystallized silicon depend on parameters such as laser power, laser intensity profile, substrate preheating temperature (the wafer is held on a heated vacuum chuck), and scanning speed.[49] Typical recrystallization conditions of a 500 nm-thick LPCVD polysilicon film deposited on a 1 μm thermal oxide grown on a silicon wafer are: spot size of 50-150 μm (defined as the laser spot diameter at $1/e^2$ intensity, TEM_{00} mode), power of 10-15 watts, scanning speed of 5-50 cm/sec, and substrate heating at 300-600°C. Several sample structures have also been studied, all aiming at improving the crystal quality of the recrystallized silicon film, eliminating grain boundaries and reducing surface roughness. Some of them will be described next, all for the case of recrystallization with an argon laser.

2.2.2. Seeding

Polysilicon films recrystallized on an amorphous SiO_2 substrate have a random crystal orientation. X-ray diffraction studies of polysilicon recrystallized with a Gaussian laser profile indicate the presence of crystallites having (111), (220), (311), (400), (331), (110), and (100) orientations.[50] This is clearly unacceptable for device fabrication, since different crystal orientations will result in different gate oxide growth rates.

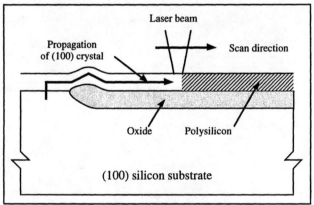

Figure 2.2.1: Principle of the lateral seeding process.

Ideally, one wants a uniform (100) orientation for all crystallites. From there comes the idea of opening a window (seeding area) in the insulator to allow contact between the silicon substrate and the polysilicon layer. Upon melting and recrystallization, lateral epitaxy can take place and the recrystallized silicon will have a uniform (100) orientation, as shown in Figure 2.2.1.[51] Unfortunately, the (100) crystal orientation can only be dragged less than 100 μm away from the seeding window, at which distance defects appear which cause loss of the (100) orientation.

2.2.3. Encapsulation

Because of some surface tension and de-wetting effects, polysilicon films have a tendency to "bead up" (form droplets) on SiO_2 upon melting if the laser power is too high. A less dramatic mass transport is observed at optimum laser power, but the flatness of the original polysilicon film is lost after laser melting and recrystallization.[52] In order to solve this problem, a capping layer of SiO_2 and/or Si_3N_4 can be used as a "cast" to improve wetting of the molten silicon and to prevent silicon delamination from the underlying insulator. Silicon dioxide caps are too soft for preventing surface waviness of the recrystallized silicon film.[53] On the other hand, nitrogen dissolves in the molten silicon if a Si_3N_4 cap is used. This nitrogen migrates towards the Si-buried insulator interface upon further processing and creates a high density of interface states at this interface.[54,55] The best results are obtained using a cap combining both silicon dioxide and silicon nitride. A surface waviness of at least 20 nm seems, however, unavoidable unless a planarization step is used after recrystallization.[56]

2.2.4. Beam shaping

The normal intensity profile of an Ar^+ laser operated in the TEM_{00} mode is Gaussian (before and after focusing). As a consequence, the molten zone produced by a stationary laser spot will be circular (or elliptical if a semi-cylindrical lens is used). When the beam is scanned across the sample, the molten zone will evolve with time as indicated in Figure 2.2.2.

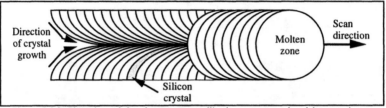

Figure 2.2.2: Plan view of the chevron recrystallization pattern produced by scanning a Gaussian laser beam.

The recrystallization of the quenched silicon proceeds along the thermal gradients which are perpendicular to the trailing edge of the molten zone. As a consequence, crystals grow from the edges of the scanned line towards its center, and in the direction of the scan. The resulting crystals present a "chevron pattern", as shown in Figure 2.2.2.[57] The silicon crystals are elongated, with a width of a few micrometers and a length of 10-20 micrometers. The grains are separated by grain boundaries which have detrimental effects on device properties.

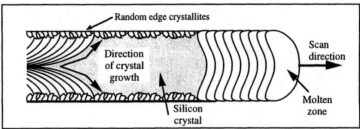

Figure 2.2.3: Growth of a large SOI crystal using a shaped laser beam.

Growth of large SOI crystals can be obtained if the trailing edge of the molten zone is concave, as presented in Figure 2.2.3. This can be achieved by masking part of the beam.[58]

In this approach, however, the available laser power is reduced. More efficient beam shaping can be achieved by merging different laser modes (doughnut-shaped beam) [59] or by recombining a split laser beam.[60]

In order to obtain not only a single large crystal, but a large single-crystal area, the laser beam has to be raster-scanned on the wafer with some overlap between the scans. Unfortunately, small random crystallites arise at the edges of the large crystals, which precludes the formation of large single-crystal areas, and grain boundaries are formed between the single-crystal stripes. The location of these grain boundaries depends on the scanning parameters and the stability of the beam. In other words, from a macroscopic point of view, the location of the boundaries is quasi-random, and the yield of large circuits made in this material will be zero. A solution to this problem is to use stripes of an anti reflecting (AR) material (SiO$_2$ and/or Si$_3$N$_4$) to obtain the photolithographically-controlled beam shaping of an otherwise Gaussian beam.[61,62,63]

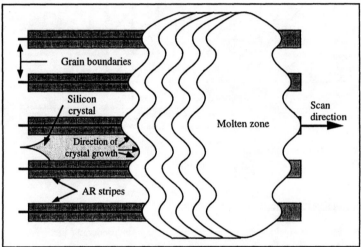

Figure 2.2.4: Recrystallization using anti reflection stripes.

This technique is called selective annealing because more energy is selectively deposited on the silicon covered by AR material. It permits the growth of large adjacent crystals with straight grain boundaries, the location of which is controlled by a lithography step (Figure 2.2.4). Although this technique imposes constraints to circuit design, it allows for placing film defects outside the active area of transistors. The technique can be used with a laser scan parallel, slanted or perpendicular to the anti reflection stripes (AR stripes).

Unfortunately, a rotation of the crystal orientation is observed, and the (100) orientation cannot be kept for more that $\cong$200 μm from the seeding window. Defects then appear in the crystals (stacking faults, microtwins or even grain boundaries) until a (110) orientation is reached.[64] This rotation can be minimized by forcing the solidification front (the interface between molten and recrystallizing silicon) to be coincident with the [111] facets of the re-solidifying silicon. This can be achieved by using AR stripes parallel to the (150) direction and a laser scan parallel to the (100) direction (Figure 2.2.5). Using this technique, chip-wide (several mm x several mm) defect-free, (100)-oriented single-crystal areas have been produced.[65]

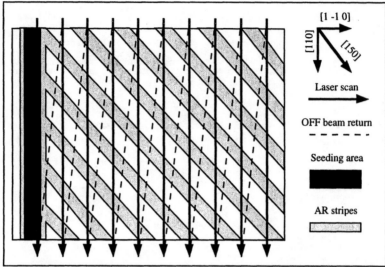

Figure 2.2.5: Recrystallization of a large SOI area using selective annealing, seeding and control of the orientation of the crystal facets in the solidification front.

Recrystallization techniques based on a patterned anti reflection coating are primarily used in Japan for the fabrication of three-dimensional integrated circuits (3-D circuits) with up to 4 active device layers.[66,67,68,69,70]

2.2.5. Silicon film patterning

Several other techniques have been developed for producing device-worthy silicon films by laser recrystallization, including the use of periodic seeding windows [71], recrystallization of patterned silicon islands [72,73,74,75] and defect filtering through the use of V-shaped, interconnected silicon islands, as illustrated in Figure 2.2.6.[76] These techniques, however, impose severe (and often unacceptable) constraints to the circuit layout since the size and shape of the active areas is fixed by laser recrystallization constraints rather than by circuit design considerations.

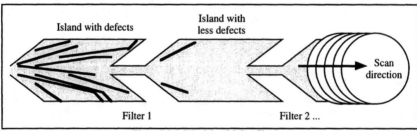

Figure 2.2.6: Principle of defect filtering.

It is also worth mentioning that several attempts have been made to create an "artificial seed", having a (100) orientation, in the patterned silicon film itself, without the need for opening a seeding window to the substrate.[77,78]

2.3. E-beam recrystallization

The recrystallization of a polysilicon film on an insulator using an electron beam (e-beam) is in many respects very similar to the recrystallization using a cw laser. Similar seeding techniques are used, and an SiO_2 (or Si_3N_4) encapsulation layer is used to prevent the melted silicon from de-wetting.[79] The use of an e-beam for recrystallizing SOI layers has some potential advantages over laser recrystallization since the scanning of the beam can be controlled by electrostatic deflection, which is far more flexible than the galvanometric deflection of mirrors. Indeed, the oscillation frequency of a laser beam scanned using galvanometer-driven mirrors is limited to a few hundred hertz, while e-beam scan frequencies of 50 MHz have been utilized.[80] The absorption of the energy deposited by the electron beam is almost the same in most materials, such that the energy absorption in a sample is quite independent of crystalline state and optical reflectivity of the different materials composing it. This improves the uniformity of the recrystallization of silicon deposited over an uneven substrate, but precludes the use of a patterned anti reflection coating. Structures with tungsten stripes have, however, been proposed to achieve differential absorption.

2.3.1. Scanning techniques

The flexibility of electron optics and electrostatic deflection permits one to explore different types of scanning configurations. A first technique (Figure 2.3.1.A) involves the simple raster scan of the focused electron beam on the sample. This technique produces a chevron recrystallization pattern with grain size up to 20 μm, and a texture which looks very much like the one which is obtained by scanning with a Gaussian laser spot.[81] A second technique makes use of a focused linear e-beam source (Figure 2.3.1.C).[82] This technique allows for the rapid recrystallization of large areas of SOI, but the control of the uniformity of the beam intensity is difficult.

The most flexible and powerful e-beam recrystallization technique consists of the synthesis of a pseudo-linear source through rapid scanning of a focused beam (Figure 2.3.1.B). A continuous, linear molten zone can be created in the silicon film if the period (1/frequency) of the scan is smaller than the thermal constant of the SOI system. This is achieved if a frequency over 2 MHz is used. If a sinusoidal scanning is used, the position of the beam is given by $y(t) = W \sin (2\pi ft)$, and the deposited energy, which is proportional to the dwell time at the position y, is significantly higher at the edges of the scanned line ($y=W$) than at its center ($y=0$). This can be compensated for by modulating the amplitude of the rapid sinusoidal scan, such that an uniform energy deposition profile can be obtained across the entire scanned line (Figure 2.3.2).

17

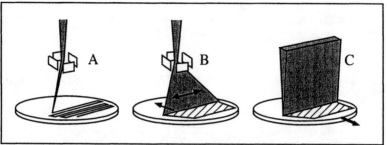

Figure 2.3.1: Beam configurations for e-beam recrystallization: (A) scanned spot, (B) synthesized line source, (C) focused line source.

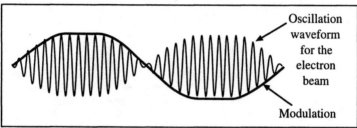

Figure 2.3.2: Electron beam oscillation waveform used to synthesize a uniform-intensity line source.[83]

2.3.2. Seeding and wafer heating

Two seeding techniques have been proved to be successful when used with e-beam recrystallization. The first one is similar to that used for laser recrystallization (Figure 2.2.1). Using this technique, multilayer SOI recrystallization has been achieved to produce 3D IC's.[84] The second technique makes use of a periodic seeding structure. The scanning can be parallel, slanted or perpendicular to the seeding windows.[85] Back-side heating of the wafers is needed in order to reduce the thermal stress across the wafer during recrystallization. Both radiation heating from halogen lamps or a glowing filament, and heating from a second electron beam, directed at the back of the wafer, can be used.[86,87]

2.4. Zone-melting recrystallization

One of the main limitations of laser recrystallization is the small molten zone produced by the focused beam, which results in a long processing time needed to recrystallize a whole wafer. Recrystallization of a polysilicon film on an insulator can also be carried out using incoherent light (visible or near IR) sources. In this case, a narrow (a few millimeters) but long molten zone can be created on the wafer. A molten zone length of the size of an entire wafer diameter can readily be obtained. As a result, full recrystallization of a wafer can be carried out in a single pass. Such a recrystallization technique is generally referred to as

Zone-Melting Recrystallization (ZMR) because of the analogy between this technique and the float-zone refining process used to produce silicon ingots. An excellent review of the ZMR mechanisms can be found in the book of E.I. Givargizov.[88] The first method which successfully achieved recrystallization of large-area samples makes use of a heated graphite strip which is scanned across the sample to be recrystallized. The set-up is called "graphite strip heater" (Figure 2.4.1). A heated graphite susceptor is used to raise the temperature of the entire sample up to within a few hundred degrees below the melting temperature of silicon. Additional heating is locally produced at the surface of the wafer using a heated graphite strip located a few millimeters above the sample and scanned across it.[89,90] A typical sample is made of a silicon wafer on which a $\cong$ 1-2 μm-thick oxide is grown. A $\cong$ 0.5-1 μm-thick layer of LPCVD amorphous or polycrystalline silicon is then deposited, and the whole structure is capped with a 2 μm-thick layer of deposited SiO_2 covered by a thin Si_3N_4 layer. The capping layer helps minimize mass transport and protects the molten silicon from contaminants (such as carbon from the strip heater). Recrystallization is carried out in a vacuum or an inert gas ambient in order to keep the graphite elements from burning.

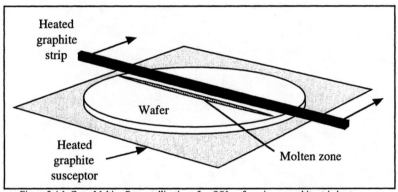

Figure 2.4.1: Zone-Melting Recrystallization of an SOI wafer using a graphite strip heater.

Both the graphite susceptor and the graphite strip can be replaced by lamps to achieve ZMR of SOI wafers. A lamp recrystallization system is composed of a bank of halogen lamps which is used to heat the wafer from the back to a high temperature (1100°C or above), and a top halogen or mercury lamp whose light is focused on the sample by means of an elliptical reflector (Figure 2.4.2).[91,92,93,94] An unpolished quartz plate may be inserted between the lamp bank and the wafer in order to homogenize the energy deposition at the back of the wafer. As in the case of strip heater recrystallization, a narrow, wafer-long molten zone is created and scanned across the wafer with a speed on the order of 0.1-1 mm/sec.

ZMR can be carried out using an elongated laser spot as well. Indeed, a linear molten zone can be created using a high-power (300 W) CW YAG:Nd laser (wavelength=1.06 μm). The circular laser beam can be transformed into a linear beam using 90°-crossed cylindrical lenses. The produced linear spot, which is focused on the sample, can exceed 10 cm in length. In order to get free carrier absorption from the silicon, the substrate has to be heated to a temperature of 1200-1300°C by a bank of halogen lamps. The laser beam melts a 0.5-1 mm-wide zone of the polysilicon layer sitting on top of an oxidized silicon substrate.

Crystallization is performed in a single pass of the molten zone across the wafer. The scanning speed can vary from 0.1 to 2 mm/s. The lateral temperature gradient is controlled by slightly focusing or defocusing the beam.[95]

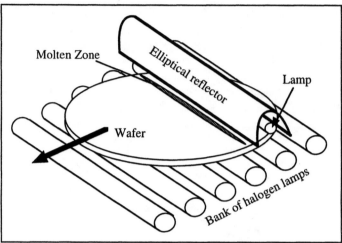

Figure 2.4.2: ZMR recrystallization of an SOI wafer using lamps.

The physics of zone-melting recrystallization is rather independent of the type of radiation source used (graphite strip heater, focused lamp system or elongated laser spot), but it is quite different from that of laser recrystallization. Indeed, the dwell time (time during which a portion of silicon is exposed to the beam) is in the order of a millisecond or less in the case of cw laser processing, while the dwell time is in the order of seconds in the case of ZMR. The thermal gradients are also quite different: in a typical laser recrystallization experiment, the substrate is heated to 300-600°C, while ZMR uses preheating temperatures between 1100 and 1350°C.

2.4.1. Zone-melting recrystallization mechanisms

2.4.1.1. Melting front

Contrary to what occurs in zone melting and recrystallization of bulk silicon, the crystalline "memory" of the silicon film is not completely erased upon melting, due to the thin-film configuration and to the presence of a stiff capping material which is capable of keeping a morphological memory.[96] As a consequence, the final quality of the crystal shows some dependence on the melting front dynamics, and instabilities in the melting front can be responsible for defects in the recrystallized film. Upon zone-melting of thin SOI films (0.5 μm or less), the front solid/liquid interface does not advance continuously at the speed of scanning but progresses in a succession of bursts. This phenomenon is known as "explosive melting".[97] This effect is attributed to the slightly different melting temperature of silicon grains having different crystal orientations and superheating in polycrystalline silicon. As a result, some solid silicon crystallites can be found in the molten silicon near the solid/liquid interface, and molten silicon droplets can be found ahead of the melting front. Sudden melting of superheated silicon grains causes the melting front to propagate in bursts (Figure 2.4.3).

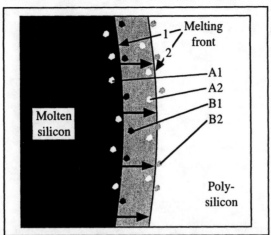

Figure 2.4.3: Evolution of melting front during ZMR processing (1 = at time 1 and 2 = at time 2 > time 1). A = solid crystallites in the molten zone, B = liquid silicon droplets ahead of the molten zone.

Because bursts of melting are very rapid, the SiO_2 cap cannot flow and accommodate for the volume decrease of melting silicon, and bubbles can be created. These bubbles can give rise to voids in the recrystallized film. This effect is inversely proportional to the silicon film thickness and is usually not observed during recrystallization of thick silicon films.[98]

2.4.1.2. Solidification front

The solidification front is even more important than the melting front in determining the crystalline properties of the final SOI film. The recrystallized film has a (100) normal orientation, even if no seeding is used. Grain boundary-like defects repeated at 100 µm to several mm intervals are found in the recrystallized films. These defects can either be real grain boundaries, subgrain boundaries (*i.e.* low angle boundaries between adjacent (100) crystals) or dislocation networks. Subgrain boundaries have a distinct wishbone pattern (Figure 2.4.4). In the best cases, the dislocation networks can be reduced to a few isolated dislocations.

There has been quite a debate concerning the origin of these defects. Two mechanisms have been proposed, based either on a cellular growth (or cellular-dendritic growth) mechanism or on a faceted growth process. Cellular growth postulates the presence of a high concentration gradient of impurities such as nitrogen or carbon in the molten silicon.[99,100,101,102] In that case, the material can be regarded as a dilute alloy which is locally constitutionally supercooled.

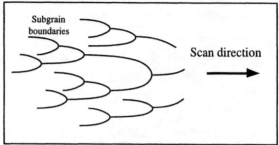

Figure 2.4.4: Pattern of subgrain boundaries in a ZMR film.

The criterion for the existence of constitutional supercooling is:

$$\frac{G}{R} < \frac{m \, C_0 \, (1-k)}{k \, D}$$

(2.4.1)

where R is the scan rate, G is the thermal gradient at the solidification interface, m is the liquidus slope of the silicon-impurity binary system [103], C_o is the impurity concentration in solid silicon, k is the solid-liquid impurity distribution coefficient ($k=7 \times 10^{-4}$ in the case of nitrogen), and D is the diffusion coefficient of the impurity in the liquid. If there is a region where the actual temperature is below the freezing temperature (constitutional supercooling), any perturbation of the interface will tend to grow as shown in Figure 2.4.5.

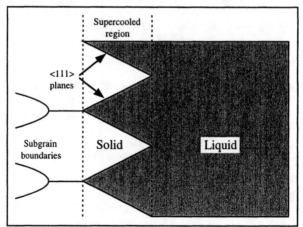

Figure 2.4.5: Cellular growth model: faceted protrusions span the undercooled region and form subgrain boundaries where they meet.[104]

Subgrain boundaries will appear at the trailing tips of the solidification interface, and their spacing will be comparable to the width of the supercooled region. This model can generally account for the spacing of the subgrain boundaries as a function of thermal gradients (but not as a function of scanning speed) and is best applicable when low thermal gradients are used (low thermal gradient regime). It also accounts for the rejection of impurities such as nitrogen towards the boundaries, which can be evidenced by techniques such as AES (Auger Electron Spectroscopy) mapping.[105] The cellular growth model, however, is limited to those cases where there is a high concentration of impurities in the silicon film and when thermal gradients are extremely low, which is not always the case in real ZMR processing.

The second model is the faceted growth model.[106,107] It is based on the three following assumptions: 1) the solid presents only <111> facets at the solid-liquid interface if the normal orientation of the film is (100) and scanning takes place in the <100> direction (<111> facets have the lowest interface energy and are the slowest growing planes), 2) the growth rate of each facet is limited by the nucleation rate of new atomic layers (which is dependent on the facet size - the smaller the facet, the larger the rate of addition of quenching atomic layers on it), and 3) a new re-entrant corner is formed as soon as a tip of the solid-liquid interface reaches the melting isotherm. Simulations of the dynamics of the ZMR solidification front have been carried out based on the three above assumptions.[108] These simulations show that the loci traced by the trailing tips of the liquid-solid interface form a branching pattern similar to the wishbone structure of the subgrain boundaries. The model accounts for the subboundary spacing as a function of the scan speed as well. The formation of subboundaries is illustrated in Figure 2.4.6 where the solidification front is depicted at three consecutive times.

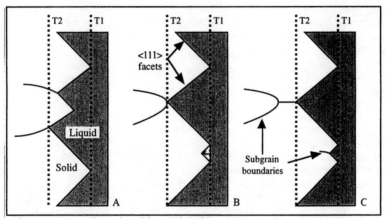

Figure 2.4.6: Faceted growth model: evolution of the solidification front at three consecutive times. Subboundaries are originated at the tips of the interface (A), the characteristic wishbone pattern is due to the coalescence of two tips (B), and a new <111> facets form at re-entrant corner of the melting isotherm, T_1 (C). T_2 is the solidification isotherm.[109]

Boundaries originate at the trailing tips of the solidification interface. Their formation is related to a concentration of stress and a pile-up of defect-generating impurities dissolved into the silicon. The silicon film has a (100) normal orientation, with <100> direction parallel to the scan motion. Growth occurs by the addition of atoms at ledges or steps which sweep rapidly across the <100> facets. Therefore, it takes a longer time for adding a new layer of atoms on a long facet than on a short one. The distance a wedge-shaped pair of facets extends into the liquid region, which determines the spacing between the subgrain boundaries originating at the interior corners, is limited by two temperatures, T_1 and T_2, shown as dotted isothermal contours. T_2 is the temperature at which the generation of new ledges occurs at such a rate that the forward advance of the <111> facets exactly matches the imposed scanning velocity. At temperatures close to T_2 the speed with which ledges sweep across a <111> facet is much greater than the rate of forward advance of the solidification front. Therefore, the rate of forward advance is determined by the rate of generation of new ledges. (The velocity of the ledges decreases as a ledge encounters higher temperatures, and, eventually, the forward velocity of the ledge matches the scanning velocity when T_1 is reached.) Subboundaries tend to be parallel to the scan direction, but small perturbations cause them to move laterally. Quite often two subboundaries coalesce in a wishbone pattern and form a single subgrain boundary (Figure 2.4.6.C). When the distance between adjacent subboundaries becomes so large that the wedge formed by a pair of facets would extend beyond the T_1 isotherm, the interface becomes flattened (Figure 2.4.6.B). This flattened <100> interface is not stable, and two new <111> facets immediately develop, giving rise to a new subboundary (Figure 2.4.6.C).

This model accounts for the subboundary spacing as a function of thermal gradient (which influences the distance between T_1 and T_2), and scan speed. The higher the thermal gradient, the smaller the spacing between subboundaries. The thermal gradient is controlled by the profile of energy deposited on the wafer by the incoherent light source. The misalignment between adjacent crystals has been found to decrease as a function of subboundary spacing, and is ruled by the following empirical law: $\Theta = 19.7 \ S^{-0.9}$, where Θ is the angle of relative misalignment (in degrees), and S is the subboundary spacing (in micrometers).[110] The faceted model is best applicable in cases where high thermal gradients are used (high-thermal gradientregime).

Actual ZMR processing involves both faceted and cellular growth (faceted growth appears to be prevalent near the melting isotherm, and cellular-dendritic growth is more important near the solidification isotherm).[111] Furthermore, more recent studies reveal that the establishment of an equilibrium between cooling of the high-reflectivity liquid silicon and heating of the lower-reflectivity solidified silicon can also explain the formation of a stable solidification interface which produces straight (unbranched) dislocation trails as only observable defect. Indeed, when optimal conditions are used (i.e. when ultra-high purity silane is used for polysilicon deposition, when backside heating of the wafer is very uniform, when low thermal gradients are used ("low thermal gradient regime"), and when the scanning speed is thoroughly controlled, a stable solidification front can be obtained, subboundaries can be eliminated, and only parallel trails of isolated dislocations (*i.e.* without wishbone patterns) are observed in the SOI layer.[112] Such an "optimized" material exhibits average defect densities lower than 5×10^4 dislocations cm^{-2}, an intrinsic doping level below 2×10^{15} cm^{-2}, junction leakage below 10^{-6} A/cm^2 and a minority carrier lifetime higher than 30 µs.[113]

2.4.2. Role of the encapsulation

During the preparation of wafers for Zone-Melting Recrystallization, it is necessary to cap the polysilicon film with a layer which prevents the molten silicon from beading up. The tendency to bead up (or: delaminate, ball up) is a consequence of the poor wetting properties of molten silicon on SiO_2.[114,115,116] Indeed, the wetting angle of molten silicon in a capillary sandwich of SiO_2 must be smaller than 90 degrees (Figure 2.4.7). A wetting angle of 87 degrees is experimentally observed [117], which renders wetting possible, but extremely unstable. As a consequence, de-wetting and balling-up of the silicon occurs in most cases where a pure SiO_2 cap is used. The wetting angle of silicon on silicon nitride is much lower (25 degrees). If a thin ($\cong$ 5 nm) Si_3N_4 layer is deposited between the polysilicon film and the oxide cap, wetting is substantially improved, and reliable recrystallization can be achieved. Unfortunately, recrystallization using this much nitride does not produce (100)-oriented SOI films. Reliable (100) recrystallization can be obtained by introducing less nitrogen at the silicon-cap interface (about one-third of a monolayer of nitrogen). This can be achieved by various methods: NH_3 annealing of the SiO_2 cap, deposition of SiN_x or Si_3N_4 on the cap, or plasma nitridation of the cap. These processes allow small amounts of nitrogen to diffuse towards the Si-SiO_2 interface during ZMR processing, which drastically improves wetting and, therefore, the quality of recrystallization.

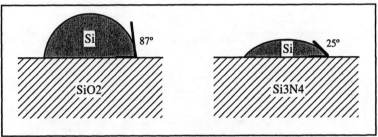

Figure 2.4.7: Wetting angle of molten silicon on SiO_2 and Si_3N_4.[118]

2.4.3. Mass transport

The thickness uniformity of an SOI layer is of crucial importance, especially for thin-film applications. Indeed, some thin SOI MOSFET parameters such as threshold voltage and subthreshold slope are extremely sensitive to film thickness variations. SOI films produced by the ZMR technique exhibit two types of thickness non-uniformity. The first one is long-range mass transport caused by the sweeping of the molten zone from one side of the wafer to the other. The silicon film is slightly thinner where the recrystallization begins and slightly thicker where it ends. The second one is a short-range waviness, perpendicular to the scan direction, and associated with the non-planar shape of the solidification front and the generation of subgrain boundaries (or trails of dislocations).

This short-range waviness is caused by the finite stiffness of the cap layer. For the "low-thermal gradient" regime (*i.e.* when non-branching, parallel and equally spaced dislocation trails are produced), the waviness is the result of the formation of a meniscus at the surface of the liquid silicon and of a surface tension equilibrium along the cap - solid silicon - liquid silicon triple line. The waviness of the recrystallized silicon film exhibits a periodical parabolic shape, the thickness reaching minima at the subgrain boundaries and maxima half way between the boundaries (Figure 2.4.8). A peak-to-peak thickness variation of $\cong$20 nm is measured on commercial ZMR wafers. Because of film thickness variations and wetting problems during ZMR processing, recrystallization of thin (< 0.3 μm) films is unpractical.

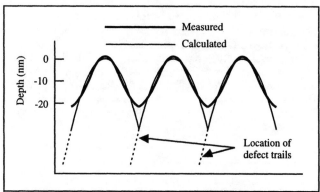

Figure 2.4.8: Amplitude of thickness variation measured perpendicular to the ZMR scan direction.[119]

2.4.4. Impurities in the ZMR film

Evidence of high contamination levels (such as SiC precipitates in subgrain boundaries) was reported in early ZMR material. Nowadays, the use of high-purity gases for polysilicon and SiO_2 deposition allow for the fabrication of electronic-grade ZMR films. The impurity found in highest concentration in the recrystallized silicon film is oxygen. The concentration of oxygen in the molten silicon during ZMR processing is 2.5×10^{18} atoms/cm^3 (oxygen solubility in liquid silicon).

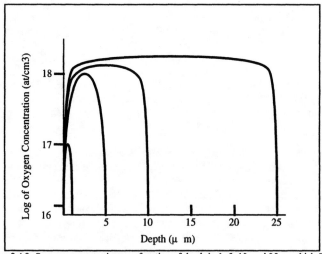

Figure 2.4.9: Oxygen concentration as a function of depth in 1, 5, 10, and 25 μm-thick ZMR silicon films.

The segregation coefficient for oxygen at the solidification interface being close to 1, a similar concentration is found in the silicon film right after solidification. During cooling down after ZMR processing, however, the oxygen dissolved in the silicon film segregates towards the upper and lower Si-SiO$_2$ interfaces, which decreases the oxygen concentration in the silicon film. The effectiveness of this segregation is limited by the diffusion coefficient of oxygen in silicon to a fraction of a micron. As a result, the final peak oxygen concentration in thin ZMR silicon film is $\cong 10^{17}$ atoms/cm^3, while it is close to 2.5×10^{18} atoms/cm^3 in thicker films (Figure 2.4.9).[120]

The concentration of other impurities (carbon, transition metals, and shallow level impurities) is generally quite low in ZMR films, with typical values below 10^{16} cm^{-3}, as indicated by SIMS (Secondary Ion Mass Spectroscopy), AES (Auger Electron Spectroscopy), HIBS (Heavy Ion Back Scattering) and spreading resistance measurements.[121,122]

Finally, one can mention that *thick* silicon films (10-100 µm) can be recrystallized over an insulator using the LEGO technique (Lateral Epitaxial Growth over Oxide).[123] This technique makes use of a bank of stationary halogen lamps which heat the front side of a wafer. A thick polysilicon film is deposited on an oxidized silicon wafer where windows have been opened in the oxide for seeding purposes. As the temperature of the wafer is raised, the top polysilicon melts across the whole wafer. The temperature is then carefully ramped down, and crystal growth proceeds epitaxially from the seeding windows over the oxide until the entire silicon film is recrystallized. Recrystallization of thick silicon films can also be carried out on non-planar substrates using a scanning lamp apparatus.[124] This last method has been successfully applied to the realization of dielectrically isolated high-voltage MOSFETs.[125]

2.5. Homoepitaxial techniques

Silicon-on-insulator can be produced by homoepitaxial growth of silicon on silicon, provided that the crystal growth can extend laterally on an insulator (SiO$_2$, typically). This can be achieved either using a classical epitaxy reactor or by lateral solid-phase crystallization of a deposited amorphous silicon layer.

2.5.1. Epitaxial lateral overgrowth

The Epitaxial Lateral Overgrowth technique (ELO) consists in the epitaxial growth of silicon from seeding windows over SiO$_2$ islands or devices capped with an insulator. It can be performed in an atmospheric or in a reduced-pressure epitaxial reactor [126-116]. The principle of ELO is illustrated in Figure 2.5.1. Typical sample preparation for ELO involves patterning windows in an oxide layer grown on a (100) silicon wafer. The edges of the windows are oriented along the <010> direction. After cleaning, the wafer is loaded into an epitaxial reactor and submitted to a high-temperature hydrogen bake to remove the native oxide from the seeding windows. Epitaxial growth is performed using *e.g.* an SiH$_2$Cl$_2$ + H$_2$ + HCl gas mixture. Unfortunately, nucleation of small silicon crystals with random orientation occurs on the oxide. These crystallites can be removed by an *in-situ* HCl etch step. Once the small nuclei are removed, a new epitaxial growth step is performed, followed by an etch step, and so on, until the oxide is covered by epitaxial silicon.

Epitaxy proceeds from the seeding windows both vertically and laterally, and the silicon crystal is limited by <100> and <101> facets (Figure 2.5.1.A). When two growth fronts, seeded from opposite sides of the oxide, join together, a continuous silicon-on-insulator film is formed, which contains a low-angle subgrain boundary where the two growth fronts meet. Because of the presence of <101> facets on the growing crystals, a groove is observed over the center of the SOI area (Figure 2.5.1.B).

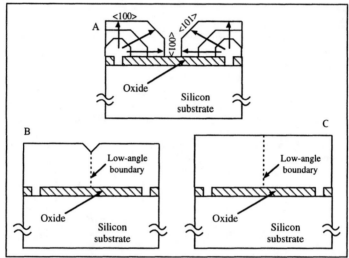

Figure 2.5.1: Epitaxial Lateral Overgrowth (ELO): growth from seeding windows (A), coalescence of adjacent crystals (B), self-planarization of the surface (C).

This groove, however, eventually disappears if additional epitaxial growth is performed (Figure 2.5.1.C).[127] Three-dimensional stacked CMOS inverters have been realized by lateral overgrowth of silicon over MOS devices.[128] The major disadvantage of the ELO technique is the nearly 1:1 lateral-to-vertical growth ratio, which means that a 10 μm-thick film must be grown to cover 20 μm-wide oxide patterns (10 μm from each side). Furthermore, 10 additional micrometers must be grown in order to get a planar surface. Thinner SOI films can, however, be obtained by polishing the wafers after the growth of a thick ELO film.[129] The ELO technique has been used to fabricate three-dimensional and dual-gate devices.[130,131]

A variation of the ELO technique, called "tunnel epitaxy", "confined lateral selective epitaxy" (CLSEG) or "pattern-constrained epitaxy" (PACE), has been reported by several groups.[132,133,134,135] In this technique, a "tunnel" of SiO_2 is created, which forces the epitaxial silicon to propagate laterally (Figure 2.5.2). With this method, a 7:1 lateral-to-vertical growth ratio has been obtained.

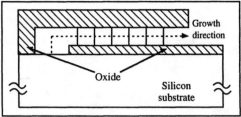

Figure 2.5.2: Principle of tunnel epitaxy.

2.5.2. Lateral solid-phase epitaxy

Lateral Solid-Phase Epitaxy (LSPE) is based on the lateral epitaxial growth of crystalline silicon through the controlled crystallization of amorphous silicon (α-Si).[136,137,138,139,140,141] A seed is needed to provide the crystalline information necessary for the growth. The thin amorphous silicon film can either be deposited or obtained by amorphizing a polysilicon film by means of a silicon ion implantation step. LSPE is performed at relatively low temperature (575-600°C) in order to obtain regrowth while minimizing random nucleation in the amorphous silicon film.

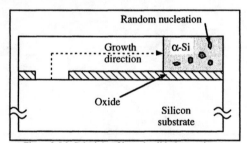

Figure 2.5.3: Principle of lateral solid-phase epitaxy.

The lateral epitaxy rate is in the order of 0.1 nm/s in undoped α-Si and 0.7 nm/s in heavily ($3\times10^{20}cm^{-3}$) phosphorous-doped α-Si. The distance over which lateral epitaxy can be performed over an oxide layer is in the order of 8 μm for undoped material, and 40 μm if a heavy phosphorus doping is used, further lateral extension of LSPE being limited by random nucleation in the α-Si film (Figure 2.5.3). An increase of the lateral extension of LSPE in undoped silicon can be obtained by alternating stripes of P-doped and undoped material, the stripes being perpendicular to the growth front (this can be performed through a mask step and ion implantation).[142] This way, the growth in the doped material "stimulates" the growth front in the undoped material. Research activity on LSPE is mainly carried out in Japan, where it is believed to be a technique of choice for the fabrication of 3D IC's using low-temperature processing.

2.6. FIPOS

The process of Full Isolation by Porous Oxidized Silicon (FIPOS) was invented in 1981.[143,144] It relies on the conversion of a layer of silicon into porous silicon and on the subsequent oxidation of this porous layer. The oxidation rate of porous silicon being orders of magnitude higher than that of monolithic silicon, a full porous silicon buried layer can be oxidized while barely growing a thin oxide on silicon islands on top of it.

The original FIPOS process is described in Figure 2.6.1. It relies on the fact that p-type silicon can readily be converted into porous silicon by electrochemical dissolution of p-type silicon in HF (the sample is immersed into an HF solution and a potential drop is applied between the sample and a platinum electrode dipped into the electrolyte). The conversion rate of n-type silicon is much lower. Porous silicon formation proceeds as follows: at first, an Si_3N_4 film is patterned over a p-type silicon wafer, and boron is implanted to control the density of the porous silicon surface layer. The N⁻ material is formed by conversion of the P⁻ silicon into N⁻ silicon by proton (H^+ ion) implantation. The p-type silicon is then converted into porous silicon by anodization in a hydrogen fluoride solution. Optimal conversion yields a 56% porosity (porosity is controlled by the HF concentration in the electrolyte, the applied potential, and the current density at the sample surface during anodization). In this way, the volume of the buried oxide formed by oxidation of the porous layer is equal to that of the porous silicon, and stress in the films can be minimized.

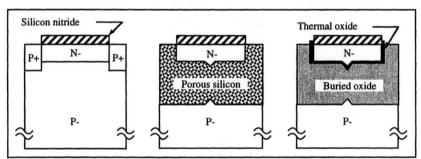

Figure 2.6.1: The original FIPOS process. From left to right: formation of N⁻ islands and P^+ current paths, formation of porous silicon, and oxidation of the porous silicon.

Porous silicon contains an intricate network of pores. As a result, the area of silicon exposed to the ambient is extremely high, and porous silicon oxidizes very rapidly. This allows one to grow a thick buried oxide (oxidation depth is comparable to the width of the N⁻ silicon islands) while growing only a thin oxide at the edges of the N⁻ silicon islands in which the devices will be made. This last oxide being a thermally grown oxide, it provides the silicon islands with a high-quality bottom interface.

The original FIPOS technique produces high-quality SOI islands, and has been used to produce devices with good electrical characteristics [145,146] but it has several limitations. Indeed, the formation of a thick oxidized porous silicon layer is needed to isolate even small islands. Such a thick oxide induces wafer warpage, especially if the islands are unevenly distributed across the wafer. A second limitation is the formation of a little cusp of unanodized silicon at the bottom-center of the silicon islands, where the two (left and right) anodization fronts meet during porous silicon formation (Figure 2.6.1). This limits the use of such a technique for thin-film SOI applications, where having a good thickness uniformity of the silicon islands is of crucial importance. Several variations of the process have been proposed, such as the formation of a buried P+ layer (over a P- substrate) located below N- silicon islands, the whole structure being produced *e.g.* by epitaxy.[147] This technique solves the problem of having to produce a very thick buried layer to isolate wide islands, since it permits to obtain an island width-to-porous silicon layer thickness larger than 50.

A different approach for the fabrication of FIPOS structures is based on the preferential anodization of the N+ layer of an N-/N+/N- structure (Figure 2.6.2).[148,149,150] With this method, the thickness of both the silicon islands and the porous silicon layer are uniform and easily controlled by the N+ doping profile (*e.g.* obtained by antimony implant on a lightly-doped, n-type wafer, and subsequent epitaxy of an N- superficial layer). Another advantage of the N-/N+/N- approach is the automatic endpoint on the island isolation. As soon as all of the easily anodized N+ layer is converted into porous silicon, the anodization current drops, and anodization stops due to a change in anodization potential threshold between the N+ layer and the N- silicon in the substrate and the islands. Such an automatic end of reaction control is not available in the N-/P+/P- approach where anodization of the P- substrate occurs as soon as the P+-layer has been converted into porous silicon, unless the anodization current is reduced to zero. After oxidation of the porous silicon layer [151], a dense buried oxide layer is obtained.

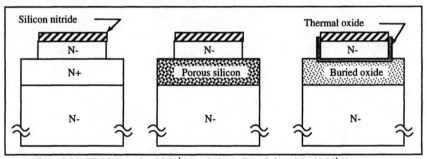

Figure 2.6.2: FIPOS formation (N-/N+/N- technique). From left to right: N-/N+/N- structure, formation of porous silicon, and oxidation of the porous silicon.

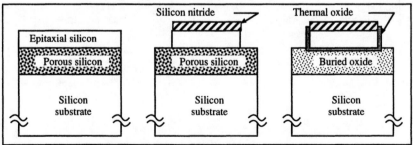

Figure 2.6.3: FIPOS formation by blanket anodization and epitaxy, trench etching and porous silicon oxidation (left to right).

It is worth noticing that porous silicon is a single crystal material, in spite of the fact that it contains many voids. A blanket porous silicon layer can, therefore , be created on a wafer, and "regular" single-crystal silicon can be grown onto it using MBE (molecular-beam epitaxy) or low-temperature PECVD (plasma-enhanced chemical vapor deposition) techniques.[152] Trenches are then etched through the dense silicon overlayer in order to reach the buried porous silicon layer, the oxidation of the porous silicon can then be performed (Figure 2.6.3). This method has the advantage of easy blanket anodization, but it has the drawback of relying on delicate epitaxial procedures (MBE or PECVD). Large-scale integrated circuits (16k and 64k SRAMs) [153,154] as well as thin-film SOI devices [155] have been demonstrated using FIPOS technology. The technique, however, suffers from the fact that the anodization and the buried oxide formation are in most cases carried out after a mask step, which means that "FIPOS substrates" cannot be commercialized by a vendor.

2.7. SIMOX

The acronym SIMOX stands for "Separation by IMplanted OXygen". The principle of SIMOX material formation is very simple (Figure 2.7.1), and consists in the formation of a buried layer of SiO_2 by implantation of oxygen ions beneath the surface of a silicon wafer. The Buried OXide layer is often referred to as "BOX".

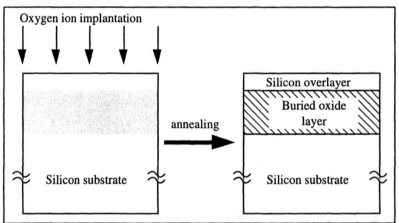

Figure 2.7.1: The principle of SIMOX: a heavy-dose oxygen implantation into silicon followed by an annealing step produces a buried layer of silicon dioxide below a thin, single-crystal silicon overlayer.

Processing conditions must be such that a single-crystal overlayer of silicon is maintained above the oxide. It is worth noticing that in conventional microelectronics, ion implantation is used to introduce atoms into silicon at the impurity level. Boron, for instance, is introduced in ppb (part per billion) quantities for threshold voltage adjust purposes, and the heaviest implants (S&D implants) introduce impurity concentrations of the order of one percent (for 99% of silicon atoms). In the case of SIMOX, ion implantation is used to synthesize a new material, namely SiO_2. This means that 2 atoms of oxygen have to be implanted for every silicon atom over a depth over which silicon dioxide has to be formed. In other words, the implanted dose required to form a BOX layer has to be $\cong$200-500 times higher than the heaviest doses commonly used for microelectronics processing. Indeed, O^+ doses of 1.8×10^{18} cm^{-2} are commonly used to produce "standard" SIMOX material, while doses of $\cong10^{16}$ cm^{-2} are the usual upper boundary for doping impurity implantations.

2.7.1. A brief history of SIMOX

Some of the important milestones indicating the progress achieved in SIMOX technology are summarized in Table 2.7.1.

The first SiO_2 layer synthesized by ion implantation of oxygen into silicon was reported in 1963, and in 1977, the first buried oxide layer was produced by the same technique. A year later, the acronym of SIMOX was proposed for this technique, and 19-stage ring oscillators were fabricated in that material. At that time, most of the SOI research worldwide was concentrating on recrystallization techniques, and the newcoming SIMOX technique was considered quite exotic; researchers were skeptical about its possible use in microelectronics. Indeed, early SIMOX material was of poor quality. The defect density (dislocations, oxide precipitates and polycrystalline silicon inclusions) was so high that epitaxial silicon had to be grown on top of the silicon overlayer in order to obtain a silicon layer good enough for device fabrication. In addition, the implantation of a mere 2x2 cm^2 area on a silicon wafer could take on the order of 24 hours, using a conventional ion implanter. Nonetheless, a 1k SRAM was fabricated by NTT in 1982, and the SOI community started to look at this material with less skeptical eyes.

In 1985, the first NV-200 high-current oxygen implanter was produced by Eaton according to NTT's specifications. This machine was designed to deliver an O^+ ion beam of up to 100 mA at energies of up to 200 keV. The availability of such a machine was the key to the successful development of SIMOX technology.

Before 1985, SIMOX material was usually annealed at 1150°C (which is the highest temperature available in conventional furnaces equipped with quartz tubes). In 1985, it was demonstrated that annealing at 1300°C in a furnace equipped with either a polysilicon or a silicon carbide tube, and even annealing at a few degrees below the melting point of silicon (1405°C) in a lamp annealing system could dramatically improve the quality of the material and produce atomically sharp silicon-oxide interfaces.

In 1987, it was discovered that the use of multiple implantation/annealing steps could be used in place of a single, high-dose implantation could dramatically decrease the dislocation density in the silicon overlayer (less than 10^3 dislocations/cm^2). Finally, it was discovered in 1991 that one could produce SIMOX using an oxygen dose as low as 4×10^{17} cm^{-2}. Improvements of the NV-200 implanter and the advent of new implanters designed for

SIMOX production yielded steadily better material, with less metal and carbon contamination such that, for instance, 16 Mb DRAMs have been fabricated in SIMOX.

More recent developments include low-dose implantation for production of thin buried oxides and silicon overlayers with a low defect density. The research for producing SIMOX material using low-dose implantation is also driven by economical reasons, since the production costs of SIMOX material are proportional to the dose and the energy used for the implantation.

Year	Name/Company	Milestone	Ref.
1963	Watanabe & Tooi	Synthesis of SiO2 by oxygen implantation	156
1978	Izumi et al. (NTT)	SIMOX acronym, First SIMOX circuits	157
1982	NTT	1kb SRAM	158
1985	EATON - NTT	NV-200 high current oxygen implanter	
1985	LETI	High-temperature annealing (1300°C)	159
1986	AT&T	High-temperature annealing (1405°C)	160
1987	IBIS	SIMOX wafers commercial (3" to 6")	
1987	British Tel./Surrey	Masked implantation	161
1987	Hewlett-Packard	Thin-film (fully depleted) CMOS circuits	162
1987	Monsanto	Multiple implantation	163
1988	Hewlett-Packard	2 GHz thin-film CMOS circuits	164
1988	Harris	High-temperature CMOS 4kb SRAM	165
1989	SPIRE	Low-energy, low-dose implants	166
1989	TI / Harris	64kb SRAM	167
1989	LETI	Thin-film 16k SRAM	168
1989	AT&T	6.2 GHz thin-film CMOS circuits	169
1989	NTT	21 ps CMOS ring oscillator	170
1990	TI	Commercial 64 SRAM	
1991	TI	256kb SRAM	171
1991	IBM	256kb fully-depleted SRAM	172
1991	Ibis	Ibis 1000 high-current oxygen implanter	
1991	Westinghouse	14 GHz f_T microwave SOI MOSFETs	173
1992	NTT	Low-dose SIMOX	174
1192	UCL	High-temperature (300°C) SOI op amps	175
1993	Westinghouse	23 GHz f_T microwave SOI MOSFETs	176
1993	TI	1Mb SRAM, 20 ns access time @ 5V	177
1993	IBM	512 kb SRAM, 3.5 ns access time @ 1V	178
1993	Mitsubishi	1 M gate array	179
1995	Samsung	16 Mb DRAM	180
1996	Mitsubishi	0.9 volt,16 Mb DRAM	181

Table 2.7.1: Some milestones of SIMOX technology.

2.7.2. Oxygen implantation

The shape and quality of SIMOX layers obtained by oxygen implantation depend on both the oxygen dose and the temperature of the wafer during the implantation process.

2.7.2.1. Oxygen dose

When a heavy dose of oxygen ions is implanted into silicon, several effects occur. These effects depend on the implanted dose. Stoichiometric SiO_2 contains 4.4×10^{22} oxygen atoms/cm³. Therefore, the implantation of 4.4×10^{17} atoms/cm² should be sufficient to produce a 100 nm-thick buried oxide layer. Unfortunately, due to the statistical nature of ion implantation, the oxygen profile in silicon does not have a box shape, but rather a skewed Gaussian profile, and the implanted atoms spread over more than 100 nm, such that SiO_2 stoichiometry is not reached (Figure 2.7.2). If the wafer is annealed after such a "low-dose" implant, oxide precipitates will be formed at a depth equal to the depth of maximum oxygen concentration, but no continuous layer of SiO_2 will be formed. Experiments show that a dose of 1.4×10^{18} cm² must be implanted in order to create a continuous buried oxide layer. The standard dose which is most common is 1.8×10^{18} cm², and produces a 400 nm-thick buried oxide layer upon annealing. Figure 2.7.2 describes the evolution of the profile of oxygen atoms implanted into silicon with an energy of 200 keV. At low doses, a skewed Gaussian profile is obtained. When the dose reaches $1.2...1.4 \times 10^{18}$ cm², stoichiometric SiO_2 is formed (66 at.% of oxygen for 33 at.% of silicon), and further implantation does not increase the peak oxygen concentration, but rather broadens the overall profile (_i.e._ the buried oxide layer becomes thicker). This is possible because the diffusivity of oxygen in SiO_2 (10^{17} cm².s⁻¹ at 500°C [182]) is high enough for the oxygen to readily diffuse to the Si-SiO_2 interface where oxidation occurs. The dose at which the buried oxide starts to form ($\cong 1.4 \times 10^{18}$ cm²) is called "critical dose".

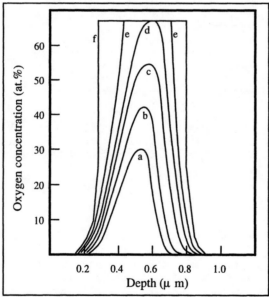

Figure 2.7.2: Evolution of the oxygen concentration profile with the implanted dose: a) 4×10^{17} cm², b) 6×10^{17} cm², c) 10^{18} cm², d) 1.2×10^{18} cm², e) 1.8×10^{18} cm², and f) 2.4×10^{18} cm². Energy is 200 keV.[183]

Doses below or above that level are called "subcritical" or "supercritical", respectively. It is important to notice that a very large amount of oxygen is introduced into the silicon. Since the volume of synthesized SiO_2 is larger than the volume of silicon consumed to form the buried SiO_2 layer, swelling of the wafer is observed. In addition, significant sputtering of the top silicon layer is observed when the implanted dose exceeds 10^{18} cm^{-2} (Figure 2.7.3). Recently it has been shown that thin BOX layers could be synthesized using subcritical doses. This material, called "low-dose SIMOX", is described in Section 2.7.6.

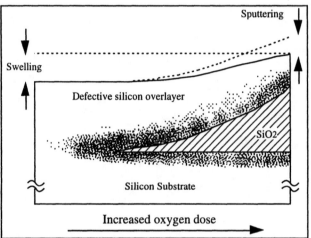

Figure 2.7.3: Evolution of the buried oxide layer formation as a function of implanted dose. From left to right: creation of a damaged layer into the silicon, amorphization of the silicon, creation of a buried oxide layer. Swelling and sputtering effects are also represented.[184]

The sputtering rate ranges between 0.1 to 0.2 sputtered Si atom per implanted oxygen atom. Sputtering can be minimized by growing an SiO_2 layer on the silicon wafer prior to the implantation step. Swelling slightly increases the thickness of the wafer, while sputtering thins it down.

2.7.2.2. Implant temperature

The temperature at which the implantation is performed is also an important parameter which influences the quality of the silicon overlayer. Indeed, the oxygen implantation step does amorphize the silicon which is located above the projected range. If the temperature of the silicon wafer during implantation is too low, the silicon overlayer gets completely amorphized, and it forms polycrystalline silicon upon further annealing, an undesirable effect. When the implantation is carried out at higher temperatures (above 500°C)-(Figure 2.7.4), the amorphization damage anneals out during the implantation process ("self annealing"), and the single-crystallinity of the top silicon layer is maintained. The silicon overlayer, however, is highly defective.

At higher implant temperatures (700-800°C), SiO_2 precipitates form in the silicon overlayer, mostly near its bottom interface, although large precipitates have been observed near the surface region of the silicon film. This effect seems to set an upper limit on the implantation temperature of approximately 700°C. Implantation temperatures most commonly used range between 600°C and 650°C. Wafer heating during implantation is automatically achieved by the energy deposited into the wafer if a high-current implanter is used (beam current of 30-50 mA). If a conventional, low-current implanter is employed, wafer heating can be provided by a bank of halogen lamps or a heated chuck.

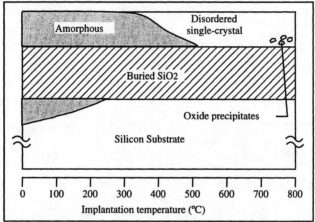

Figure 2.7.4: Evolution of the crystallinity of the SIMOX structure as a function of implantation temperature (dose = 1.8×10^{18} cm^{-2}, energy = 200 keV).[185]

2.7.3. Annealing parameters

After the oxygen ion implantation has been performed, a thermal annealing step is necessary to form a device-worthy SIMOX structure. Generally speaking, the higher the annealing temperature, the better the quality of the material. Until 1985, temperatures of 1150°C were used. The resulting material was of rather poor quality (it had $\cong 10^9$ dislocations per cm^2). Later, furnace annealing at temperatures of 1250°C and 1300°C was achieved, which greatly improved the quality of the silicon overlayer.[186] Annealings were even carried out in lamp annealing systems at temperatures up to 1405°C.[187] Six-hour annealings at 1300-1350°C in furnaces equipped with either polysilicon or silicon carbide tubes are now considered standard. Annealing is carried out in a nitrogen or argon ambient with 2% oxygen added. The oxygen allows for the growth of some oxide on the superficial silicon layer which protects the silicon from the pitting phenomenon which occurs when silicon is annealed at high temperature in pure nitrogen. A pure nitrogen ambient can be used as well, provided that a protecting CVD oxide layer has been deposited on the wafer prior to annealing. In addition, annealing can be performed in an argon gas ambient. Better material qualities seem to be obtained when the annealing is carried out in argon (rather than in nitrogen). The kinetics of the annealing of the SIMOX structure is quite complicated and has been investigated by several research groups. Therefore, we shall describe only the results which are of immediate practical interest.

The evolution of the SIMOX structure after different annealing steps is presented in Figure 2.7.5. We will take the example of the implantation of 1.5×10^{18} oxygen ions/cm^2 at 200 keV into silicon.[188] Three regions can be distinguished in the as-implanted sample: 1) a 420 nm-thick, highly disordered but nonetheless single-crystal top silicon layer containing SiO$_2$ precipitates, the size of which increases from the top surface to the Si-SiO$_2$ interface, 2) a 180 nm-thick amorphous BOX, and 3) a heavily damaged silicon layer extending ...450 nm... into the substrate (Figure 2.7.5.A).

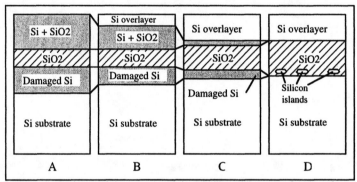

Figure 2.7.5: Evolution of the structure of the SIMOX structure as a function of post-implantation annealing temperature (implant dose=1.5 10^{18} cm^{-2}, energy=200 keV). A: As-implanted; B: 2-hour annealing at 1150°C; C: 6-hour annealing at 1185°C; and; D: 6-hour annealing at 1300°C.

After annealing at 1150°C, the first 80 nm of the top of the silicon overlayer are denuded from SiO$_2$ precipitates (Figure 2.7.5.B). The lower part of the silicon layer is highly defective and contains a large density of oxide precipitates where dislocations are pinned. The mean diameter of these oxide precipitates is approximately 25 nm. The BOX - silicon substrate interface region displays a lamellar structure (layered mixtures of Si and SiO$_2$). After annealing at 1185°C, the top 200 nm of the silicon layer are free from oxide precipitates (Figure 2.7.5.C), but large SiO$_2$ precipitates are found closer to the Si layer-BOX interface. These precipitates are fewer but larger than in the case of the 1150°C annealing. The lower SiO$_2$-Si substrate interface still shows a lamellar structure. Upon annealing at 1300°C, the entire silicon overlayer is denuded from oxide precipitates, and the top Si/BOX interface is atomically sharp (Figure 2.7.5.D). Small silicon islands (remnants of the lamellar structure) are found within the BOX at a distance of 25 nm from the bottom interface. These islands have the same crystal orientation as the substrate, and they are 30 nm thick and 30 to 200 nm long.

The evolution of the SIMOX structure can be explained by thermodynamical considerations. During annealing, both dissolution of the oxide precipitates and precipitation of the dissolved oxide take place in the oxygen-rich silicon layers. In order to minimize the total surface energy of the SiO$_2$ precipitates, thus creating a more stable system, small precipitates dissolve into silicon, and large precipitates grow from the dissolved oxygen. At any given temperature (and for a given concentration of oxygen in the silicon), there exists a critical precipitate radius, r_c, below which a precipitate will disappear, and above which it will be stable. It can be shown that $r_c = - \dfrac{2\sigma}{\Delta H_v} \dfrac{T_E}{T_E - T}$, where T is the temperature, T_E is the temperature of equilibrium between the two phases (solid precipitate or dissolved oxide), ΔH_v is the volume enthalpy of formation of the SiO_2 phase, and σ is the surface energy of the precipitates. The critical radius increases with temperature and becomes essentially infinite at very high temperatures (*i.e.* when $T \rightarrow T_E$), such that the only stable precipitate is the BOX itself, which has an infinite radius of curvature. This phenomenon of growth of the large precipitates at the expense of smaller ones is known as the "Ostwald ripening". At relatively low annealing temperatures, the dissolution of small precipitates near the surface will contribute to the formation of a denuded zone at the top of the silicon overlayer, while larger precipitates will remain stable deeper into the silicon film (Figure 2.7.5.B). The complete dissolution of all precipitates upon high temperature annealing ($\geq 1300°C$) and the diffusion of the dissolved oxygen towards the buried layer explain the thickness increase of the BOX when high-temperature annealing is performed.

The use of a nitrogen ambient during annealing can induce the formation of silicon oxynitride around the oxide precipitates and inhibit their dissolution into the silicon matrix. Therefore, the use of an inert gas, such as argon, is preferred to nitrogen for the high-temperature annealing step used in the SIMOX formation process.[189,190]

2.7.4. Multiple implants

The SIMOX implantation process introduces damage and stress into the silicon overlayer, creating crystalline defects (dislocations). Unlike oxide precipitates, dislocations are not eliminated by high-temperature annealing. It can be shown that the creation of some of these micro-defects can be related to the introduction of impurities such as carbon into the silicon film during implantation [191], and with the improvements of the cleanliness and the design of oxygen implanters a steady reduction of the dislocation densities has been reported.[192] In 1987, it was observed that the creation of defects were drastically minimized when the implanted dose remained below a threshold value of approximately 4×10^{17} ions cm^{-2} at 150 keV. This threshold is unfortunately subcritical, such that no continuous buried oxide layer can be formed upon annealing. Oxide precipitates are formed, however, in the vicinity of the projected range of the oxygen ions (*i.e.* at a depth corresponding to the peaks of the Gaussian profiles of Figure 2.7.2). If the implant and annealing processes are repeated a second and a third time (such that a total dose of 1.2×10^{18} atoms/cm^2 is implanted), high-quality SIMOX material can be produced, without generating significant amounts of dislocations in the silicon overlayer; dislocation densities as low as 10^3 cm^{-2} have been demonstrated. In addition to the low defect density, atomically sharp $Si-SiO_2$ interfaces are produced, and the bottom of the BOX is free of silicon islands.[193]

2.7.5. Low-energy implantation

Recent SIMOX material research is oriented towards the fabrication of thin silicon films on thin buried oxides using low-energy, low-dose implantation. The drive for such a development is three-fold: first, the total-dose hardness of thin buried oxides is expected to be better than that of thicker ones. Secondly, direct fabrication of thin silicon films is attractive for thin-film device applications (instead of thinning the film down after thicker material fabrication). Finally, the production cost of a SIMOX wafer is proportional to both the beam energy and to the implanted dose. It has been shown that the implantation of subcritical doses at energies ranging between 30 to 80 keV, followed by a high-temperature annealing step can produce thin and continuous buried oxide layers as well as thin silicon overlayers with low defect densities.[194] For instance, the implantation of 1.5×10^{17} O^+ ions cm-2 at 30 keV produces a 57 nm-thick silicon layer on top of a 47 nm-thick buried oxide layer. A potential additional benefit from this technique is the reduction of contamination of the wafers, since the introduction of impurities (carbon, heavy metals) is proportional to the implanted oxygen dose.

2.7.6. Low-dose SIMOX

The "low-dose" SIMOX is obtained by implanting O^+ ions at a specific dose located in a very narrow window around 4×10^{17} cm-2. With a single implantation and a 6 hour anneal at 1320°C, a continuous BOX having a thickness of 80-nm is formed. Figure 2.7.6 presents the structure of the buried oxide versus dose around the process window for an implant energy of 120 keV. [195] At a doses of 3×10^{17} cm-2, isolated oxide precipitates are formed. For a dose of 5×10^{17} cm-2, silicon precipitated form in the BOX. Only doses within a very narrow process window around 4×10^{17} cm-2 produce a continuous, precipitate-free BOX.

The choice of the implantation energy is a critical parameter. At 190 keV, which is the standard SIMOX energy, some SiO_2 islands are found in the silicon overlayer. Indeed, when the peak of implant defect generation and the projected range of the oxygen ions are distinct, two precipitation sites can occur and oxide precipitates can form at both the oxygen projected range and the peak of defect generation. A reduction of the implant energy to 120 keV is sufficient to merge the two precipitation sites and obtain a single and continuous BOX (Figure 2.7.7). The use of a low implantation dose does significantly reduce the defect density. Indeed, dislocation densities on the order of 300 cm-2 are found on low-dose SIMOX material. From an economical point of view, the use of low-dose implantation is obviously advantageous, since the throughput of the oxygen implanter is inversely proportional to the implanted dose.

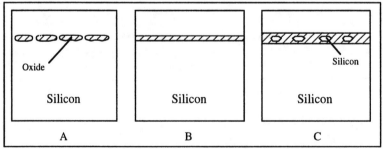

Figure 2.7.6: Evolution of the buried oxide structure for a dose of A: 3×10^{17}; B: 4×10^{17}; C: 5×10^{17} O^+ cm-2 and an energy of 120 keV. [196]

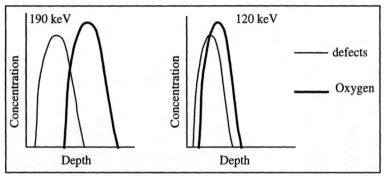

Figure 2.7.7: Defect and oxygen ion concentration produced by 190 and 120 keV implants.

It is possible to increase the thickness of the BOX produced by low-dose oxygen implantation. Indeed, high-temperature (1350°C) oxidation of a low-dose SIMOX wafer causes the increase of the thickness of the buried oxide (Figure 2.7.8). This phenomenon is called high-temperature internal oxidation (ITOX). As long as the thermal oxide grown on the silicon overlayer is thinner than 500 nm, there exists a linear relationship between the thickness of this oxide layer (t_{ox}) and the thickness increase of the buried oxide (Δt_{BOX}): $\Delta t_{BOX} = 0.06\ t_{ox}$. The internal oxide grows at the expense of the bottom of the silicon overlayer. High-temperature internal oxidation has been shown to significantly improve the roughness of interface between the silicon overlayer and the BOX.

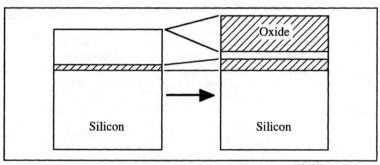

Figure 2.7.8: Principle of internal thermal oxidation (ITOX).[197,198]

2.7.7. Material quality

The silicon film and buried oxide thicknesses are quite uniform in SOI wafers produced by the SIMOX technique. It is interesting to note that the thickness of the layers depends on the dose of oxygen locally introduced in the substrate, and that the thinner the silicon film, the thicker the BOX (Figure 2.7.9).

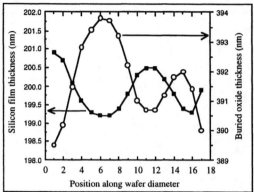

Figure 2.7.9: Silicon overlayer thickness and BOX thickness of a typical "standard" SIMOX wafer, measured at 17 positions along a diameter of the wafer.[199]

2.7.7.1. Silicon overlayer

The evolution of the dislocation density with time in SIMOX is reported in Figure 2.7.10 (adapted from [200]). The improvements are attributed to the better control of the implant conditions (cleanliness, stability of the beam, optimization of scanning techniques, and uniformity of wafer heating), and, more recently, to the introduction of multiple implant techniques and low-dose SIMOX formation.

Early SIMOX material contained high concentrations of light metals such as aluminum and of heavy metals, such as iron, chromium and copper. These where sputtered from the implanter walls or the rotating drum on which the wafers were placed onto the wafers and thermally driven into the silicon by the wafer heating and the subsequent thermal annealing step. Such metallic impurities increase the leakage currents of junctions made in SIMOX material and degrade the radiation hardness of SIMOX MOSFETs. Great efforts have been made to reduce the metal contamination sources, and silicon or silicon dioxide shields have been placed in the implanters such that the beam never "sees" any metallic part. The design of high-temperature furnaces used to anneal the SIMOX material has also been optimized in order to minimize the diffusion of metallic contaminants from the heating elements into the furnace tubes.

One defect is typical of SIMOX silicon layers. It is called "HF defect" since it is revealed when a SIMOX is dipped for a certain time into HF. Normally, HF does not etch silicon, unless some oxidizing agent is present. When HF defects are present, HF etches small holes in the silicon overlayer and reaches the buried oxide, where it etches a cavity which can easily be observed by optical microscopy. HF defects are due to the presence of metallic silicide and silicate compounds, such as $CaSi_2$, $CaSiO_3$, Ca_2SiO_4, $FeSi$, $FeSiO_3$, ... which react with HF and allow for the formation of micro-holes in the silicon overlayer.[201]

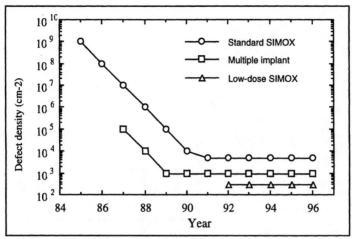

<u>Figure 2.7.10</u>: Evolution of the dislocation densities in SIMOX with time.

2.7.7.2. Buried oxide

The electrical characteristics of the SIMOX buried oxide are inferior to those of thermally-grown SiO_2.[202] The most important difference between the SIMOX buried oxide (BOX) and thermally grown oxide is that the former usually contains excess silicon in some form.[203] In this sense the SIMOX BOX is similar to deposited SiO_2 with deliberately introduced excess silicon. The electrical conduction in the BOX has two components: a "bulk" component, which is area-dependent, and a localized component, which is due to defects, and which is not area dependent.

◊ *Bulk conduction*: The buried oxide of SIMOX structures exhibit a well-defined and reproducible bulk conduction which, in contrast to defect conduction, is area dependent. [204] This bulk conduction is quasi ohmic and time dependent up to an applied voltage of 80 V (for a 400 nm-thick BOX). The temperature dependence of the bulk conduction has an activation energy of 0.3 eV and can be associated with trapping defects. At higher applied voltages the current-voltage characteristics are super-linear and resemble those observed for deposited SiO_2 films that contain excess silicon in the form of Si clusters. The high-field conduction depends very little on temperature, indicating that the conduction is controlled by tunneling between silicon clusters or O_3Si-SiO_3 bonds (E' centers). The average size of the clusters can be estimated to be approximately 0.5 nm (*i.e.* three silicon atoms), and their density to be $2x10^{19}$ cm^{-3}, which means that the concentration of excess silicon is on the order of $6x10^{19}$ cm^{-3}. The bulk conduction depends on polarity. This effect is probably associated with the asymmetry in the distribution of the excess silicon in the BOX. Supplemental oxygen implantation eliminates the trapping-controlled conduction at low field and renders the high-field conduction somewhat similar to that characteristic of thermally grown oxide, but without being identical to it (Figure 2.7.11). Supplemental oxygen implantation and annealing can be used to match the density of the BOX to that of a thermal oxide.[205] This process has been demonstrated to improve the total-dose hardness of SIMOX buried oxides.[206,207]

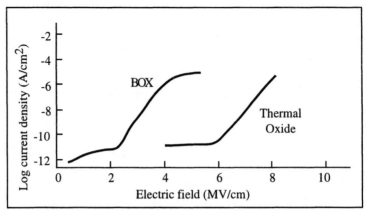

Figure 2.7.11: Oxide "bulk" current density as a function of the electric field.

◊ *Defect conduction*: The BOX layers exhibit localized, defect-induced conduction superimposed on the background (bulk) conduction. Some of this conduction can be associated with the presence of vertical silicon filaments (or: silicon pipes) which run across the buried oxide from the silicon wafer to the silicon overlayer.[208] These silicon pipes find their origin in the micro-masking effect caused by particles on the surface of the wafer during the oxygen implantation step. They present a high conductivity of ohmic nature until the current density becomes high enough, at which point the filament "blows" like a fuse. The conductivity then returns back to the background "bulk" conductivity. Another type of "defect" conductivity can be attributed to silicon filaments which do not extend all the way from the silicon wafer up to the silicon overlayer. These defect give rise to a very low current, until the applied voltage reaches several volts. Then the current suddenly rises, probably due to a field-emission mechanism, and a blow-out mechanism similar to that found in silicon pipes occurs at higher applied voltages (Figure 2.7.12).

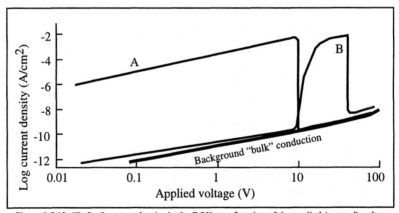

Figure 2.7.12: "Defect" current density in the BOX as a function of the applied (ramped) voltage ($t_{BOX} \cong 400$ nm). A: Current due to silicon pipes; B: Field-emission current.

Some of the characteristics of modern SIMOX material are listed in Table 2.7.2, as well as these which appear to be within reach in a near future.

Parameter	Current	Future	Unit
Wafer diameter	150 to 200	200	mm
Silicon film thickness	100 to 200	50	nm
Silicon film thickness uniformity	< 10	< 5	nm
Buried oxide (BOX) thickness	400	400 or 80	nm
Buried oxide (BOX) thickness uniformity	< 20	< 4	nm
Roughness	0.3	0.2	nm
Dislocation density	$< 10^5$*; $< 300^{\lozenge}$	< 100	cm^{-2}
HF defect density	< 1	< 1	cm^{-2}
BOX pipe density	< 0.2	< 0.2	cm^{-2}
Metallic contamination	$< 10^{11}$	$< 5 \times 10^{10}$	cm^{-2}

Table 2.7.2: Characteristics of modern SIMOX material (*= standard SIMOX; $\lozenge$= low-dose SIMOX).[209]

2.8. SIMNI and SIMON

In analogy to the SIMOX acronym, SIMNI and SIMON mean "Separation by IMplanted NItrogen" and "Separation by IMplanted Oxygen and Nitrogen", respectively.

2.8.1. SIMNI

Just as buried oxide can be synthesized by oxygen ion implantation, a buried silicon nitride layer (Si_3N_4) can be obtained by implanting nitrogen into silicon. The critical dose for forming a buried nitride layer is 1.1×10^{18} cm^{-2} at 200 keV [210], but lower doses can be employed if the implant energy is lower. The obtained buried nitride and silicon overlayer thicknesses are 190 nm and 215 nm, respectively, in the case of a 7.5×10^{17} N^+ ions cm^{-2} implantation at 160 keV followed by a 1200°C anneal.[211] As a result of the lower dose needed to form a buried layer than in the case of SIMOX (standard, single implant SIMOX), lower dislocation densities are observed ($< 10^7$ cm-2). As in the case of SIMOX, the profiles of nitrogen atoms implanted into silicon can be modeled.[212] The most significant difference between oxygen and nitrogen profiles is that the peak of the nitrogen distribution does not saturate once stoichiometry is attained. This is due to the low diffusion coefficient of nitrogen in Si_3N_4 (10^{-28} cm^2 s^{-1} at a temperature of 500°C compared with 10^{-17} cm^2 s^{-1} for O_2 in SiO_2). The result of this low diffusion coefficient is that both nitrogen bonded to silicon and "unbonded" nitrogen are found in the buried layer if supercritical doses are implanted. During implantations, nuclei of crystalline silicon nitride are formed. Upon annealing (*e.g.* at 1200°C), nitride crystallites grow radially outwards from the nucleation sites, in a dendritic manner, by gettering nitrogen from the surrounding silicon, and polycrystalline α-Si_3N_4 is formed. As in the case of SIMOX, low-dose, low-energy implants (60 keV, 7×10^{17} cm^{-2}) can be used to produce thinner buried dielectric layers.[213]

The formation of buried nitride was presenting advantages over early SIMOX material: use of lower implant doses, non-corrosion of hot-filament implanter sources, and lower defect densities in the silicon overlayer. Unfortunately, nitride buried layers are polycrystalline (buried oxide layers are amorphous), and the grain boundaries between the Si_3N_4 crystallites are at the origin of leakage currents between devices made in the silicon overlayer and the underlying silicon substrate. Furthermore, Si-Si_3N_4 interfaces are known to have a higher density of surface states than Si-SiO_2 interfaces. MOS device fabrication has, however, been demonstrated in SIMNI material.[214]

2.8.2. SIMON

The implantation of both oxygen and nitrogen ions into silicon has been achieved by several research groups. These were attempting to combine the advantages of both SIMNI (formation of a buried layer by implantation of a relatively low dose and low defect generation) with those of SIMOX (formation of an amorphous rather than polycrystalline buried layer and good Si-dielectric interfaces).[215] Buried oxynitride layers may also present better radiation hardness performances than pure SIMOX material.

Buried oxynitride layers have been formed by implantation of different doses of both nitrogen and oxygen into silicon. Different implant schemes have been proposed (oxygen can be implanted before nitrogen, or vice-versa). The kinetics of synthesis of oxynitrides by ion implantation is more complicated than that of pure SIMOX or pure SIMNI materials. In some instances, gas (nitrogen) bubbles can be formed within the buried layer. It is, however, possible to synthesize buried oxynitride layers which are stable and remain amorphous after annealing at 1200°C. The resistivity of the buried oxynitride layers can reach up to 10^{15} Ω.cm, which is comparable to that of oxynitride layers formed by other means.

2.9. Wafer bonding and etch-back

The principle of the bonding and etch-back technique is very simple: two oxidized wafers are "mated" -or- "bonded" together. One of the wafers is subsequently polished or etched down to a thickness suitable for SOI applications. The other wafer serves as a mechanical substrate, and is called "handle wafer" (Figure 2.9.1).

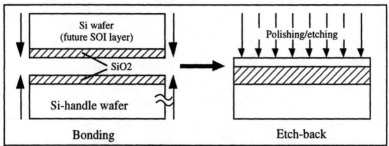

Figure 2.9.1: Bonding of two oxidized silicon wafers (left), and polishing/etching back of one of the wafers.

2.9.1. Mechanisms of bonding

When two flat, hydrophilic surfaces such as oxidized silicon wafers are placed against one another, bonding naturally occurs, even at room temperature. The contacting forces are believed to be caused by the attraction of hydroxyl groups (OH)⁻ and H_2O molecules adsorbed on the two surfaces. This attraction can be significant enough to cause spontaneous formation of hydrogen bonds across the gap between the two wafers.[216] This attraction propagates from a first site of contact across the whole wafer in the form of a "contacting wave" with a speed of several cm/s (Figure 2.9.2).[217,218]

After the wafers have been bonded at room temperature, it is customary to anneal the structure to strengthen the bonding. The bond strength increases with annealing temperature, and three phases of bond strengthening occurring at different temperatures can be distinguished. Firstly, at temperaures above 200°C the adsorbed water separates from the Si-OH groups and forms tetramer water clusters. The second phase occurs at temperatures greater than 700°C when the bonds between hydroxyl groups start to be replaced by Si-O-Si bonds. Finally, at higher temperatures (1100°C and above), the viscous flow of the oxide allows for complete "welding" of the wafers. The quality of bonding depends on the roughness of surfaces in contact. For example, it has been shown that much more uniform bonding is obtained if the roughness of the bonded wafers is less than 0.5 nm than if it is larger than that value.[219]

When the wafers are annealed at 700°C and above, the following reaction takes place : Si-OH + OH-Si → Si-O-Si + H_2O. As a result of water vapor formation during this reaction, voids appear between the bonded wafers. These voids are called "intrinsic voids" and they disappear at higher temperatures, as the water reacts and oxidizes the silicon (as in a thermal wet oxidation process). The hydrogen produced by the latter reaction diffuses easily out of the wafers.

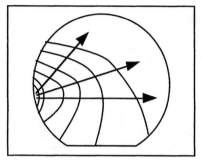

Figure 2.9.2: Propagation of a contacting wave.

Another type of voids, called "extrinsic voids" are caused by the presence of particles between the wafers (Figure 2.9.3). Voids with a diameter of several millimeters are readily created by particles 1 μm or less in size. Because of the usually organic nature of the particulates, the extrinsic voids cannot be eliminated by high-temperature annealing. The only way of obtaining void-free bonding is, therefore, to clean the wafers with ultra-pure chemicals and water and to carry out all bonding operations in an ultra-clean environment. The presence of voids is easily revealed by infrared or ultrasonic imaging, as well as by X-ray tomography. The presence of voids is, of course, undesirable, since they can lead to delamination of devices located in imperfectly bonded areas during device processing.

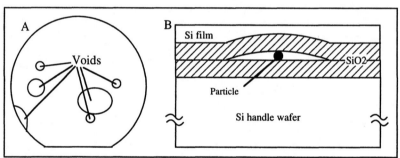

Figure 2.9.3: Extrinsic voids, as seen by infrared, ultrasonic or X-ray imaging of bonded wafers (A), and cross-section of an extrinsic void caused by a particle (B).

The early problems of dust particles between the wafers leading to unbonded areas have now been solved satisfactorily. Either a sufficiently good clean room has to be used or the so-called micro-clean room concept. This concept involves bonding the wafers in a particle-free enclosure. Deionized and filtered water is flushed between the two closely-spaced wafers. The wafers are then dried by spinning under an infrared lamp and finally mated.[220,221]

2.9.2. Etch-back techniques

After bonding of the wafers has been carried out, the top wafer has to be thinned down from a thickness of ...600 μm... to a few micrometers or less in order to be useful for SOI device applications. Two basic thinning approaches can be used: grinding followed by chemico-mechanical polishing, and grinding followed by selective etch-back. The grinding operation is a rather crude but rapid step which is used to remove all but the last several micrometers of the (top) bonded wafer. The thinning method using chemico-mechanical polishing is cheap, but its use is, so far, limited to the fabrication of rather thick SOI films because of the absence of an etch stop. Much more accurate are the techniques using, after initial grinding, a chemical etch-back procedure with etch stop(s). The etch stop is obtained by creating doping concentration gradients at the surface (*i.e.* right next to the oxide layer used for bonding) of the top wafer. For instance, in the double etch-stop technique, a lightly doped wafer is used, and a P^{++} layer is created at its surface by ion implantation. Then, a lowly doped epitaxial layer is grown onto it. This epitaxial layer will be the SOI layer at the end of the process. After grinding, two chemical etch steps are used. The first one, either an ethylenediamine-pyrocatechol-water (EPW) etch [222] or a potassium hydroxide solution, etches the substrate and stops on the P^{++} layer. Then, a 1:3:8 $HF:HNO_3:CH_3COOH$ etch is used to remove the P^{++} layer. The combined selectivity of the etches is better than 10,000:1. The final thickness uniformity of the SOI layer depends on the uniformity of the silicon thickness grown epitaxially, as well as on the uniformity of the P^{++} layer formation, but thickness standard deviations better than 12 nm can be obtained.

More elaborate etch-stop techniques have been recently developed,[223] such as the use of a P^{++} Si-Ge-B etch stop. In this approach, a low-doped (boron, 10^{16} cm^{-3}) epitaxial silicon layer is grown on the epi P^{++} Si-Ge-B layer. The structure is etched back in a double etch-back process. The etching of the Si-Ge-B layer is carried out in a 1:3:12 volume $HF:HNO_3:CH_3COOH$ solution with H_2O_2 added, which provides an etch selectivity of 40. At the end of the process, the SOI layer is oxidized and the oxide is stripped, which lowers the boron concentration in the SOI film to 10^{15} atoms/cm^3.[224]

A special device has been developed to accurately control the thickness uniformity of the top silicon film.[225] It consists of a small-area, computer-controlled polishing tool with a continuous film thickness measurement feedback to the computer. The technique has been perfected with the ÅcuThin™ process [226], where the SOI film is thinned using a scanning small-size plasma head. This technique, called plasma-assisted chemical etching (PACE) is used to planarize the silicon film and can produce 300 nm-thick SOI material with an rms thickness variation better than 2 nm (σ=1...1.6 nm).[227,228,229]

Another technique that combines the formation of a porous silicon layer and wafer bonding to produce BESOI material with good film thickness uniformity, is called "epitaxial layer transfer by bonding and etch back of porous silicon" (ELTRAN®).[230] As presented in Figure 2.6.3, it is possible to grow a high-quality silicon epitaxial film on porous silicon.[231] The porous silicon wafer with the epitaxial film (which is the SOI layer-to-be) can then be oxidized and bonded to another wafer. Polishing is then used to thin the SOI wafer down to the porous silicon layer, and chemical etching is then used to remove the latter. A mixture of

49

HF and H_2O_2 is best used, since it provides an etch selectivity as high as 10^5 between the porous silicon and the epitaxial layer. At the end of the process only the epi layer remains on top of the oxide layer and the handle wafer. The ELTRAN® process has been shown to produce 470 nm-thick SOI layers with a 14 nm thickness uniformity.

The crystal quality of the SOI material obtained by wafer bonding and etch-back is, in principle, as good as that of the starting silicon wafer. The dislocation density, however, is increased in the case where epi is used on a P^{++} buried layer as an etch stop. The dislocation density, however, is still below 10^3 cm^{-2}. The minority carrier lifetime is comparable to bulk silicon ($\cong 20$ μs).

2.9.3. UNIBOND material

The UNIBOND® material is produced by the so-called Smart-Cut® process.[232,233] It combines ion implantation technology and wafer bonding techniques to transfer a thin surface layer from a wafer onto another wafer or an insulating substrate. The basic process steps of this technique are the following:

◊ Ion implantation of hydrogen ions into an oxidized silicon wafer (wafer A). The implanted dose is on the order of 5×10^{16} cm^{-2}. At this stage micro cavities and micro bubbles are formed at a depth equal to the implantation range (R_p).[234] The wafer is preferably capped with thermally grown SiO_2 prior to implantation. This oxide layer will become the buried oxide of the SOI structure at the end of the process. During implantation the oxide becomes less hydrophilic due to some surface carbon contamination. It is, therefore, necessary to carefully clean the wafers before bonding to restore their hydrophilicity.[235]

◊ Hydrophilic bonding of wafer A to a handle wafer (wafer B) is performed. Wafer B can either be bare or oxidized, depending on the final desired buried oxide thickness.

◊ A two-phase heat treatment of the bonded wafers is then carried out. During the first phase, which takes places at a temperature of 500°C, the implanted wafer A splits into two parts: a thin layer of monocrystalline silicon which remains bonded onto wafer B, and the remainder of wafer A which can be recycled for use as another wafer B. The basic mechanism of the wafer splitting upon hydrogen implantation and thermal treatment is similar to the surface flaking and blistering of materials exposed to helium or proton bombardment.[236,237] During the anneal, the average size of the microcavities increases. This size increase takes place along a <100> direction (*i.e.* parallel to the wafer surface) and an interaction between cavities is observed, which eventually results into the propagation of a crack across the whole wafer. This crack is quite parallel to the bonding wafer. The second heat treatment takes place at a higher temperature (1100°C) and is aimed at strengthening the bond between the handle wafer and the SOI film.[238]

◊ Finally, chemo-mechanical polishing is performed on the SOI film to give it the desired mirror-like surface. Indeed, this layer exhibits significant micro-roughness after wafer A splitting, such that a final touch-polish step is necessary. This polishing step reduces the surface roughness to less than 0.15 nm and consumes a few hundred angströms of the SOI film.

The typical silicon film thickness provided by this process is 200 nm. The silicon film roughness after wafer splitting is better than 4 nm. It is better than 2 nm after the final polishing step. The film thickness uniformity is better than 10 nm (max-min, 200-mm wafer), and the dislocation density is lower than 10^2 cm^{-2}. The metal contamination is lower than 5×10^{10} cm^{-2}. Unlike in SIMOX, the buried oxide of the UNIBOND® material is pipe free. The minority carrier lifetime in the UNIBOND® material is on the order of 100 μs, *i.e.* ten times larger than in SIMOX.[239]

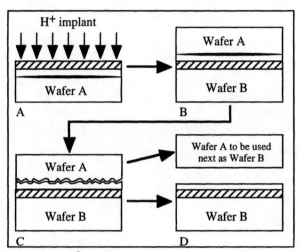

Figure 2.9.4: The Smart-Cut® process. A: Hydrogen implant; B: Wafer bonding; C: Splitting of wafer A; D: Polishing of both wafers. Wafer A is recycled as a future handle wafer.

From an economical point of view, the Smart-Cut® process is significantly better than classical BESOI processes. Indeed, the Smart-Cut® process requires only $N+1$ starting wafers to produce N SOI wafers, while other BESOI processes require $2N$ wafers to produce N SOI wafers. In addition, this process can potentially be used to transfer a thin layer of *any* semiconductor material on top of an insulator. It is worthwhile noting that patterns etched on a silicon wafer can be transferred onto another wafer using the Smart-Cut® process.[240]

2.10. Silicon on diamond

Diamond can be used as the buried insulator in SOI structures. The main motivation for the use of diamond is its high thermal conductivity which can be helpful to solve thermal problems encountered in some SOI power devices.

In the temperature range between 100 and 600K undoped diamond is a highly insulating material. A resistivity above 10^{13} Ω.cm and a breakdown electric field above 10^7 V/cm have been reported.[241] At the same time the heat conductivity of diamond is approximately 10 times higher than that of silicon and 1000 times higher than that of SiO_2.

Silicon-on diamond (SOD) can be fabricated by combining chemical vapor deposition (CVD) of diamond and wafer bonding or epitaxial lateral overgrowth of silicon.[242,243] The diamond layer can also be deposited by the hot-filament method using an ambient of 1-2% of CH_4 in H_2 at a temperature of 650-750°C and under a pressure of 30 to 50 Torr.[244] The efficiency of the good heat evacuation in SOD devices has been demonstrated.[245,246] Compared with bulk silicon, the thermal conductivity of the silicon-on-diamond structure composed of 1 μm of silicon on 300 μm of diamond is increased by 850%.[247]

It is also worth mentioning that aluminum nitride can be used as an alternative to diamond.[248,249]

References

1   I. Golecki, in "Comparison of thin-film transistor and SOI technologies", Ed. by H.W. Lam and M.J. Thompson, Mat. Res. Soc. Symp. Proc., Vol. 33, p. 3, 1984

2   T. Sato, J. Iwamura, H. Tango, and K. Doi, in "Comparison of thin-film transistor and SOI technologies", Ed. by H.W. Lam and M.J. Thompson, Mat. Res. Soc. Symp. Proc., Vol. 33, p. 25, 1984

3   H.M. Manasevit and W.I. Simpson, J. Appl. Phys, Vol.35, p. 1349, 1964

4   RCA part number MWS-5101

5   B.E. Forbes, Hewlett-Packard Journal, p. 2, 1977

6   R.J. Hollingsworth, A.C. Ipri, and C.S. Kim, IEEE J. Solid-State Circuits, Vol. SC-13, p. 664, 1978

7   A.G.F. Dingwall, R.G. Stewart, B.C. Leung, and R.E. Stricker, Proc. IEDM, p. 193, 1978

8   D. Adams, D. Uehara, D. Wheeler, and D. Williams, Proc. of the IEEE SOS/SOI Technology Workshop, p. 58, 1987

9   A.E. Schmitz, R.H. Walden, M. Montes, D. M. Courtney, and E. Stevens, Tech. Digest of Symposium on VLSI Technology, p. 67, 1988

10   D.J. Dumin, S. Dabral, M.H. Freytag, P.J. Robertson, G.P. Carver, and D.B. Novoty, IEEE Trans. on Electron. Dev., Vol. 36, p. 596, 1989

11   R.A. Johnson, C.E. Chang, P.R. de la Houssaye, G.A. Garcia, I. Lagnado, and P.M. Asbeck, Proceedings of the IEEE International SOI Conference, p. 18, 1995

12   H.M. Manasevit, I. Golecki, L.A. Moudi, J.J. Yang, and J.E. Mee, J. Electrochem. Soc., Vol. 130, p. 1752, 1983

13   G.D. Robertson, Jr., P.K. Vasudev, R.G. Wilson, and V.R. Deline, Appl. Surf. Sci., Vol. 14, p. 128, 1982/83

14   T. Sato, J. Iwamura, H. Tango, and K. Doi, in "Comparison of thin-film transistor and SOI technologies", Ed. by H.W. Lam and M.J. Thompson, Mat. Res. Soc. Symp. Proc., Vol. 33, p. 25, 1984

15   I. Golecki, in "Comparison of thin-film transistor and SOI technologies", Ed. by H.W. Lam and M.J. Thompson, Mat. Res. Soc. Symp. Proc., Vol. 33, p. 3, 1984

16   G. Yaron and L.D. Hess, Solid-State Electronics, Vol. 23, IEEE Trans. on Electron Devices, Vol. 27, p. 573, 1980

17   S.S. Lau, S. Matteson, J.W. Mayer, P. Revesz, J. Gyulai, J.Roth, T.W. Sigmond, and T. Cass, Appl. Phys. Lett, Vol. 34, p. 76, 1979

18   J. Amano and K. Carey, Appl. Phys. Lett., Vol. 39, July 1980

19   P.K. Vasudev and D.C. Mayer, in "Comparison of thin-film transistor and SOI technologies", Ed. by H.W. Lam and M.J. Thompson, Mat. Res. Soc. Symp. Proc., Vol. 33, p. 35, 1984

20   M.E. Roulet, P. Schwob, I. Golecki, and M.A. Nicolet, Electronics Letters, Vol. 15, p. 527, 1979

21   P.K. Vasudev and D.C. Mayer, in "Comparison of thin-film transistor and SOI technologies", Ed. by H.W. Lam and M.J. Thompson, Mat. Res. Soc. Symp. Proc., Vol. 33, p. 35, 1984

22   D.J. Dumin, S. Dabral, M.H. Freytag, P.J. Robertson, G.P. Carver, and D.B. Novoty, IEEE Trans. on Electron. Dev., Vol. 36, p. 596, 1989

23   A.C. Ipri, Applied Solid-State Sciences, Supplement 2, Silicon Integrated Circuits, Part A, Ed. by. D. Kahng, Academic Press, p. 253, 1981

24   I. Golecki, H.M. Manasevit, L.A. Moudy, J.J. Yang, and J.E. Mee, Appl. Phys. Letters, Vol. 42, p. 501, 1983

25   I. Golecki, in "Comparison of thin-film transistor and SOI technologies", Ed. by H.W. Lam and M.J. Thompson, Mat. Res. Soc. Symp. Proc., Vol. 33, p. 3, 1984

26   I. Golecki, R.L. Maddox, H.L. Glass, A.L. Lin, and H.M. Manasevit, presented at the 26th electronic Materials Conf., Santa Barbara, CA, June 1984

27   M. Ihara, Y.Arimoto, M. Jifiku, T. Kimura, S. Kodama, H. Yamawaki, and T. Yamaoka, J. Electrochem. Soc., Vol. 129, p. 2569, 1982

28   Y. Hokari, M. Mikami, K. Egami, H. Tsuya, and M. Kanamori, Techn. Digest of IEDM, p. 368, 1983

29   K. Ikeda, H. Yamawaki, T. Kimura, M. Ihara, and M. Ozeki, Ext. Abstr. 5th International Workshop on Future Electron Devices, Miyagi-Zao, Japan, p. 225, 1988

30   Asano and H. Ishiwara, Jpn. J. Appl. Phys., Vol. 21, suppl. 21-1, p. 187, 1982

31   T.R. Harrison, P.M. Mankiewich, and Dayem, Appl. Phys. Lett., Vol. 41, p. 1102, 1982

32   H. Onoda, M. Sasaki, T. Katoh, and N. Hirashita, IEEE Trans. Electron Dev., Vol. ED-34, p. 2280, 1987

33   S. Sugiura, T. Yoshida, and K. Shono, Jpn. J. Appl. Phys, Vol. 22, p. L426, 1983

34   T. Takenaka and K. Shono, Jpn. J. Appl. Phys, Vol. 13, p. 1211, 1974

35   M. Morita, S. Isogai, N. Shimizu, K. Tsubouchi, and N. Mikoshiba, Jpn. J. Appl. Phys, Vol. 20, p. L173, 1981

36   D.M. Jackson, Jr. and R.W. Howard, Trans. Met. Soc. AIME, Vol. 223, p. 468, 1965

37   H.M. Manasevit, D.H. Forbes, and I.B. Cardoff, Trans Met. Soc. AIME, Vol. 236, p. 275, 1966

38   G. Shimaoka and S.C. Chang, J. Vac. Sci. Technol., Vol. 9, p. 235, 1972

39   C.H. Fa and T.T. Jew, IEEE Trans. on Electron Devices, Vol. ED-13, p. 290, 1966

40   T.I.Kamins, Solid-State Electronics, Vol. 15, p. 789, 1972

41   S.W. Depp, B.G. Huth, A. Juliana, and R.W. Koepcke, in "Grain Boundaries in Semiconductors",Ed. by H.J. Leamy, G.E. Pike, and C.H. Seager, Mat. Res. Soc. Symp. Proceedings, Vol. 5, p. 297, 1982

42   T.I. Kamins and P.J. Marcoux, IEEE Electron Device Letters, Vol. EDL-1, p. 159, 1980

43   H. Shichiro, S.D.S. Malhi, P.K. Chatterjee, A.H. Shah, G.P. Pollack, W.H. Richardson, R.R. Shah, M.A. Douglas, and H.W. Lam, in "Comparison of thin-film transistor and SOI technologies", Ed. by H.W. Lam and M.J. Thompson, Mat. Res. Soc. Symp. Proceedings, North-Holland, Vol. 33, p. 193, 1984

44   S.D.S. Malhi, in "Comparison of thin-film transistor and SOI technologies", Ed. by H.W. Lam and M.J. Thompson, Mat. Res. Soc. Symp. Proceedings, North-Holland, Vol. 33, p. 147, 1984

45   G.K Celler, H.J. Leamy, L.E. Trimble, and T.T. Sheng, Appl. Phys Lett., Vol. 39, p. 425, 1981

46   C. Hill, in"Laser and Electron-Beam Solid Interactions and Material Processing", Ed. by J.F. Gibbons, L.D. Hess, and T.W. Sigmon, Mat. Res. Soc. Symp. Proceedings, Vol. 1, p. 361, 1981

47   N.M. Johnson, D.K. Biegelsen, H.C. Tuan, M.D. Moyer, and E. Fennel, IEEE Electron Dev. Lett, Vol. EDL-3, p. 369, 1982

48   E.I. Givargizov, V.A. Loukin, and A.B. Limanov, in "Physical and Technical Problems of SOI Structures and Devices", Kluwer Academc Publishers, NATO ASI Series - High Technology, Vol. 4, Ed. by J.P. Colinge, V.S. Lysenko and A.N. Nazarov, p. 27, 1995

49   G.J. Willems, J.J. Poortmans, and H.E. Maes, J. Appl. Phys., Vol. 62, p. 3408, 1987

50   T.I. Kamins, M.M. Mandurah, and K.C. Saraswat, J. Electrochem. Soc., Vol. 125, p. 927, 1978

51   H.W. Lam, R.F. Pinizotto and A.F. Tasch, J. Electrochem. Soc., vol 128, p. 1981, 1981

52   K.F. Lee, T.J. Stultz, and J.F. Gibbons, Semiconductors and Semimetals, Vol. 17, cw Processing of Silicon and other Semiconductors, Academic Press, p.227, 1984

53   M. Ohkura, K. Kusukawa, I. Yoshida, M. Miyao, and T. Tokuyama, Ext. Abstr. of the 15th Conf. on Solid-State Devices and Materials, Japan Society of Appl. Physics, p. 43, 1983

54   H.P. Le and H.W. Lam, Ext. Abstr. of Electrochem. Soc. Spring Meeting, Vol. 82-1, p. 240, 1982

55   D.J. Wouters and H.E. Maes, J. Appl. Phys, Vol. 66, p. 900, 1989

56   J.P. Colinge, H.K. Hu, and S. Peng, Electronics Letters, Vol. 21, p. 1102, 1985

57   P.Seegebrecht, Ext. Abstracts of 5th Internat. Workshop on Future Electron Devices, Miyagi-Zao, Japan, p. 19, 1988

58   J.M. Hodé, J.P. Joly, and P. Jeuch, Ext. Abstr. of Electrochem. Soc. Spring Meeting, Vol. 82-1, p. 232, 1982

59   S. Kawamura, J. Sakurai, M. Nakano, and M. Tagaki, Appl. Phys. Lett., Vol. 40, p. 232, 1982

60   N.A. Aizaki, Appl. Phys. Lett., Vol. 44, p. 686, 1984

61   J.P. Colinge, E. Demoulin, D. Bensahel, and G. Auvert, Appl. Phys. Lett., Vol. 41, p. 346, 1982

62   J.P. Colinge, Ext. Abstracts of 2nd Internat. Workshop on Future Electron Devices, Shuzenji, Japan, p. 13, 1985

63   T. Morishita, T. Miyajima, J. Kudo, M. Koba, and K. Awane, Ext. Abstracts of 2nd Internat. Workshop on Future Electron Devices, Shuzenji, Japan, p. 35, 1985

64   E. Fujii, K. Senda, F. Emoto, and Y. Hiroshima, Appl. Phys. Lett., Vol. 63, p. 2633, 1988

65   K. Sugahara, S. Kusunoki, Y. Inoue, T. Nishimura, and Y. Akasaka, J. Appl. Phys., Vol. 62, p. 4178, 1987

66   T. Nishimura, Y. Inoue, K. Sugahara, S. Kusunoki, T. Kumamoto, S. Nakagawa, M. Nakaya, Y. Horiba, and Y. Akasaka, Proceedings of International Electron Device Meeting, p. 111, 1987

67   T. Kunio, K. Omaya, Y. Hayashi, and M. Morimoto, Proceedings of International Electron Device Meeting, p. 837, 1989

68   Y. Akasaka, Ext. Abstracts of 8th Internat. Workshop on Future Electron Devices, Kochi, Japan, p. 9, 1990

69   Y. Inoue, T. Ipposhi, T. Wada, K. Ichinose, T. Nishimura, and Y. Akasaka, Proceedings Symp. VLSI Technology, p. 39, 1989

70   S. Toyoyama, K. Kioi, K. Shirakawa, T. Shinozaki, and Ohtake, Ext. Abstracts of the 8th Internat. Workshop on Future Electron Devices - Three-Dimensional ICs and Nanometer Functional Devices, Kochi, Japan, p. 109, 1990

71   D. Bensahel, Ext. Abstracts of 5th Internat. Workshop on Future Electron Devices, Miyagi-Zao, Japan, p. 9, 1988

72   D.K. Biegelsen, N.M. Johnson, D.J. Bartelink, and M.D. Moyer, Appl. Phys. Lett., Vol. 38, p. 150, 1981

73   S. Akiyama, N. Noshii, K. Yamazaki, M. Yoneda, S. Ogawa, and Y. Terui, Ext. Abstracts of 2nd Internat. Workshop on Future Electron Devices, Shuzenji, Japan, p. 41, 1985

74   R. Mukai, N. Sasaki, T. Iwai, S. Kawamura, and M. Nakano, Proceedings of International Electron Device Meeting, p. 360, 1983

75   A.J. Auberton-Hervé, J.P. Joly, P. Jeuch, J. Gautier, and J.M. Hodé, Proceedings of International Electron Device Meeting, p. 808, 1984

76   S. A. Lyon, R. J. Namanich, N.M. Johnson, and D.K. Biegelsen, Appl. Phys. Lett., Vol. 40, p. 316, 1982

77   D.K. Biegelsen, Ext. Abstracts of 2nd Internat. Workshop on Future Electron Devices, Shuzenji, Japan, p. 31, 1985

78   K. Egami and M. Kimura, Ext. Abstracts of 2nd Internat. Workshop on Future Electron Devices, Shuzenji, Japan, p. 105, 1985

79   R.C. McMahon, Microelectronic Engineering, Vol. 8, p. 255, 1988

80   T. Hamasaki, T. Inoue, I. Higashinakagawa, T. Yoshii, and H. Tango, J. Appl. Phys., Vol. 59, p. 2971, 1986

81   K. Shibata, T. Inoue, T. Takigawa, and S. Yoshii, Appl. Phys. Lett., Vol. 39, p. 645, 1981

82   J.A. Knapp and S.T. Picraux, J. Crystal Growth, Vol. 63, p. 445, 1983

83   T. Yoshii, T. Hamasaki, K. Suguro, T. Inoue, M. Yoshimi, K. Taniguchi, M. Kashiwagi, and H. Tango, Ext. Abstracts of 2nd Internat. Workshop on Future Electron Devices, Shuzenji, Japan, p. 51, 1985

84   M. Yoshimi, H. Hazama, M. Takahashi, S. Kambayashi, M. Kemmochi, and H. Tango, Ext. Abstracts of 5th Internat. Workshop on Future Electron Devices, Miyagi-Zao, Japan, p. 143, 1988

85   R.C. McMahon, Microelectronic Engineering, Vol. 8, p. 255, 1988

86   T. Hamasaki, T. Inoue, I. Higashinakagawa, T. Yoshii, and H. Tango, J. Appl. Phys., Vol. 59, p. 2971, 1986

87   J.R. Davis, R.A. McMahon, and H. Ahmed, Electronics Letters, Vol. Vol. 18, p. 163, 1982

88   E.I. Givargizov, Oriented Crystallization on Amorphous Substrates, Plenum Press, New York, 1991

89   J.C.C. Fan, M.W. Geis, and B.Y. Tsaur, Appl. Phys. Lett., Vol. 38, p. 365, 1981

90   M.W. Geis, H.I. Smith, B.Y. Tsaur, J.C.C. Fan, D.J. Silversmith, R.W. Mountain, and R.L. Chapman, in "Laser-Solid Interactions and Transient Thermal Processing of Materials", Narayan, Brown and Lemons Eds., (North-Holland), MRS Symposium Proceedings, Vol. 13, p. 477, 1983

91   A. Kamgar and E. Labate, Mat. Letters, Vol. 1, p. 91, 1982

92   T.J Stultz, Appl. Phys. Lett., Vol. 41, p. 824, 1982, and
      T.J. Stultz, J.C. Sturm, and J.F. Gibbons, in "Laser-Solid Interactions and Transient Thermal Processing of Materials", Narayan, Brown and Lemons Eds., (North-Holland), MRS Symposium Proceedings, Vol. 13, p. 463, 1983

93   D.P. Vu, M. Haond, D. Bensahel, and M. Dupuy, J. Appl. Phys., Vol. 54, p. 437, 1983

94   P.W. Mertens, D.J. Wouters, H.E. Maes, A. De Veirman, and J. Van Landuyt, J. Appl. Phys., Vol. 63, p. 2660, 1988

95   E.I. Givargizov, V.A. Loukin, and A.B. Limanov, in "Physical and Technical Problems of SOI Structures and Devices", Kluwer Academc Publishers, NATO ASI Series - High Technology, Vol. 4, Ed. by J.P. Colinge, V.S. Lysenko and A.N. Nazarov, p. 27, 1995

96   D. Dutartre, in "Silicon-On-Insulator and Buried Metals in Semiconductors", Sturm, Chen, Pfeiffer and Hemment Eds., (North-Holland), MRS Symposium Proceedings, Vol. 107, p. 157, 1988, and
      D. Dutartre, M. Haond, and D. Bensahel, in "Semiconductor-On-Insulator and Thin Film Transistor Technology", Chiang, Geis and Pfeiffer Eds., (North-Holland), MRS Symposium Proceedings, Vol. 53, p. 89, 1986

97   D. Dutartre, Appl. Phys. Lett., Vol. 48, p. 350, 1986

98   D. Dutartre, in "Silicon-On-Insulator and Buried Metals in Semiconductors", Sturm, Chen, Pfeiffer and Hemment Eds., (North-Holland), MRS Symposium Proceedings, Vol. 107, p. 157, 1988, and
      D. Dutartre, M. Haond, and D. Bensahel, in "Semiconductor-On-Insulator and Thin Film Transistor Technology", Chiang, Geis and Pfeiffer Eds., (North-Holland), MRS Symposium Proceedings, Vol. 53, p. 89, 1986

99   H.A. Atwater, Jr., VLSI Memo 83-149, Dept. of Electr. Eng. and Computer Science, MIT, Cambridge, MA, p. 23, 1983

100   H.J. Leamy, C.C. Chang, H. Baumgard, R.A. Lemons, and J. Cheng, Mat. Lett., Vol. 1, p. 33, 1982

101   R.A. Lemons, M.A. Bosch, and D. Herbst, in "Laser-Solid Interactions and Transient Thermal Processing of Materials", Narayan, Brown and Lemons Eds., (North-Holland), MRS Symposium Proceedings, Vol. 13, p. 581, 1983

102   M.W. Geis, H.I. Smith, B.Y. Tsaur, J.C.C. Fan, D.J. Silversmith, and R.W. Mountain, J. Electrochem. Soc., Vol. 129, p. 2812, 1982

103   H.F. Wolf, "Semiconductor Data", Pergamon Press, Oxford, 1969

104   R.A. Lemons, M.A. Bosch, and D. Herbst, in "Laser-Solid Interactions and Transient Thermal Processing of Materials", Narayan, Brown and Lemons Eds., (North-Holland), MRS Symposium Proceedings, Vol. 13, p. 581, 1983

105   P.W. Mertens, D.J. Wouters, H.E. Maes, A. De Veirman, and J. Van Landuyt, J. Appl. Phys., Vol. 63, p. 2660, 1988

106   M.W. Geis, H.I. Smith, B.Y. Tsaur, J.C.C. Fan, D.J. Silversmith, R.W. Mountain, and R.L. Chapman, in "Laser-Solid Interactions and Transient Thermal Processing of Materials", Narayan, Brown and Lemons Eds., (North-Holland), MRS Symposium Proceedings, Vol. 13, p. 477, 1983

107   M.W. Geis, H.I. Smith, D.J. Silversmith, R.W. Mountain, and C.V. Thomson, J. Electrochem. Soc., Vol. 130, p. 1178, 1983

108   L. Pfeiffer, S. Paine, G.H. Gilmer, W. Saarloos, and K.W. West, Phys. Rev. Lett., Vol. 54, p. 1944, 1985

109   M.W. Geis, H.I. Smith, B.Y. Tsaur, J.C.C. Fan, D.J. Silversmith, R.W. Mountain, and R.L. Chapman, in "Laser-Solid Interactions and Transient Thermal Processing of Materials", Narayan, Brown and Lemons Eds., (North-Holland), MRS Symposium Proceedings, Vol. 13, p. 477, 1983

110   L. Pfeiffer, K.W. West, D.C. Joy, J.M. Gibson, and A.E. Gelman, in "Semiconductor-On-Insulator and Thin Film Transistor Technology", Chiang, Geis and Pfeiffer Eds., (North-Holland), MRS Symposium Proceedings, Vol. 53, p. 29, 1986

111   J.S. Im, C.K. Chen, C.V. Thompson, M.W. Geis, and H. Tomita, in "Silicon-On-Insulator and Buried Metals in Semiconductors", Sturm, Chen, Pfeiffer and Hemment Eds., (North-Holland), MRS Symposium Proceedings, Vol. 107, p. 169, 1988

112   M.W. Geis, C.K. Chen, H.T. Smith, P.M. Nitishin, B.Y. Tsaur, and R.W. Mountain, in "Semiconductor-On-Insulator and Thin Film Transistor Technology", Chiang, Geis and Pfeiffer Eds., (North-Holland), MRS Symposium Proceedings, Vol. 53, p. 39, 1986

113   P.M. Zavaracky, D.P. Vu, L. Allen, W. Henderson, H. Guckel, J.J. Sniegowski, T.P. Ford, and J.C.C. Fan, in "Silicon-On-Insulator and Buried Metals in Semiconductors", Sturm, Chen, Pfeiffer and Hemment Eds., (North-Holland), MRS Symposium Proceedings, Vol. 107, p. 213, 1988

114   E.W. Maby, M.W. Geis, Y.L. LeCoz, D.J. Silversmith, R.W. Mountain, and D.A. Antoniadis, Electron Dev. Lett., Vol. 2, p. 241, 1981

115   C.K. Chen, L. Pfeiffer, K.W. West, M.W. Geis, S. Darack, G. Achaibar, R.W. Mountain, and B.Y. Tsaur, in "Semiconductor-On-Insulator and Thin Film Transistor Technology", Chiang, Geis and Pfeiffer Eds., (North-Holland), MRS Symposium Proceedings, Vol. 53, p. 53, 1986

116   M. Haond, D. Dutartre, R. Pantel, A. Straboni, and B. Vuillermoz, in "Semiconductor-On-Insulator and Thin Film Transistor Technology", Chiang, Geis and Pfeiffer Eds., (North-Holland), MRS Symposium Proceedings, Vol. 53, p. 59, 1986

117   E. Yablonovitch and T. Gmitter, J. Electrochem. Soc., Vol. 131, p. 2625, 1984

118   C.K. Chen, L. Pfeiffer, K.W. West, M.W. Geis, S. Darack, G. Achaibar, R.W. Mountain, and B.Y. Tsaur, in "Semiconductor-On-Insulator and Thin Film Transistor Technology", Chiang, Geis and Pfeiffer Eds., (North-Holland), MRS Symposium Proceedings, Vol. 53, p. 53, 1986

119   P.W. Mertens and H.E. Maes, in "Beam-Solid Interactions: Physical Phenomena", Knapp, Børgesen and Zuhr Eds., (North-Holland), Mat. Res. Soc. Symposium Proceedings, Vol. 157, p. 461, 1990

120 P.W. Mertens, J. Leclair, H.E. Maes and W. Vandervorst, J. Appl. Phys., Vol. 67, p. 7337, 1990

121 J. Knapp, L.P. Allen, and P.M. Zavaracky, Proc. of the IEEE SOS/SOI Technology Conference, p. 80, 1989

122 C. Karulkar, R.J. Hillard, and P. Rai-Choudhury, Proc. of the IEEE SOS/SOI Technology Conference, p. 97, 1989

123 G.K. Celler and L.E. Trimble, in "Energy Beam-Solid Interactions and Transient Thermal Processing", Fan and Johnson Eds., (North-Holland), MRS Symposium Proceedings, Vol. 23, p. 567, 1984

124 B. Tillack, K. Hoeppner, H.H. Richter, and R. Banisch, Materials Science and Engineering (Elsevier Sequoia), Vol. B4, p. 237, 1989

125 B. Tillack, R. Banisch, H.H. Richter, K. Hoeppner, O. Joachim, J. Knopke, and U. Retzlaf, Proceedings IEEE SOS/SOI Technology Conference, p. 121, 1990

126 L. Jastrzebski, J.F. Corboy, J.T. McGinn, and R. Pagliaro, Jr., J. Electrochem Soc., Vol. 130, p. 1571, 1983

127 L. Jastrzebski, J.F. Corboy, J.T. McGinn, and R. Pagliaro, Jr., J. Electrochem Soc., Vol. 130, p. 1571, 1983

128 R.P. Zingg, B. Höfflinger, and G.W. Neudeck, IEDM Technical Digest, p. 909, 1989

129 R.P. Zingg, H.G. Graf, W. Appel, P. Vöhringer, and B. Höfflinger, Proc. IEEE SOS/SOI Technology Workshop, p. 52, 1988

130 J.P. Denton and G.W. Neudeck, Proceedings of the IEEE International SOI Conference, p. 135, 1995

131 J.C. Chang, J.P. Denton, and G.W. Neudeck, Proceedings of the IEEE International SOI Conference, p. 88, 1996

132 A. Ogura and Y. Fujimoto, Ext. Abstracts of 8th Internat. Workshop on Future Electron Devices, Kochi, Japan, p. 73, 1990, and
A. Ogura and Y. Fujimoto, Appl. Phys. Lett., Vol. 55, p. 2205, 1989

133 P.J. Schubert and G.W. Neudeck, IEEE Electron Device Letters, Vol. 11, p. 181, 1990

134 H.-S. Wong, K. Chan, Y. Lee, P. Roper, and Y. Taur, Symposium on VLSI Technology, Digest of Technical Papers, p. 94, 1996

135 S. Venkatesan, C. Subramanian, G.W. Neudeck, and J.P. Denton, Proceedings of the IEEE International SOI Conference, p. 76, 1993

136 Y.Kunii, M. Tabe, and K. Kajiyama, J. Appl. Phys., Vol. 54, p. 2847, 1983, and
Y.Kunii, M. Tabe, and K. Kajiyama, J. Appl. Phys., Vol. 56, p. 279, 1984

137 H. Ishiwara, H. Yamamoto, S. Furukawa, M. Tamura, and T. Tokuyama, Appl. Phys. Lett., Vol. 43, p. 1028, 1983

138 J.A Roth, G.L. Olson, and L.D. Hess, in "Energy Beam-Solid Interactions and Transient Thermal Processing", Fan and Johnson Eds., (North-Holland), MRS Symposium Proceedings, Vol. 23, p. 431, 1984

139 H. Ishiwara, H. Yamamoto, and S. Furukawa, Ext. Abstracts of 2nd Internat. Workshop on Future Electron Devices, Shuzenji, Japan, p. 63, 1985

140 Y. Kunii and M. Tabe, Ext. Abstracts of 2nd Internat. Workshop on Future Electron Devices, Shuzenji, Japan, p. 69, 1985

141 T. Dan, H. Ishiwara, and S. Furukawa, Ext. Abstracts of 5th Internat. Workshop on Future Electron Devices, Miyagi-Zao, Japan, p. 189, 1988

142   M. Miyao, M. Moniwa, K. Kusukawa, and S. Furukawa, J. Appl. Phys., Vol. 64, p. 3018, 1988

143   K. Imai, Solid-State Electron., Vol. 24, p.59, 1981

144   S.S. Tsao, IEEE Circuits and Devices Magazine, Vol. 3, p. 3, 1987

145   K. Ansai, F. Otoi, M. Ohnishi, and H. Kitabayashi, Proc. IEDM, p. 796, 1984

146   S. Muramoto, H. Unno, and K. Ehara, Proc. Electrochem. Soc. Meeting, Boston, Vol. 86, p. 124, May 1986

147   L. Nesbit, Tech. Digest of IEDM, p. 800, 1984

148   R.P. Holmstrom and J.Y. Chi, Appl. Phys. Lett., Vol. 42, p. 386, 1983

149   K. Barla, G. Bomchil, R. Herino, and. A. Monroy, IEEE Circuits and Devices Magazine, Vol. 3, p. 11, 1987

150   E.J. Zorinsky, D. B. Spratt, and R. L. Vinkus, Tech. Digest IEDM, p. 431, 1986

151   K. Barla, G. Bomchil, R. Herino, J.C. Pfister, and J. Baruchel, J. Cryst. Growth, Vol. 68, p. 721, 1984, and
K. Barla, R. Herino, and G. Bomchil, J. Appl. Phys, Vol. 59, p. 439, 1986

152   M.I.J. Beale, N.G. Chew, A.G. Cullis, D.B. Garson, R.W. Hardeman, D.J. Robbins, and I.M. Young, J. Vac. Sci. Technol. B, p. 732, 1985

153   H. Takai and T. Itoh, J. Electronic Materials, Vol. 12, p.293, 1983

154   T.Mano, T. Baba, H. Sawada, and K. Imai, Tech. Digest Symposium on VLSI Technology, p. 12, 1982

155   K. Ehara, H. Unno, and S. Muramoto, Electrochem. Soc. Extended Abstracts, Vol. 85-2, p. 457, 1985

156   M.Watanabe and A. Tooi, Japan. J. Appl. Phys., Vol. 5, p. 737, 1966

157   K. Izumi, M. Doken, and H. Ariyoshi, Electronics Letters, Vol. 14, p. 593, 1978

158   K. Izumi, Y. Omura, M. Ishikawa, and E. Sano, Techn. Digest of the Symposium on VLSI Technology, p. 10, 1982

159   J. Stoemenos, C. Jaussaud, M. Bruel, and J. Margail, J. Crystal Growth, Vol. 73, p.546, 1985

160   G.K. Celler, P.L.F. Hemment, K.W. West, and J.M. Gibson, Appl. Phys. Lett., Vol.48, p. 532, 1986

161   J.R. Davis, A. Robinson, K. Reeson, and P.L.F. Hemment, Proc. IEEE SOS/SOI Technology Workshop, p. 71, 1987

162   J.P. Colinge and T.I. Kamins, Proc. IEEE SOS/SOI Technology Workshop, p. 69, 1987

163   D. Hill, P. Fraundorf, and G. Fraundorf, Proc. IEEE SOS/SOI Technology Workshop, p. 29, 1987

164   J.P. Colinge, J. Kang, W. McFarland, C. Stout, and R. Walker, Proc. IEEE SOS/SOI Technology Workshop, p. 68, 1988

165   W.A. Krull and J.C. Lee, Proc. IEEE SOS/SOI Technology Workshop, p. 69, 1988

166   F. Namavar, E. Cortesi, B. Buchanan, and P. Sioshansi, Proc. IEEE SOS/SOI Technology Workshop, p. 117, 1989

167   T. W. Houston, H. Lu, P. Mei, T.G.W. Blake, L.R. Hite, R. Sundaresan, M. Matloubian, W.E. Bailey, J. Liu, A. Peterson, and G. Pollack, Proc. IEEE SOS/SOI Technology Workshop, p. 137, 1989

168 A.J. Auberton-Hervé, B. Giffard, and M. Bruel, Proceedings IEEE SOS/SOI Technology Workshop, p. 169, 1989

169 G.K. Celler, A. Kamgar, H.I. Cong, R.L. Field, S.J. Hillenius, W.S. Lindenberger, L.E. Trimble, and J.C. Sturm, Proc. IEEE SOS/SOI Technology Workshop, p. 139, 1989

170 H. Miki, Y. Omura, T. Ohmameuda, M. Kumon, K. Asada, K. Izumi, T. Asai, and T. Sugano, Techn. Digest of IEEE Internat. Electron Device Meeting (IEDM), p. 906, 1989

171 W.E. Bayley, H. Lu, T.G.W. Blake, L.R. Hite, P. Mei, D. Hurta, T.W. Houston, and G.P. Pollack, Proceedings of the IEEE International SOI Conference, p. 134, 1991

172 L.K. Wang, J. Seliskar, T. Bucelot, A. Edenfeld, and N. Haddad, Techn. Digest of IEEE Internat. Electron Device Meeting (IEDM), p. 679, 1991

173 A.K. Agrawal, M.C. Driver, M.H. Hanes, H.M. Hobgood, P.G. McMullin, H.C. Nathanson, T.W. O'Keeffe, T.J. Smith, J.R. Szedon, and R.N. Thomas, Techn. Digest of IEDM, p. 687, 1991

174 S. Nakashima, Y. Omura, and K. Izumi, Proceedings 5th International Symposium on Silicon-on-Insulator Technologies, The Electrochemical Society, Vol. 92-13, p. 358, 1992

175 P. Francis, A. Terao, B. Gentinne, D. Flandre, and JP Colinge, Techn. Digest of IEDM, p. 353, 1992

176 M.H. Hanes, A.K. Agrawal, T.W. O'Keeffe, H.M. Hobgood, J.R. Szedon, T.J. Smith, R.R. Siergiej, P.G. McMullin, H.C. Nathanson, M.C. Driver, andR.N. Thomas, IEEE Electron Device Letters, Vol. EDL 14, p. 219, 1993

177 H. Lu, E. Yee, L. Hite, T. Houston, Y.D. Sheu, R. Rajgopal, C.C. CHen, J.M. Hwang, and G. Pollack, Proceedings of IEEE International SOI Conference, p. 182, 1993

178 G.G. Shahidi, T.H. Ning, T.I. Chappell, J.H. Comfort, B.A. Chappell, R. Franch, C.J. Anderson, P.W. Cook, S.E. Schuster, M.G. Rosenfield, M.R. Polcari, R.H. Dennard, and B. Davari, Techn. Digest of IEDM, p. 813, 1993

179 T. Iwamatsu, Y. Yamaguchi, Y. Inoue, T. Nishimura, and N. Tsubouchi, Techn. Digest of IEDM, p. 475, 1993

180 H.-S. Kim, S.-B. Lee, D.-U. Choi, J.-H. Shim, K.-C. Lee, K.-P. Lee, K.-N. Kim, and J.-W. Park, Digest of Technical Papers of the Symposium on VLSI Technology, p. 143, 1995

181 T. Oashi, T. Eimori, F. Morishita, T. Iwamatsu, Y. Yamaguchi, F. Okuda, K. Shimomura, H. Shimano, S. Sakashita, K. Arimoto, Y. Inoue, S. Komori, M. Inuishi, T. Nishimura, and H. Miyoshi, Technical Digest of IEDM, p. 609, 1996

182 P.L.F. Hemment, in "Semiconductor-On-Insulator and Thin Film Transistor Technology", Chiang, Geis and Pfeiffer Eds., (North-Holland), MRS Symposium Proceedings, Vol. 53, p. 207, 1986

183 E.A. Mayell-Ondruz and I.H. Wilson, Thin Solid Films, Vol. 114, p. 357, 1984

184 P.L.F. Hemment, K.J. Reeson, J.A. Kilner, R.J. Chater, C. Maesh, G.R. Booker, J.R. Davis, and G.K. Celler, Nucl. Instr. Meth. Phys. Res., Vol. B21, p. 129, 1987

185 P.L.F. Hemment, in "Semiconductor-On-Insulator and Thin Film Transistor Technology", Chiang, Geis and Pfeiffer Eds., (North-Holland), MRS Symposium Proceedings, Vol. 53, p. 207, 1986

186　C. Jaussaud, J. Margail, J. Stoemenos, and M. Bruel, in " Silicon-On-Insulator and Buried Metals in Semiconductors", Sturm, Chen, Pfeiffer, and Hemment Eds, (North-Holland), MRS Symposium Proceedings, Vol. 107, p. 17, 1988

187　G.K. Celler, P.L.F. Hemment, K.W. West, and J.M. Gibson, Appl. Phys. Lett., Vol.48, p. 532, 1986

188　C. Jaussaud, J. Margail, J. Stoemenos, and M. Bruel, in " Silicon-On-Insulator and Buried Metals in Semiconductors", Sturm, Chen, Pfeiffer, and Hemment Eds, (North-Holland), MRS Symposium Proceedings, Vol. 107, p. 17, 1988

189　S.J. Krause and S. Visitserngtrakul, Proceedings of the IEEE SOS/SOI Technology Conference, p. 47, 1990

190　S. Cristoloveanu, A. Ionescu, T. Wetteroth, H. Shin, D. Munteanu, P. Gentil, S. Hong, and S. Wilson, Journal of the Electrochemical Society, Vol. 144, No. 4, p. 1468, 1977

191　L. Jastrzebski, J.F. Corboy, C.W. Magee, J.H. Thomas, A.C. Ipri, D.A. Peters, G.W. Cullen, and H. Friedman, J. Electrochem. Soc., Vol. 135, p. 1746, 1988

192　M.A. Guerra, Proc. of the 4th International Symposium on Silicon-on-Insulator Technology and Devices, Ed. by D. Schmidt, the Electrochemical Society, Vol. 90-6, p. 21, 1990

193　D. Hill, P. Fraundorf, and G. Fraundorf, Proc. IEEE SOS/SOI Technology Workshop, p. 29, 1987, and
D. Hill, P. Fraundorf, and G. Fraundorf, J. Appl. Phys., Vol. 63, p. 4933, 1988

194　F. Namavar, E. Cortesi, B. Buchanan, and P. Sioshansi, Proc. IEEE SOS/SOI Technology Workshop, p. 117, 1989

195　B. Aspar, C. Pudda, A.M. Papon, A.J. Auberton-Hervé, and J.M. Lamure, in "Silicon-on-Insulator Technology and devices", Ed. by. S. Cristoloveanu, The Electrochemical Society, Proceedings Vol. 94-11, p. 62, Abstract 541, 1994

196　A.J. Auberton-Hervé, B. Aspar and J.L. Pelloie, in "Physical and Technical Problems of SOI Structures and Devices", Kluwer Academc Publishers, NATO ASI Series - High Technology, Vol. 4, Ed. by J.P. Colinge, V.S. Lysenko and A.N. Nazarov, p. 3, 1995

197　S. Nakashima, T. Katayama, Y. Miyamura, A. Matsuzaki, M. Imai, K. Izumi, and N. Ohwada, Proceedings of the IEEE International SOI Conference, p. 71, 1994

198　M. Tachimori, S. Masui, T. Nakajima, K. Kawamura, I. Hamaguchi, T. Yano, and Y. Nagatake, in "Silicon-on-Insulator Technology and Devices VII", Ed. by. P.L.F. Hemment, S. Cristoloveanu, K. Izumi, T. Houston, and S. Wilson, Proceedings of the Electrochemical Society, Vol. 96-3, p. 53, 1996

199　J.P. Colinge, "SOI technology", in <u>Low-power HF microelectronics: a unified approach</u>, edited by G.A.S. Machado, IEE circuits and systems series 8, the Institution of Electrical Engineers, p. 139, 1996

200　G.K. Celler, presented at the 6th International Symposium on Silicon Materials Science and Technology, Semiconductor Silicon, Ed. by H. Huff, K. Barraclough and J.-I. Chikawa, the Electrochemical Society, Vol. 90-7, p. 472, 1990

201　H.J. Hovel, Proceedings of the IEEE International SOI Conference, p. 1, 1996

202　S. Nakashima, M. Harada, and T. Tsuchiya, Proceedings of the IEEE International SOI Conference, p. 14, 1993

203   A.G. Revesz and H.L. Hughes, in "Physical and Technical Problems of SOI Structures and Devices", Kluwer Academc Publishers, NATO ASI Series - High Technology, Vol. 4, Ed. by J.P. Colinge, V.S. Lysenko and A.N. Nazarov, p. 133, 1995

204   A.G. Revesz, G.A. Brown, and H.L. Hughes, J. Electrochemical Society, Vol. 140, No 11, p. 3222, 1993

205   B.J. Mrstik, P.J. McMarr, R.K. Lawrence, and H.L. Hughes, IEEE Transactions on Nuclear Science, Vol. 41, No. 6, p. 2277, 1994

206   R. Stahlbush, H. Hughes, and W. Krull, IEEE Transactions on Nuclear Science, Vol. 40, No. 6, p. 1740, 1993

207   M.E. Zvannt, C. Benefield, H.L. Hughes, and R.K. Lawrence, IEEE Transactions on Nuclear Science, Vol. 41, No. 6, p. 2284, 1994

208   G.A. Brown and A.G. Revesz, Proceedings of the IEEE International Conference, p. 174, 1991

209   A.J. Auberton-Hervé, J.M Lamure, T. Barge, M. Bruel, B. Aspar, and J.L. Pelloie, Semiconductor International, October 1995

210   P.L.F. Hemment, in "Semiconductor-On-Insulator and Thin Film Transistor Technology", Chiang, Geis and Pfeiffer Eds., (North-Holland), MRS Symposium Proceedings, Vol. 53, p. 207, 1986

211   L. Nesbit, S. Stiffler, G. Slusser, and H. Vinton, J. Electrochem. Soc., Vol. 132, p. 2713, 1985

212   E. Sobeslavsky and W. Skorupa, Phys. Stat. Sol. (A), Vol. 144, p. 135, 1989

213   W. Skorupa, K. Wollschläger, R. Grötzschel, J. Schöneich, E Hentschel, R. Kotte, F. Stary, H. Bartsch, and G. Götz, Nucl. Instr. and Meth. in Phys. Research, Vol. B32, p. 440, 1988

214   G. Zimmer and H. Vogt, IEEE Trans. on Electron Devices, Vol. 30, p. 1515, 1983

215   L. Nesbit, G. Slusser, R. Frenette, and R. Halbach, J. Electrochem. Soc., Vol. 133, p. 1186, 1986

216   W.P. Maszara, Proc. of the 4th International Symposium on Silicon-on-Insulator Technology and Devices, Ed. by D. Schmidt, the Electrochemical Society, Vol. 90-6, p. 199, 1990

217   W.P. Maszara, G. Goetz, A. Caviglia, and J.B. McKitterick, J. Appl. Phys., Vol. 64, p. 4943, 1988

218   R. Stengl, T. Tan, and U. Gösele, Japanese Journal of Applied Physics, Vol. 28, No.. 10, p. 1735, 1989

219   T. Abe, M. Nakano, and T. Itoh, Proc. of the 4th International Symposium on Silicon-on-Insulator Technology and Devices, Ed. by D. Schmidt, the Electrochemical Society, Vol. 90-6, p. 61, 1990

220   R. Stengl, K.Y. Ahn, and U. Gösele, Japanese Journa of Applied Physics, Vol. 27, p. L2367, 1988

221   U. Gösele, M. Reich, and Q.Y. Tong, Microelectronic Engineering, Vol. 28, No. 1-4, p. 391, 1995

222   N.F. Raley, Y. Sugiyama, and T. Van Duzer, J. Electrochem. Soc., Vol. 131, p. 161, 1984

223   C.E. Hunt and C.A. Desmond, in "Semiconductor wafer bonding", Ed. by. U. Goesele, T. Abe, J. Haisman and M. Schmidt, Proceedings of the Electrochemical Society, Vol. 92-7, p. 165, 1992

224   C.A. Desmond, C.E. Hunt, and S.D. Collins, in "Silicon-on-Insulator Technology and devices", Ed. by. S. Cristoloveanu, The Electrochemical Society, Proceedings Vol. 94-11, p. 111, Abstract 541, 1994

225   A. Yamada, O. Okabayashi, T. Nakamura, E. Kanda, and M. Kawashima, Ext. Abstr. 5th International Workshop on Future Electron Devices, Miyagi-Zao, Japan, p. 201, 1988

226   See for example: Solid-State Technology, p. 155, June 1994, or back cover page of European Semiconductor, May 1994

227   A. M. Ledger and P.J. Clapis, Proceedings of the IEEE International SOI Conference, p. 64, 1993

228   P.B. Mumola and G.J. Gardopee, Extended abstracts of the International Conference on Solid-State Devices and Materials, Yokohama, Japan, p. 256, 1994

229   P.B. Mumola, G.J. Gardopee, T. Feng, A.M. Ledger, P.J. Clapis, and E.P. Miller, Proceedings of the Second International Symposium on "Semiconductor Wafer Bonding: Science, Technology, and Applications", Ed. by M.A. Schmidt, C.E. Hunt, T. Abe, and H. Baumgart, The Electrochemical Society, Proceedings Vol. 93-29, p. 410, 1993

230   T. Yonehara, K. Sakaguchi, and N. Sato, Applied Physics Letters, Vol. 64, p. 2108, 1994

231   T. Unagami and M. Seki, Journal of the Electrochemical Society, Vol. 125, p. 1339, 1978

232   M. Bruel, Electronics Letters, Vol. 31, p. 1201, 1995

233   M. Bruel, Nuclear Instruments and Methods in Physics Research B, Vol. 108, p. 313, 1996

234   A.J. Auberton-Hervé, Technical Digest of IEDM, p. 3, 1996

235   C. Maleville, B. Aspar, T. Poumeyrol, H. Moriceau, M. Bruel, A.J. Auberton-Hervé, T. Barge, and F. Metral, in "Silicon-on-Insulator Technology and Devices VII", Ed. by. P.L.F. Hemment, S. Cristoloveanu, K. Izumi, T. Houston, and S. Wilson, Proceedings of the Electrochemical Society, Vol. 96-3, p. 53, 1996

236   W.J. Choyke, R.B. Irwin, J.N. McGruwer, J.R. Townsend, N.J. Doyle, B.O. Hall, J.A. Spitznagell, and S. Wood, Nuclear Instruments and Methods, Vol. 209/210, p. 407, 1983

237   W.K. Chu, R.H. Kastl, R.F. Lever, S. Mader, and B.J. Masters, Phys. Rev. B, Vol. 16 (9), p. 3851, 1977

238   B. Aspar, M. Bruel, H. Moriceau, C. Maleville, T. Poumeyrol, A.M. Papon, A. Claverie, G. Benassayag, A.J. Auberton-Hervé, and T. Barge, Microelectronic Engineering, Vol. 36, No. 1-4, p. 233, 1997

239   C. Munteanu, C. Maleville, S. Cristoloveanu, H. Moriceau, B. Aspar, C. Raynaud, O. Faynot, J.L. Pelloie, and A.J. Auberton-Hervé, Microelectronic Engineering, Vol. 36, No. 1-4, p. 395, 1997

240   B. Aspar, M. Bruel, M. Zussy, and A.M. Cartier, Electronics Letters, Vol. 32 (21), p. 1985, 1996

241 A.T. Collins, Properties and growth of diamond, Ed. by G. Davies, INSPEC, the Institution of Electrical Engineers, London, p. 288, 1994

242 N.K. Annamalai, J. Sawyer, P. Karulkar, W. Maszara, and M. Landstrass, IEEE Transactions on Nuclear Science, Vol. 40, p. 1780, 1993

243 N.K. Annamalai, P. Fechner, and J. Sawyer, Proceedings of the IEEE International SOI Conference, p 64, 1992

244 M. Matsumoto, Y. Sato, M. Kamo, and N. Setaka, Japanese Journal of Applied Physics, Vol. 21, p. L182, 1982

245 B. Edholm, A. Söderbärg, J. Olsson, and E. Johansson, Japanese Journal of Applied Physics, Vol. 34, Part 1, No. 9A, p. 4706, 1995

246 B. Edholm, A. Söderbärg, and S. Bengtsson, J. Electrochem. Soc., Vol. 143, p. 1326, 1996

247 C.Z. Gu, Z.S. Jin, X.Y. Lu, G.T. Zou, J.X. Lu, D. Yao, J.F. Zhang, and R.C. Fang, Chinese Physics Letters, Vol. 13, No. 8, p.610, 1966

248 S. Bengtsson, M. Bergh, M. Choumas, C. Olesen, and K.O. Jeppson, Japanese Journal of Applied Physics, Vol. 38, No. 8, p. 4175, 1996

249 S. Bengtsson, M. Choumas, W.P. Maszara, M. Bergh, C. Olesen, U. Södervall, and A. Litwin, Proceedings of the IEEE International SOI Conference, p. 35, 1994

CHAPTER 3 - SOI Materials Characterization

Once Silicon-On-Insulator material has been produced, it is important to characterize it in terms of quality. The most critical parameters are the defect density, the thickness of both the top silicon layer and the buried insulator, the concentration of impurities, the carrier lifetime, and the quality of the silicon-insulator interfaces.

Parameter characterized	Technique	Destructive (D) or not (ND)	Physical or Electrical
Film thickness	Visible reflectance	ND	P
	Spectroscopic ellipsom.	ND	P
	IR reflectance	ND	P
	RBS	D	P
	Stylus profilemeter	D	P
	dV_{TH}/dV_{G2} technique	D	E
Si crystal quality	TEM	D	P
	RBS	D	P
	Chemical decoration	D	P
	UV reflectance	ND	P
Stress in Si layer	Raman spectroscopy	ND	P
Impurity content	SIMS	D	P
	Spark-source mass spectroscopy	D	P
	Bulk decomposition	D	P
Carrier lifetime	Surface photovoltage	ND	P
	Ψ-MOSFET	D	E
	Measurements on devices	D	E
Si-SiO2 interfaces	MOS capacitor	D	E
	Ψ-MOSFET	D	E
	Surface photovoltage	ND	P

Table 3.1: Some SOI material characterization techniques.

Some of the techniques used for characterization of SOI materials are very powerful, but destructive (*e.g.*: transmission electron microscopy - TEM). These are tools of choice for thorough examination of the materials, but can hardly be employed on a routine basis in a production environment. Some other techniques may be less sensitive, but are non destructive and provide information in a matter of seconds. These may be used to assess the quality of large amounts of SOI wafers. Physical techniques, such as optical film thickness measurement techniques, can be employed on virgin SOI wafers, while some others need the fabrication of devices in the SOI material, and are, therefore, destructive by definition. We will now describe in more detail some of the characterization techniques used to assess the quality of SOI materials (Table 3.1). An extensive review of the characterization techniques developed for SOI materials and devices can be found in the book of S. Cristoloveanu and S.S. Li.[1]

3.1. Film thickness measurement

Accurate determination of the thickness of both the silicon overlayer and the buried insulator layer is essential for device processing as well as for the evaluation of parameters such as the threshold voltage. General-purpose film thickness measurement techniques can, of course, be utilized to evaluate SOI structures. These include the etching of steps in the material whose thickness is to be determined and the use of a stylus profilemeter to measure the height of the steps. Rutherford backscattering (RBS) can also be used to evaluate the thickness of a silicon film on SiO_2, but is not practical for the measurement of the buried oxide thickness, since oxygen atoms are lighter than silicon atoms. Cross-sectional transmission electron microscopy (XTEM) is probably the most powerful film thickness measurement technique, and atom lattice imaging can provide a built-in ruler for accurate distance measurements. All the above techniques are destructive and time consuming. Routine inspection and thickness mapping of SOI wafers necessitate contactless (non-destructive) methods. Both spectroscopic reflectometry and spectroscopic ellipsometry satisfy this requirement. Electrical film thickness measurement can also be performed on manufactured fully-depleted devices. This measurement is based on the body effect (variation of front threshold voltage with back-gate bias) and can be used to verify that the thickness of the silicon film in which the devices are made is indeed what is expected from processing.

3.1.1. Spectroscopic reflectometry

The thickness of both the silicon overlayer and the buried oxide can be evaluated in a non-destructive way by measuring the reflectivity spectrum of an SOI wafer. Visible light (λ ranging from 400 to 800 nm) is usually utilized. The spectrum can then readily be compared with the theoretical reflectivity spectrum which can be calculated from the theory of thin-film optics.

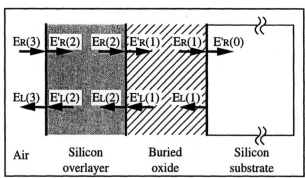

Figure 3.1.1: Multilayer structure and light-induced electric fields.

For SOI samples with well-defined Si-SiO$_2$ interfaces, a three-layer model can be used (silicon on oxide on silicon), while more complicated multilayer structures have to be employed if the interfaces are not well defined (in the case of as-implanted SIMOX material, for example [2]). Let us consider the case of an SOI structure with sharp Si-SiO$_2$ interfaces (Figure 3.1.1) and use the following notation: the semi-infinite silicon substrate, the buried oxide, the silicon overlayer and the air ambient are numbered 0,1,2, and 3, respectively. The light-induced electric fields at the different interfaces between the n-th and the (n+1)-th layer are expressed by the following equation [3,4,5]:

$$\begin{pmatrix} E_R(n+1) \\ E_L(n+1) \end{pmatrix} = W_{n+1,n} \begin{pmatrix} E'_R(n) \\ E'_L(n) \end{pmatrix} \tag{3.1.1}$$

where $E_R(n+1)$ and $E'_R(n)$ are the electric fields of the right-going wave at the left and right of the n-th interface, respectively, and $E_L(n+1)$ and $E'_L(n)$ are the fields of the left-going wave at the left and the right of the n-th interface, respectively. The interface matrix, W, is given by:

$$W_{(n+1,n)} = \frac{1}{2} \begin{pmatrix} 1 + \dfrac{y_n}{y_{n+1}} & 1 - \dfrac{y_n}{y_{n+1}} \\ 1 - \dfrac{y_n}{y_{n+1}} & 1 + \dfrac{y_n}{y_{n+1}} \end{pmatrix} \tag{3.1.2}$$

where y_n is the refractive index of the n-th layer. y_n is a complex number composed of a real (non-absorbing) and an imaginary (absorbing) part: $y_n = N_n + jK_n$. For the wavelengths under consideration, the refractive indices of dielectric materials such as SiO$_2$ and Si$_3$N$_4$ can be considered as real and constant, while the refractive index of silicon is complex and wavelength dependent.[6] The phase change and the absorption occurring within the thickness of a layer (n-th layer) is expressed by the following relationship:

$$\begin{pmatrix} E'_R(n) \\ E'_L(n) \end{pmatrix} = \Phi_n \begin{pmatrix} E_R(n) \\ E_L(n) \end{pmatrix} \tag{3.1.3}$$

where the "phase matrix", Φ, is given by: $\Phi_n = \begin{pmatrix} e^{j\Phi_n} & 0 \\ 0 & e^{-j\Phi_n} \end{pmatrix}$

where $\Phi_n = \dfrac{2\pi}{\lambda} y_n t_n$, t_n being the thickness of the n-th layer, and λ being the wavelength. Considering the full SOI structure, we finally obtain the following expression:

$$\begin{pmatrix} E_R(3) \\ E_L(3) \end{pmatrix} = W_{3,2}\, \Phi_2\, W_{2,1}\, \Phi_1\, W_{1,0} \begin{pmatrix} E'_R(0) \\ E'_L(0) \end{pmatrix} \tag{3.1.4}$$

We will note that the silicon substrate is considered as semi-infinite. Therefore, there is no returning wave from the substrate in the above equation (E'$_L$(0) = 0). The reflectivity of the structure is given by the ratio of reflected power to incident power:

$$R(\lambda) \equiv \frac{[E_L(3)]^2}{[E_R(3)]^2} \tag{3.1.5}$$

The reflectivity is expressed in % of incident power, and E'$_R$(0) is easily eliminated from equation (3.1.5). Indeed, E'$_R$(0) can be viewed as a reference value and can be considered to be equal to unity (E'$_R$(0)=1). Equation (3.1.4) then yields the values for E$_R$(3) and E$_L$(3) from which the ratio R(λ) can be computed for each wavelength. The above relationships were established for normal incidence of the light on the sample. Angles of incidence up to ...15 degrees... can, however, be used without introducing any significant error in the measurement. An example of reflectivity spectrum of a SIMOX wafer annealed at high temperature (*i.e.*: with sharp Si-SiO$_2$ interfaces) is given in Figure 3.1.2. The determination of the thickness of both the silicon film and the buried layer is based on the comparison between the measured spectrum and the calculated one. The film thickness values can be obtained iteratively through minimizing the difference between R$_{measured}$ and R$_{calculated}$ at selected wavelengths. This iterative technique is unfortunately rather time consuming.

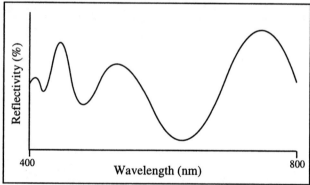

Figure 3.1.2: Example of reflectivity spectrum of a SIMOX sample.

There exists, however, another comparison technique which is significantly faster. It is based on the following observation: the reflectivity spectra present a succession of extrema (minima and maxima), and that the position of these extrema is a function of the thickness of both the silicon overlayer and the buried oxide.

Figure 3.1.3 presents a nomogram of the position of the reflectivity minima (or absorbance maxima in ref. [7]) of a SIMOX structure having a 352 nm-thick buried oxide, as a function of the silicon overlayer thickness. It can be observed that both the number of minima and the position (the wavelengths at which they occur) is a function of the silicon film thickness).

A database containing the position of the reflectivity minima for a wide range of silicon and buried oxide thicknesses can readily be computed and stored in a computer memory. Accurate estimation of film thickness can then be obtained by merely comparing the measured reflectivity minima with the values stored in the database.

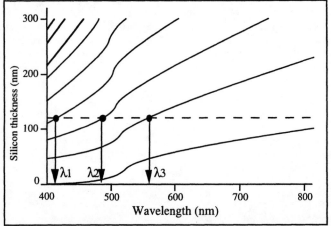

Figure 3.1.3: Nomogram of the wavelengths at which reflectivity minima occur (x-axis) as a function of silicon overlayer thickness (y-axis). The buried oxide thickness is 352 nm. For a silicon thickness of 120 nm, for instance, three reflectivity minima are found (at $\lambda = \lambda_1$, λ_2, and λ_3).

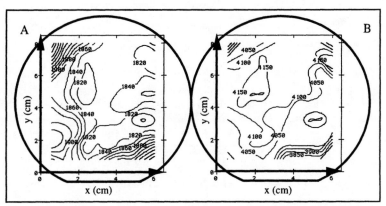

Figure 3.1.4: Mapping of the thickness of the silicon overlayer (A) and the buried oxide (B) of a 125 mm-diameter SIMOX wafer. Thickness is indicated in angströms (Å). (1Å = 0.1 nm).

Figure 3.1.4 shows an example of thickness mapping made on a SIMOX wafer using this technique.[8] Measurements based on reflectivity measurements in the visible spectrum are efficient for silicon films thinner than 500nm. Indeed, thicker SOI layers are no longer sufficiently transparent to visible light to allow for optical interferences to produce clear, well-defined reflectivity extrema and, in addition, thick films produce a large number of

71

extrema, which slows down the interpretation of the measured spectra. An alternative technique uses S-polarized light to measure the thickness of the layer (dual-beam S-polarized reflectance (DBSPR)), which has been used to produce complete film thickness mapping.[9]

In the case of thick SOI films, longer wavelengths (infrared, 1 μm $< \lambda < 5$ μm) can be used. These can penetrate through the silicon film and produce interference patterns which can be used to determine the film thickness of the SOI structure. Silicon-on insulator can be modeled as a double layer structure on a substrate. Under normal incidence conditions, the reflectance can be expressed by [10,11,12]:

$$R(\lambda) = \frac{A + f(\theta_1, \theta_2)}{F + f(\theta_1, \theta_2)}$$

where

$$f(\theta_1, \theta_2) = a_1\cos 2\theta_1 + a_2\cos 2\theta_2 + a_3\cos 2(\theta_1 + \theta_2) + a_4\cos 2(\theta_1 - \theta_2)$$
$$A = r_1^2 + r_2^2 + r_3^2 + r_1^2 r_2^2 r_3^2$$
$$a_1 = 2r_1 r_2(1 + r_3^2)$$
$$a_2 = 2r_2 r_3(1 + r_1^2)$$
$$a_3 = 2r_1 r_3$$
$$a_4 = 2r_1 r_2^2 r_3$$
$$F = 1 + r_1^2 r_2^2 + r_1^2 r_3^2 + r_2^2 r_3^2$$

The r's are the individual Fresnel reflectances of the three interfaces (air-silicon overlayer, silicon overlayer-buried oxide, and buried oxide-silicon substrate) and are given by: $r_j = \frac{n_{j-1} - n_j}{n_{j-1} + n_j}$, j=1,2,3. θ_1 and θ_2 are the phase thicknesses of the silicon and the buried oxide layers, respectively, and are given by: $\theta_j = \frac{2\pi}{\lambda} n_j t_j$, (j=1,2). The n_j are the real refractive indices of the different layers (being transparent at the wavelengths under consideration, both silicon and silicon dioxide have real refractive indices). The t_j are the layer thicknesses. By expanding the denominator of (3.1.6) in a power series, one finds:

$$R(\lambda) = \frac{A + f(\theta_1, \theta_2)}{F + f(\theta_1, \theta_2)} = \frac{A}{F} \frac{1 + f(\theta_1, \theta_2)/A}{1 + f(\theta_1, \theta_2)/F}$$

Using $\frac{1}{1 + f(\theta_1, \theta_2)/F} \cong \left(1 - (f(\theta_1, \theta_2)/F) + (f(\theta_1, \theta_2)/F)^2 + ...\right)$ for $(f(\theta_1, \theta_2)/F)^2 < 1$, one obtains:

$$R(\lambda) \cong \frac{A}{F}(1 + f(\theta_1, \theta_2)/A)\left(1 - f(\theta_1, \theta_2)/F + (f(\theta_1, \theta_2)/F)^2 - ...\right)$$

or

$$R(\lambda) \cong \frac{A}{F}\left(1 + (1/A - 1/F)\, f(\theta_1, \theta_2) + (1/F^2 - 1/AF)\, f^2(\theta_1, \theta_2) + ...\right)$$

or, more generally,

$$R(\lambda) = R_0 \sum_{n=0}^{\infty} k_n f^n(\theta_1, \theta_2)$$

where the k_n are functions of the individual Fresnel reflectances of the three interfaces. Let us consider, for instance, the first term in the expression for $f(\theta_1,\theta_2)$: $a_1 \cos 2\theta_1 = a_1 \cos \left(2\pi \cdot 2n_1 t_1 \cdot \dfrac{1}{\lambda}\right)$ where $2n_1 t_1$ is the optical thickness for a wave traversing twice the layer of refractive index n_1. As expressed above, the optical thickness is analogous to frequency, and the inverse wavelength is analogous to time in the parlance of conventional Fourier analysis. The reflectance spectrum contains several "frequency" components corresponding to $2n_1 t_1$, $2n_2 t_2$, $2(n_1 t_1 + n_2 t_2)$, $2(n_1 t_1 - n_2 t_2)$, and multiples thereof. By applying the so-called "bilinear transform" $B\{R(\lambda)\} = \dfrac{1 + R(\lambda)}{1 - R(\lambda)}$, the number of frequency components can be reduced dramatically.[13,14,15] The effects of the additional frequency modulation as a result of the wavelength dependence of the refractive indices are minimized by analyzing only data in the region between approximately 1 μm and 5 μm where the refractive indexes of both Si and SiO_2 are constant. Under these conditions, only three frequency components are dominant, and the Fourier analysis of $B\{R(\lambda)\}$ yields peaks at the following wave numbers: $V_{Si} = 2n_1 t_1$, $V_{SiO2} = 2n_2 t_2$, and $V_{total(Si+SiO2)} = 2(n_1 t_1 + n_2 t_2)$, which are the optical thicknesses of the silicon layer, the buried oxide layer, and the Si/SiO_2 structure, respectively. Conversion from wave numbers at which peaks occur (V_{Si} and V_{SiO2}) into layer thicknesses can be done using the following expressions: $t_{Si} = \dfrac{10^7 V_{Si}}{2n_{Si}}$ and $t_{SiO2} = \dfrac{10^7 V_{SiO2}}{2n_{SiO2}}$, where V is measured in cm and t in nm. This technique is complementary to the visible reflectance technique since it can be used to measure the thickness of relatively thick SOI layers, owing to the longer wavelengths used here (infrared), but it can also be used for thinner films (down to 100 nm).

3.1.2. Spectroscopic ellipsometry

Ellipsometry is based on the measurement of the change of polarization of a light beam reflected by a sample. In order to maximize the sensitivity of the measurement, ellipsometry is usually carried out at large incidence angles (75°, close to the Brewster angle). The change of polarization can be derived from equations (3.1.1)-(3.1.4) where corrections for non-normal incidence have to be introduced. The complete theory of ellipsometry is quite complicated and is outside the scope of this book. The interested reader can, however, refer to [16] for more information. Classical ellipsometry is performed at a single wavelength (usually emitted by an He-Ne laser). After passing through a polarizing filter, the beam is reflected on the sample and is directed into an analyzer, composed of a rotating polarization filter and a photodiode. The polarization direction of the reflected beam is given by the rotation angle of the latter filter when the amplitude of the light collected by the photodiode is maximum. In spectroscopic ellipsometry, the same analysis is repeated for a large number of wavelengths within the visible spectrum and the near ultraviolet. Typically, about a hundred equidistant measurements are performed at wavelengths ranging between 300 and 850 nm.

The output of the measurement consists into the spectra of $\tan\Psi$ and $\cos\delta$, which angles are defined, at each wavelength, by the relationship:

$$\tan\Psi \; e^{j\delta} = \frac{r_p}{r_s}$$

where r_p and r_s are the complex reflection coefficients r for light polarized parallel (p), and perpendicular (s - from *senkrecht*, in German) to the plane of incidence. An example of $\tan\Psi$ and $\cos\delta$ spectra is given in Figure 3.1.5.

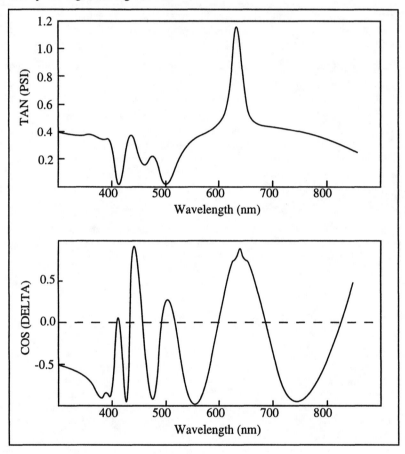

Figure 3.1.5: Example of the wavelength dependence of the ellipsometry parameters $\tan\Psi$ and $\cos\delta$ in a SIMOX structure.

The interpretation of the measured spectra is based on a simulation and regression program which minimizes the difference between the measured data and spectra calculated

74

using thin-film optics theory. Spectroscopic ellipsometry spectra contain much more information than spectroscopic reflectometry spectra. Its sensitivity is such that it can be used to measure complicated multilayer structures such as imperfect SIMOX structures which contain silicon inclusions within the buried oxide and oxide precipitates within the silicon overlayer. It can even measure the thickness of a native oxide layer on top of an SOI wafer.

Spectroscopic ellipsometry is not an analytical technique which stands of its own. Indeed, parameters such as the number of layers to be taken into account as well as their composition have to be fed into the simulator which will endeavor to tune each parameter in order to reproduce the measured data. For example, a SIMOX structure with oxide precipitates at the bottom of the silicon overlayer and silicon precipitates at the bottom of the buried oxide will need the following input parameters: estimation of the thickness of the following layers: the native oxide, the silicon overlayer without precipitates, the silicon overlayer with precipitates, the buried oxide above the silicon precipitates, the buried oxide with silicon precipitates, and the oxide below the precipitates. An estimation of the composition of the mixed layers (*e.g.*: silicon precipitates in oxide) must also be given. The refractive indices of mixed layers are calculated using the Bruggeman approximation (or effective medium approximation).[17] This approximation assume an homogeneous isotropic mixture of spheres of both components (*e.g.*: Si and SiO_2) with varying radius in order to obtain complete filling of the layer. In this approximation, the mixed layer is isotropic, and its dielectric constant, ε, is given by:

$$\varepsilon = 0.25 \left(E + \sqrt{E^2 + 8\varepsilon_1\varepsilon_2} \right) \quad \text{with} \quad E = (3c\text{-}1)(\varepsilon_2\text{-}\varepsilon_1) + \varepsilon_1$$

where c is the fraction of component 2 within the mixed layer, and ε_1 and ε_2 are the dielectric constants of component 1 and component 2, respectively. ε, ε_1, ε_2 and are complex numbers from which the expression of the refractive index $n = \sqrt{\varepsilon}$ can be extracted. The regression analysis during which measured and calculated spectra are compared, endeavors to minimize

the error function $G = \sum_{i=1}^{n} \left[(D_i^c - D_i^m)^2 + (W_i^c - W_i^m)^2 \right]$ where n is the number of different

wavelengths at which the measurement is performed. The superscripts *m* and *c* stand for "measured" and "calculated", respectively. The parameters W and D are given by the following expressions: $W_i = \dfrac{\tan^2\Psi_i - 1}{\tan^2\Psi_i + 1}$ and $D_i = \dfrac{2\cos\delta_i \tan\Psi_i}{\tan^2\Psi_i + 1}$ or $D_i = \cos\delta_i$ if $\tan\Psi_i$ is very small.

Spectroscopic ellipsometry is a very sensitive measurement technique. When the layers and interfaces of the SOI structure are well defined, a three-layer model can be used (native oxide/silicon/buried insulator), and convergence of the thickness-finding algorithm can be obtained in a matter of minutes. If the composition of the interfaces is not well defined (*e.g.*: as-implanted or low-temperature annealed SIMOX), such that mixed layers have to be simulated, many parameters have to be optimized simultaneously, and convergence can take up to several hours.

3.1.3. Electrical thickness measurement

The silicon film thickness, t_{si}, is an important parameter in thin-film, fully-depleted SOI MOSFETs. Indeed, t_{si} influences all the electrical parameters of thin-film devices (threshold voltage, drain saturation voltage, subthreshold slope, ...). Therefore, it is of interest to measure the thickness of the silicon film after device processing for debugging purposes and to check whether threshold voltage non-uniformities, for example, are due to film thickness variations or not, and whether the final targeted silicon thickness has been reached after device processing.

The dependence of the front threshold voltage, V_{th1}, of a fully-depleted n-channel SOI MOSFET on back-gate voltage, V_{G2}, is given by equation (5.3.13) of Section 5.3 (body effect):

$$\frac{dV_{th1}}{dV_{G2}} = - \frac{C_{si} \, C_{ox2}}{C_{ox1} \, (C_{si} + C_{ox2})} = \frac{-\epsilon_{si} \, C_{ox2}}{C_{ox1} \, (t_{si} \, C_{ox2} + \epsilon_{si})}$$

from which one can easily derive the following relationship [18]:

$$t_{si} = - \frac{1}{C_{ox2}} \left(\frac{\epsilon_{si} \, C_{ox2}}{C_{ox1}} - \epsilon_{si} \right) \left(\frac{dV_{th1}}{dV_{G2}} \right)^{-1}$$

where C_{ox1}, C_{ox2}, and ϵ_{si} are the gate oxide capacitance, the buried oxide capacitance, and the silicon permittivity, respectively. It is important to note that the above expression is independent of the doping concentration in the silicon film, as long as the device is fully depleted. This measurement technique assumes that C_{ox1} and C_{ox2} (or t_{ox1} and t_{ox2}) are known. The value of t_{ox1} can be obtained from an independent measurement made on bulk silicon wafers undergoing the same gate oxidation gate process as the SOI wafers, and the buried oxide thickness (which is not affected by device processing) can be measured by spectroscopic reflectometry mapping techniques before processing. It is also worth noting that the dV_{th1}/dV_{g2} measurement must be carried out for back-gate voltages for which the back-gate interface is depleted(*i.e.* neither inverted nor accumulated) (refer to Figure 5.3.4).

3.2. Crystal quality

Although all SOI material producing techniques aim at the fabrication of a perfect single-crystal silicon layer, they never completely achieve their goal, and crystal imperfections are found within the silicon. This Section describes some of the various techniques used to assess the quality of the silicon layers (crystal orientation, crystallinity, and crystal defects).

3.2.1. Crystal orientation

All SOI techniques are designed to produce silicon films with (100) normal orientation. This orientation is automatically obtained when the silicon film is produced by separation of a superficial silicon layer from a (100) silicon substrate by the formation of a buried insulator (SIMOX, SIMNI, FIPOS or wafer bonding) or when the silicon film is epitaxially grown from a single-crystal substrate having normal lattice parameters equal or close to those of (100) silicon (SOS, ELO, LSPE,...). In those cases where the silicon film is recrystallized from the melt over an amorphous insulator (laser and e-beam recrystallization, ZMR), the control of the crystal orientation is more difficult, and substantial deviation from the (100) orientation can be observed. Furthermore, even if the normal orientation is (100), in-plane misorientation can occur. The different single-crystals are then connected by subgrain boundaries.

The crystal orientation can be determined by classical techniques such as X-ray diffraction and electron diffraction in a TEM. More frequently, the electron channeling pattern (ECP) technique (or: pseudo-Kikuchi technique) is employed since it can be carried out in an SEM without any special preparation of the sample. In this technique, a stationary, defocused electron beam is incident onto the sample. The reflection of the electrons depends on the local crystal orientation relative to the incident beam (*i.e.*: it depends on the degree of channeling of the incident electrons in the silicon). The reflected electrons form a pattern which is indicative of the normal crystal orientation (a cross-shaped pattern is obtained for (100) silicon). The direction of the arms of the cross pattern can be used to determine the in-plane orientation, and a distortion of the pattern indicates a spatial rotation of the crystal axes.[19]

One of the most popular methods used to assess the crystal orientation of SOI films is the etch-pit grid technique.[20] This technique has been widely used to optimize the ZMR process. In this technique, a layer of oxide is grown or deposited on the silicon layer. Using a mask step and HF etch, circular holes are opened in the oxide layers. The holes have a 2-3 µm diameter and are repeated across the entire sample in a grid array configuration, with a pitch of 20 µm, typically. The resist is then stripped, and the holes in the oxide are used as a mask for silicon etching. The silicon is etched in a KOH solution (a mixture of 250g KOH, 800 ml deionized water and 200 ml isopropyl alcohol). The KOH solution etches silicon much more rapidly in the <100> direction than in the <111> direction. As a result, a pattern having the shape of a section of a pyramid is etched in the silicon. This pattern has a square shape if the normal orientation is (100), and the sides of the pits are <111> planes. The pit diagonals indicate [100] directions (Figure 3.2.1).

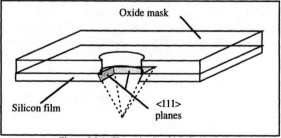

Figure 3.2.1: The etch-pit grid technique.

If the film orientation is not (100), distorted polyhedral patterns are produced by the intersection of the silicon film plane with the <111> silicon planes octahedron. The change of shape between two adjacent patterns indicates the presence of grain boundaries or the rotation of the crystalline axes. A misalignment of the diagonal directions between two patterns without modification of shape is indicative of the presence of subgrain boundaries.

3.2.2. Degree of crystallinity

In those cases where the silicon overlayer is damaged during the SOI formation process (SIMOX, SIMNI), and when epitaxial growth of silicon is performed, it is sometimes interesting to check the single-crystallinity of the formed layer. Two major techniques may help us to achieve that goal: Rutherford backscattering (RBS) and UV reflectance.

RBS is a destructive technique based on the impingement of light ions (usually He[+]) with mass M_1 on a sample consisting of atoms with mass M_2. The ions are accelerated to an energy E_0 (*e.g.*: 2 MeV) before reaching the target. These ions loose energy through nuclear and electronic interactions with the target atoms. Most of them will come to rest in the target, but a small fraction of these light ions will be backscattered over an angle Θ with an energy $E_1 = K\,E_0$, with K being the kinetic factor [21]:

$$K = \frac{\sqrt{M_2^2 - M_1^2}\,\sin\Theta + M_1\cos\Theta}{M_1 + M_2}$$

The detection of the energy of the recoiling atoms can thus be used to determine the mass M_2 of the target atoms. The probability for an elastic collision to occur and to result in a scattering event at a certain angle Θ is expressed by the differential scattering cross-section:

$\dfrac{d\sigma}{d\Omega} \propto \left(\dfrac{Z_1\,Z_2\,q^2}{4\,E_0}\right)^2 \dfrac{1}{\sin^4\Theta}$, with Ω being the detector angle. The average scattering cross-

section is then given by: $\sigma = \dfrac{1}{\Omega}\displaystyle\int_\Omega \dfrac{d\sigma}{d\Omega}\,d\Omega$. The stopping cross-section $\varepsilon = \dfrac{1}{N}\dfrac{dE}{dx}$ (E) finally

accounts for the energy loss of the particle penetrating the target due to electronic collisions or to small-angle collisions with nuclei; x is the depth below the target surface, and N is the volumic density of the target atoms. The detected signal is processed by a multichannel analyzer The output of a measurement session is a series of counts, called the backscattering yield, in every channel. To interpret the measured data, one has to convert the channel numbers into an energy scale and, therefore, to determine the energy interval **E** corresponding to a channel. Actually, the useful information is the depth scale, and **E** must be correlated with a slab i of thickness τ_i at a depth x_i. It can be shown that the energy difference between a particle backscattered at the surface (E=KE$_0$) and another one backscattered at a depth x and emerging from the target (E=E$_x$) is given by:

$$\Delta E = KE_0 - E_x = \left(\frac{K}{\cos\Theta_1}\left.\frac{dE}{dx}\right|_{in} + \frac{K}{\cos\Theta_2}\left.\frac{dE}{dx}\right|_{out}\right)x = [\varepsilon]\,N\,x$$

where Θ_1 and Θ_2 are the angles (with respect to normal) of the track of the particle before and after scattering in the target. The latter relationship assumes that dE/dx is constant along each path taken by the particle. This assumption yields a linear relationship between the energy difference, ΔE, and the depth at which scattering occurs. $[\epsilon]$ is the stopping cross-section factor. It is worthwhile noting that the analysis of RBS spectra can give information on the composition of compound materials (such as SiO_2 or, more generally, Si_xO_y).

An RBS spectrum measurement can be performed in two different ways: a crystal direction can be parallel to the incident ion beam, or it can be randomly oriented. In the former case, the ions can penetrate deeper in the crystal by channeling through the lattice, and an "aligned spectrum" is obtained. If the sample is amorphous or randomly oriented, no channeling can occur, and a "non-aligned" spectrum is obtained. Aligned spectra have lower backscattering yield because the ions penetrate deeper in the sample and have a lower probability to escape out of it after a collision. Similarly, the presence of crystalline imperfections (point defects, impurities,...) increases the backscattering yield of a crystalline, aligned target. The minimum backscattering yield, χ_{min}, is, therefore, a measure of the lattice disorder. The lower χ_{min}, the better the crystallinity. Single-crystal (100) bulk silicon has a value of χ_{min} equal to 3-4%.

Figure 3.2.2 shows typical RBS spectra obtained from a SIMOX sample. The most useful information come from the layers nearest to the surface, where the energy of the backscattered ions is highest. Part (a) of Figure 3.2.2 corresponds to the silicon overlayer. The non-aligned spectrum shows a high yield and provides information on the thickness of the silicon layer.

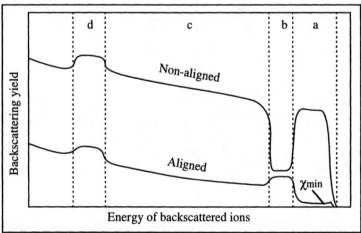

Figure 3.2.2: Aligned and non-aligned RBS spectra of a SIMOX wafer. From right to left: signal from the surface and the silicon overlayer (a), the buried oxide (b), the substrate (c), and influence of the buried oxide on the ions backscattered from the substrate (d).

The aligned spectrum gives information on its crystal quality, through χ_{min}. Part (b) corresponds to the buried oxide. Both aligned and non-aligned spectra contain information on the thickness of the layer. The non-aligned spectrum provides information on the composition of the layer (the ratio of oxygen to silicon atoms). The yield of the non-aligned spectrum in part (b) than in parts (a) and (c) because the concentration of silicon is lower in the SiO_2 buried layer than in the silicon overlayer or in the substrate. The aligned spectrum has a higher yield for SiO_2 than for Si because SiO_2 is amorphous, and no channeling can take place in the buried oxide layer. Part (c) corresponds to the silicon substrate, and the peak which can be observed in part (d) is the signal due to the oxygen present in the buried layer. The energy of the ions backscattered by the oxygen atoms is relatively low because oxygen is a light element, compared to silicon.

UV reflectance is another technique which can be used to assess the crystallinity of SOI samples. Contrarily to RBS, it is non-destructive. It has extensively been used to characterize SOS wafers, and both the microtwin density in the film and the fabrication yield of SOS circuits can be correlated to UV reflectance parameters.[22] The quality of the silicon overlayer of SIMOX wafers can also be assessed using the UV reflectance technique.[23] In the case of SOS, the methods employs a measurement of the reflectance at λ=280 nm and, if necessary, a second measurement at λ=400 nm for reference. There are two prominent maxima in the UV reflectance spectrum of single-crystal silicon, at λ=280 nm and λ=365 nm. They are caused by the optical interband transitions at the X point and along the Γ-L axis of the Brillouin zone, respectively. At short wavelengths, in particular at 280 nm, the reflectance is largely determined by the high value of the absorption coefficient $K > 10^6$ cm^{-1} corresponding to a penetration depth of less than 10 nm. Imperfect crystallinity in the near-surface region cause a broadening of the reflectance peak and a reduction of its maximum value. In the case of SIMOX wafers, more wavelengths have to be taken into consideration in order to obtain useful information about the quality of the silicon overlayer.

UV measurement of SIMOX wafers has been show to provide information on three morphological features of the material. Firstly, the overall reflectivity reduction (compared to a bulk silicon reference sample) is related to the presence of contamination in the film. This contamination can be due to the presence of carbon or SiO_x. Secondly, Rayleigh scattering caused by surface roughness shows a decrease of the reflectivity as a function of $B\lambda^{-4}$ for the shortest wavelengths (200 nm $< \lambda <$ 250 nm), where B is a constant depending on the rms roughness of the surface. Thirdly, some amorphization of the silicon overlayer reduces the intensity of the reflectance peaks at 280 and 367 nm. The sharpness and intensity of these maxima gives a measure of the degree of crystallinity within the specimen. Hence, semi-quantitative information about contamination, surface roughness and crystallinity can be obtained from UV reflectance measurements.

3.2.3. Crystal defects

Transmission electron microscopy (TEM) is one most powerful technique for the analysis of crystal defects. It is nevertheless limited by the size of the samples which can be analyzed. In cross-section TEM (XTEM), the dimensions of a sample which can be observed at once are approximately limited to a width of 20 μm and a depth of 0.7 μm. This means that the maximum observable area is on the order of 10^{-7} cm^{-2} and, consequently, that the minimum measurable defect density is approximately 10^7 defects/cm^2. Plane-view TEM allows one to observe larger sample areas. Areas of the size of the sample holder (7 mm^2 grid) can indeed be analyzed. In practice, it becomes difficult to observe dislocations with a magnification lower than 10,000 and it is more realistic to consider that an observation session yields 10 micrographs, each with a 10.000X magnification. In that case, the observed area is equal to 10^{-5} cm^2, and the minimum observable defect density is 10^5 defects/cm^{-2}. TEM observation often necessitates a lengthy and delicate sample preparation. Defect decoration techniques, combined with optical microscope observation are, therefore, preferred to TEM if the nature of the sample and the defects allow it. The main defects found in SOI materials are listed in Table 3.2.1. (Micro)twins and stacking faults are dominant in heteroepitaxial materials, while (sub)grain boundaries are found in silicon films recrystallized from the melt. Dislocations are the main crystal defect in SIMOX, FIPOS and material produced by wafer bonding.

Material	Type of defect	Concentration
SOS	Twins, stacking faults	H
SOZ	Twins, stacking faults	H
CaF2	Twins, stacking faults	H
Laser	Grain boundaries, stacking faults	H, but localized
E-beam	Grain boundaries, stacking faults	H
ZMR	Subgrain boundaries, dislocations	M
ELO	Stacking faults, dislocations	H
LSPE	Dislocations, stacking faults	H
FIPOS	Dislocations	L
SIMOX	Dislocations	L
Bur. Nitride	Dislocations	M
Bonding	Dislocations	L

Table 3.2.1: Main types of crystal defects present in the different SOI materials. H="high", M="medium", L="low".

The most common etch mixtures used for SOI defect decoration are listed in Table 3.2.2. Except for the last one (electrochemical etch), all are based on the mixture of HF with an oxidizing agent (CrO_3, $K_2Cr_2O_7$ or HNO_3). Defect decoration stems on the preferential etch (higher etch rate) of the defects with respect to silicon.

Etch name	Composition	Reference
Dash	$HF:HNO_3:CH_3COOH$ 1:3:10	[24]
Schimmel	HF:1M CrO_3 2:1	[25]
Secco	HF:0.15M $K_2Cr_2O_7$ 2:1	[26]
Stirl	HF:5M CrO_3 1:1	[27]
Wright	60ml HF:30 ml HNO_3:30 ml 5M CrO_3: 2 grams $Cu(NO_3)_2$:60 ml H_2O	[28]
Electrochemical	5% wt HF	[29,30]

Table 3.2.2: Decoration solutions used to reveal defects in SOI films.

Decoration is most effective for high-disorder defects such as grain and subgrain boundaries. The etch rate of silicon is approximately 1 μm/min for most mixtures (Dash, Secco, Stirl, Wright) - (the Schimmel etch rate is substantially lower). The decoration of dislocations in thin-film SOI material is almost impossible using classical etch mixtures, since all the silicon is removed before efficient decoration of the defects is achieved. Therefore, a new decoration technique, based on the electrochemical etching of silicon in diluted (5%) HF has been developed to reveal crystal defects in thin SOI films without etching off the silicon overlayer itself.[31] This technique necessitates the use of n-type ($N_d \cong 10^{15}$ cm^{-3}) doped silicon overlayers, and an ohmic contact must be provided to both the front side of the sample (*i.e.* to the SOI layer) and to the back of it. Electrochemical etching is performed for 10-30 minutes in 5 wt.% HF using a three-electrode configuration with the silicon controlled at +3 volts *vs.* a Cu/CuF$_2$ reference electrode. This decoration technique does nor etch defect-free silicon. Defects such as dislocations, metal contamination-related defects, and oxidation-induced stacking faults (OSF) produce pits in the silicon film during this electrochemical etch procedure, and optical microscopy is used to observe and count the pits after decoration.[32]

The incidence of a crystal defect on the electrical properties of a device depends on the nature and the geometry of the defect. We will now briefly describe the influence of some major crystal defects on the electrical properties of MOS devices made in SOI films.

3.2.3.1. (Sub)grain boundaries

A grain boundary (GB) is found when two grains with different crystal orientations meet. GBs are typical of laser and e-beam recrystallized SOI materials. The presence of a grain boundary in the channel region of a device gives rise to different effects, depending on the location of the boundary. If a GB runs from source to drain (the GB is parallel to the current direction), it acts as an enhanced-diffusion path for source and drain doping impurities during S&D reoxidation. As a consequence, source-to-drain leakage is observed in transistors with relatively short channel lengths (Figure 3.2.3.A).[33]

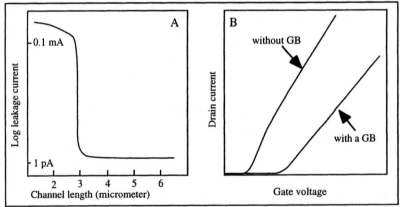

Figure 3.2.3: Influence of a grain boundary located in the channel area of SOI MOSFETs. The GB is either parallel (A) or perpendicular (B) to the current flow direction.

If the grain boundary is perpendicular to the current flow in the channel, the high density of interface traps in the grain boundary ($N_{it} \cong$ several 10^{12} cm^{-2}) retards the appearance of an inversion layer as gate voltage is increased. As a result, a dramatic increase of threshold voltage is observed (Figure 3.2.3.B). A significant reduction of the mobility in the inversion layer is also observed.[34,35]

Subgrain boundaries are found in ZMR material and are caused by the presence of two adjacent crystals with same normal crystal orientation, but with slightly different in-plane crystalline directions. A subgrain boundary (SGB), which can be viewed as a mere dislocation network, has much less electrical activity than a grain boundary. SGBs do not appear to either enhance doping diffusion or to significantly degrade threshold voltage or mobility characteristics.

3.2.3.2. Dislocations

Dislocations are the main defect found in many SOI materials. In the case of SIMOX, the dislocations are threading dislocations running vertically from the Si/buried oxide interface up to the surface of the silicon overlayer. The presence of such dislocations may pose yield and reliability hazard problems. Indeed, metallic impurities readily diffuse to dislocations upon annealing, and dislocations decorated with heavy metal impurities can cause weak points in gate oxides, so that low breakdown voltage is observed. A 1987 study shows, nevertheless, that the integrity of gate oxides grown on SIMOX is comparable to that of oxides grown on bulk silicon.[36] In addition, SIMOX technology has since then brought about steady improvement of both the dislocation density and the heavy metal contamination level.

3.2.4. Stress

The stress induced in the silicon film by the SOI fabrication process or by device processing can be measured using the Raman microprobe technique.[37] In this technique an argon laser (λ=457.9 nm) beam is focused onto the sample. The measured area can be as small as 0.6 μm.[38] The spectrum of the reflected beam is analyzed and compared to the spectrum provided by a virgin bulk silicon reference. The shift of the spectrum peak, as well as the value of the full width half maximum (FWHM) of the spectrum provide information on the stress in the silicon film. [39]

3.3. Silicon film contamination

Several types of contaminants can be introduced in the silicon overlayer during the SOI fabrication process. In the case of SIMOX, for instance, heavy metals and carbon can be sputtered from implanter parts and implanted into the wafer. Most forms of contamination can be analyzed by SIMS (secondary ion mass spectroscopy), but this technique is destructive and necessitates a rather heavy investment. Heavy metal contamination monitoring can be carried out using the SPV technique, which will be described in the next Section. Carbon and oxygen contamination are also important in SOI layers, and can contribute to stress and defect generation as well as to a decrease of the silicon overlayer resistivity.

3.3.1. Carbon contamination

Carbon contamination of SIMOX material can occur during the oxygen implantation step, most likely by interaction between the oxygen beam and either graphite implanter parts or the residual hydrocarbons present in the vacuum of the accelerator column. Carbon is known to form nucleation sites for oxygen precipitation and dislocations, such that a high ($>10^{18}$ cm^{-3}) carbon concentration can contribute to the generation of large amounts of dislocations in the silicon overlayer upon annealing of the SIMOX structure. Strong correlation has been established between the carbon concentration (measured by SIMS) and the dislocation density in the silicon overlayer of SIMOX wafers (Figure 3.3.1).[40]

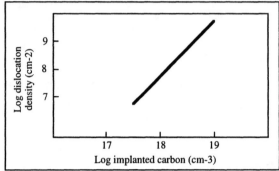

Figure 3.3.1: Dislocation density in SIMOX annealed at 1300°C as a function of implanted carbon concentration.

3.3.2. Oxygen contamination

The silicon overlayer of SIMOX material annealed at relatively low temperature (1150°C) contains large amounts of oxygen. Even moderate concentrations of oxygen (10^{17} cm^{-3}) can induce undesirable modifications of the silicon resistivity. Indeed, electrons are generated upon annealing of oxygen-containing silicon. These carriers are commonly referred to as thermal donors or new donors, depending on the temperature range in which they are activated (around 450°C for thermal donors and 750°C for new donors).

It is worth noting that 400-450°C is the usual temperature used for metal sintering, which is the last thermal operation in an integrated circuit fabrication process. Consequently, the thermal donors generated during this step will be fully activated in the finished product. The generation rate of thermal donors in 1150°C-annealed SIMOX films is 10^{11} cm^{-3} s^{-1}, and it is 10^{12} cm^{-3} s^{-1} for new donors. Thermal donor formation is explained by the formation of oxygen complexes, while the origin of new donors remains unclear but may be related to the formation of surface states between the silicon overlayer matrix and carbon-oxygen complex precipitates. Figure 3.3.2 presents the resistivity of 1150°C-annealed SIMOX material as a function of subsequent annealing temperature. The annealing time was 30 minutes in all cases, and the sheet resistivity was measured by four-point probe and spreading resistance techniques.[41]

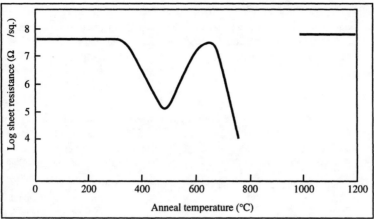

Figure 3.3.2: Sheet resistance of the silicon overlayer as a function of annealing temperature.

3.3.3. Metallic contamination

The levels of metallic contamination are usually very low and must be measured either by spark source mass spectrometry, TXRF, SIMS, using carrier lifetime measurement techniques or DLTS. It is, however, possible to use a chemical bulk decomposition technique to measure low levels of metallic contaminants. A droplet of either a 5% HF - 5% H_2O_2 or 38% HF - 68% HNO_3 solution is either placed on the SOI film or sandwiched between the SOI film and a Teflon plate. Once the chemical reaction has taken place, the etching solution is dried up and analyzed by graphite furnace atomic absorption spectroscopy (GF-AAS). Using this technique, levels of metallic contaminants down to 10^9 atoms/cm^2 have been detected.[42]

3.4. Carrier lifetime and surface recombination

The lifetime of minority carriers is a measure of the quality of the silicon films. The lifetime is affected by the presence of both crystal defects and metallic impurities. There exists no non-destructive method as such for measuring the lifetime in the silicon overlayer of an SOI structure. The measurement techniques rely either on the measurement of the lifetime in devices such as MOSFETs realized in the SOI material or on the measurement of the lifetime in the underlying silicon substrate and on the correlation between the lifetime in the substrate and the level of metal contamination within the silicon overlayer.

3.4.1. Surface photovoltage

The surface photovoltage (SPV) technique relies on the generation of a voltage at the surface of a silicon sample upon illumination. SPV uses a chopped beam of monochromatic light of photon energy $h\nu$ slightly larger than the band gap E_G of silicon. Electron-hole pairs are produced by the absorbed photons. Some of these pairs diffuse to the illuminated surface where they are separated by the electric field of the surface space-charge region whose thickness is w, thereby producing a surface photovoltage ΔV. A portion of ΔV is capacitively coupled to a transparent conducting electrode adjacent to the illuminated surface (Figure 3.4.1). This signal is then amplified to produce a quasi-dc analog output which is proportional to ΔV.

The value of ΔV is a function of the excess minority carrier (holes in the case of n-type silicon) density $\Delta p(0)$ at the edge of the surface space-charge region. This density $\Delta p(0)$ is in turn dependent on the incident light flux I_0, the optical absorption coefficient α, the optical reflectance at the illuminated surface ρ, the recombination velocity at the illuminated surface S, as well as the diffusion length L. A steady-state solution of the one-dimensional diffusion equation for the excess carrier density is given by $\Delta p(0) = \dfrac{I_0(1-\rho)}{D/L + s} \dfrac{\alpha L}{1+\alpha L}$ [43] if one assumes $\alpha w \ll 1$, $w \ll L$ and $\Delta p \ll n_0$, where n_0 is the majority carrier density. D is the minority carrier diffusion coefficient. One also assumes that $\alpha W \gg 1$, where W is the thickness of the silicon wafer.

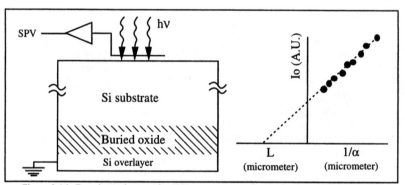

Figure 3.4.1: Experimental set-up for SPV measurement of an as-implanted SIMOX wafer (left) and graphical method for finding the carrier diffusion length (right).

A series of different wavelength values is selected to yield different values of α. At each wavelength, I_0 is adjusted to give the same value (constant magnitude) of ΔV. As a consequence, the value of $\Delta p(0)$ is constant as well. If ρ is essentially constant over the wavelength region of interest, the above equation may be rewritten: $I_0 = C[1+(\alpha L)^{-1}]$, where C is a constant. If I_0 is plotted against α^{-1} for each constant-magnitude ΔV point, the result is a linear graph whose extrapolated intercept on the negative α^{-1} axis is L (Figure 3.4.1).[44] The carrier lifetime, τ_p, can be deduced from the relationship $L^2 = D\tau$, with $D = \dfrac{kT}{q}\mu$, where μ is the carrier mobility.

The SPV technique cannot be directly used to measure the carrier recombination lifetime in the silicon overlayer of an SOI sample because the silicon film is too thin for the condition $\alpha W \gg 1$ to be met. SPV can nevertheless be employed as an indirect measurement of the heavy metal contamination of SOI material. An excellent correlation is observed between the metallic impurity concentration (Fe, Cr, Ni and Cu) in the silicon overlayer of as-implanted SIMOX wafers, measured by SIMS or by spark source mass spectrometry, and the SPV diffusion length measured in the silicon substrate. The actual relationship between the lifetime L and the heavy metal concentration C is in the form $L \cong 1/C$. The relationship between the metal concentration and the lifetime in the substrate can be explained as follows. Metal impurities are introduced in the wafer during the oxygen implantation process. In the experiment reported in ref. [45], the implantation took 6 hours, and was carried out at a wafer temperature of 580°C. In 6 hours, and at this temperature, iron, nickel, chromium and copper can diffuse in silicon at distances of 650, 3000, 200 and 6500 micrometers, respectively. Therefore, one can assume that the metal concentration is constant throughout the depth of the substrate, and that the concentration in the silicon film is correlated to that in the substrate. SPV is used to routinely monitor the heavy metal impurity level in SIMOX wafers.[46] The surface photovoltage technique can be used to monitor the interface charges at the silicon-buried insulator interface as well.[47] Other optical techniques, such as the measurement of the decay of free carriers generated by pulsed optical excitation, can be found in the literature.[48,49]

3.4.2. Lifetime measurement in devices

The minority carrier lifetime is an important parameter which has significant impact on device characteristics. Indeed, junction leakage appears if the generation lifetime is short (see *e.g.* Section 7.2.1). A high recombination lifetime, on the other hand, enhances parasitic bipolar problems (see Section 5.7). Minority carrier lifetime is affected by device processing, and its value after device fabrication may be significantly different from its value before processing.

3.4.2.1. Generation lifetime and surface generation

The minority carrier generation lifetime can be extracted from leakage current measurements [50,51] and Zerbst-like techniques.[52,53] A first method is based on the combined measurement of the leakage current and the capacitance of a junction as a function of the diode reverse bias. This method was demonstrated in thick SOI with non-reach-through junctions [54] (Figure 3.4.2).

The current flowing through a reverse-biased $(P^+\text{-}N)$ junction consists of several components, which include the diffusion current, I_{diff}, the bulk generation current I_{gen}^b, the surface generation current I_{gen}^s, and the field-enhanced generation current, I_{fe}. The last term involves field-dependent current not included in the other three current components.

If the applied reverse bias V_r is such that the depletion region underneath the junction does not reach through the buried oxide, the total reverse current, I_r, is given by:

$$I_r(V_r) = A_s q n_i S_0 + A_b q \int_0^{W_i(V_r)} \frac{n_i}{\tau_g(x)} \, dx + I_{diff}(V_r) + I_{fe}(V_r) \qquad (3.4.1)$$

where S_0 is the surface generation velocity, A_b is the area of the space-charge region in the bulk of the silicon, A_s is the area of the diode junction, n_i is the intrinsic carrier concentration, and W_i is the depletion depth in which generation takes place.

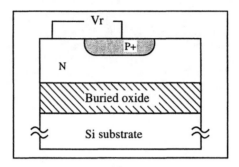

Figure 3.4.2: P$^+$-N diode for lifetime measurement in thick SOI film.

W_i remains smaller than the depletion depth in the N-type silicon film, W, up to voltages so high that I_{fe} is no longer negligible. It can be expressed by [55]:

$$W_i = \sqrt{\frac{2\varepsilon_{si}kT}{q^2 N_d}} \left(\sqrt{\ln\frac{N_d}{n_i} + \frac{qV_r}{kT}} - \sqrt{\ln\frac{N_d}{n_i}} \right) \qquad (3.4.2)$$

where N_d is the n-type doping concentration in the silicon film. If V_r is not too large, I_{diff} and I_{fe} are negligible.

I_{gen}^s is independent of V_r. Using $W = \sqrt{\frac{2\varepsilon_{si}}{q N_d}(\phi + V_r)}$, $C = A_b \varepsilon_{si}/W$ and thus $dW/dV_r = (\varepsilon_{si}A_b/C^2).dC/dV_r$, where C is the junction capacitance, W is the depletion depth, and ϕ is the junction built-in potential, equation (3.4.1) can be expressed in a differential form:

$$\tau_g(W_i) = \frac{\varepsilon_{si}q n_i A_b^2}{C^2} \frac{dW_i}{dW} \frac{dC/dV_r}{dI_r/dV_r} \qquad (3.4.3)$$

where $\dfrac{dW_i}{dW} = \left(\dfrac{V_r + \dfrac{kT}{q}\ln\dfrac{N_a N_d}{n_i^2}}{V_r + \dfrac{kT}{q}\ln\dfrac{N_d}{n_i}} \right)^{1/2}$ is a correction factor which takes into account the

difference between W_i and W. It is worthwhile noting that even though W_i may differ from W significantly, $dW_i/dW \cong 1$ for a wide range of bias voltages and doping concentrations. N_a is the doping concentration of the P^+ diffusion.

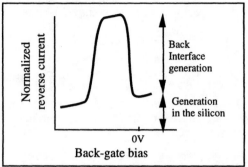

Figure 3.4.3: Reverse diode current as a function of back-gate bias.

The generation lifetime can be obtained from (3.4.3) through performing measurement of both reverse leakage current and junction capacitance and can be plotted as a function of depth in the silicon film. It is experimentally observed that the measured lifetime drops significantly within a Debye length ($\cong 100$ nm) of the silicon-buried oxide interface, due to the contribution of the surface generation. Study of the surface generation velocity can be performed by performing the measurement for different values of back-gate voltage. If the back interface is either in accumulation or inversion, only those generation centers which are within the silicon film contribute to the generation current. If, on the other hand, the back interface is depleted, generation at the interface provides yet another contribution to the total generation current, which results in a peak in $I_r(V_{G2})$ characteristics (Figure 3.4.3).[56]

Another method is more appropriate for the measurement of the minority carrier generation lifetime in thin silicon films. It is based on the measurement of the time required to create an inversion layer in a deep-depletion (or accumulation-mode) MOSFET. The minority carriers needed to create the inversion layer can only come from generation in the bulk of the silicon film or at the Si-SiO$_2$ interfaces. The dc characteristics of such a device are described in Section 5.8. We will consider the case of an n-channel (N^+-N^--N^+) device.[57,58] From equation (5.8.7), and taking into account both the presence of interfaces traps and the formation of an inversion layer (for large negative gate biases), one finds:

$$V_G\text{-}V_{FB} = \frac{qN_dx_{depl}^2}{2\varepsilon_{Si}} + \frac{qN_dx_{depl}}{C_{ox}} + \frac{Q_i + Q_{it}}{C_{ox}} \qquad (3.4.4)$$

where Q_i is the inversion charge and Q_{it} is the surface-state charge. N_d is the n-type doping density in the channel region.
One can also write:

$$x_{depl}(t) = t_{si} - x_a = t_{si} - \frac{L}{\sigma\, W\, V_{DS}} I_D(t) = t_{si} - B\, I_D(t) \qquad (3.4.5)$$

89

where $B = \dfrac{L}{\sigma\, W\, V_{DS}}$ if the device is operating in the linear regime (V_{DS} is small), and where x_a is the width of the non-depleted portion of the silicon film, and σ is the conductivity of the N⁻ silicon. L is the gate length, and W is the device width. Equation (3.4.4) can be rewritten:

$$Q_i(t) + Q_{it}(t) = C_{ox}\,(V_G\text{-}V_{FB}) - \frac{qN_dC_{ox}}{2\varepsilon_{Si}}\,(x_{depl}(t))^2 - qN_d x_{depl}(t)$$

or

$$Q_i(t) + Q_{it}(t) = C_{ox}\,(V_G\text{-}V_{FB}) - \frac{qN_dC_{ox}}{2\varepsilon_{Si}}\left(\left(x_{depl}(t) + \frac{\varepsilon_{Si}}{C_{ox}}\right)^2 - \left(\frac{\varepsilon_{Si}}{C_{ox}}\right)^2\right) \quad (3.4.6)$$

Using the classical model for generation of carriers in a semiconductor and the definition of surface generation velocity, we can write [59]:

$$\frac{d(Q_i(t) + Q_{it}(t))}{dt} = \frac{q\,n_i}{\tau_{gen}}\left(x_{depl}(t) - x_{depl}(t{=}\infty)\right) + q\,n_i\,S_0 \quad (3.4.7)$$

where τ_{gen} is the generation lifetime in the silicon film, and S_0 is the surface generation velocity. Taking the derivative of (3.4.6) and combining with (3.4.7) and (3.4.5), one obtains:

$$F(t) = \frac{B}{\tau_{gen}}\left(I_D(t{=}\infty) - I_D(t)\right) + S_0 \quad (3.4.8)$$

where

$$F(t) = -\frac{N_d\,C_{ox}}{2\,n_i\,\varepsilon_{Si}}\,\frac{d}{dt}\left(\left(t_{Si} + \frac{\varepsilon_{Si}}{C_{ox}}\right) - B\,I_D(t)\right)^2 \quad (3.4.9)$$

Plotting F(t) as a function of $\left(I_D(t{=}\infty) - I_D(t)\right)$ yields the values of both τ_{gen} (slope of the dotted line) and S_0 (y-axis intercept) (Figure 3.4.4.B).

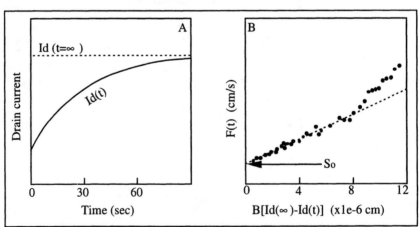

Figure 3.4.4: Drain current as a function of time in a depletion SOI MOSFET pulsed into inversion (A) and extraction of τ_{gen} and S_0 from the slope and the intercept of equation (3.4.9) (B).

90

This model assumes that the surface generation velocity is constant, which is true only when the silicon interface underneath the gate is in inversion, which condition yields low values of S_0. A better value of the surface generation is obtained by taking S_0 equal to the difference between the measured data and the extrapolated line of Figure 3.4.4.B. A graph of $S_0(t)$ can then be produced Figure (3.4.5.A).

The surface generation velocity is highest right after the application of the gate bias (no inversion layer is present to act as a screen between the interface and the depletion layer). Its value then decreases with time, as the inversion layer forms (Figure 3.4.5.A). Similarly, the effective generation lifetime can be plotted as a function of the depth in the silicon film. Indeed, the depth of the depletion zones varies with the magnitude of the applied gate bias. Small gate biases allow one to measure the effective lifetime near the surface (τ_{gen} is low because of the presence of the front interface - Figure 3.4.5.B), larger biases measure it deeper in the silicon film (τ_{gen} increases), and yet larger biases can be used to probe the back interface (τ_{gen} decreases again).

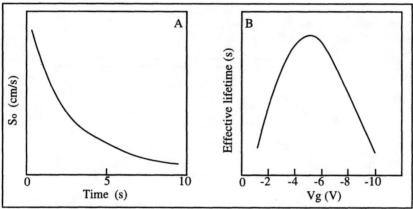

Figure 3.4.5: Surface generation as a function of time after gate biasing (A) and effective generation lifetime as a function of gate bias (B).

The above relationships were developed for edgeless devices. If a device with edges is measured, generation at the edges has to be taken into account, and equation (3.4.8) must be written [60,61]:

$$F(t) = \frac{B}{\tau_{gen}} \left(I_D(t=\infty) - I_D(t) \right) + \frac{2 L S_{edge}}{L W} B \left(I_D(t=\infty) - I_D(t) \right) + S_0$$

or

$$F(t) = \frac{B}{\tau_{eff}} \left(I_D(t=\infty) - I_D(t) \right) + S_0 \qquad (3.4.10)$$

such that one obtains:

$$\frac{1}{\tau_{eff}} = \frac{1}{\tau_{gen}} + \frac{2S_{edge}}{W} \qquad (3.4.11)$$

where W is the width of the device and S_{edge} is the surface generation velocity at the edges of the device.

3.4.2.2. Recombination lifetime

The recombination lifetime, or more exactly the effective recombination lifetime (which includes the influence of interface recombination), is a critical parameter for all characteristics involving parasitic bipolar effects. Its value can be estimated through the measurement of the gain β of lateral bipolar transistors with different base widths. Indeed, based on the bipolar transistor theory, one can write (equation 5.7.2): $\beta \cong 2\,(L_n/L_B)^2 - 1$, where L_B is the base width, which can be assumed, in first approximation, to be equal to the effective channel length, L_{eff}, and L_n is the electron diffusion length (we consider here the case of a NPN (n-channel) device). From the relationship $L_n^2 = D_n\,\tau_n$, where D_n and τ_n are the diffusion coefficient and the effective recombination lifetime of the minority carriers (electrons) in the base, respectively.

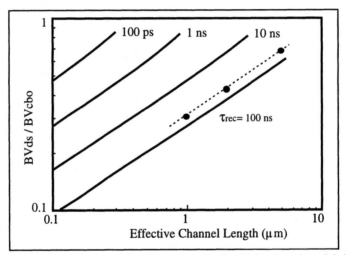

Figure 3.4.6: Determination of the effective recombination lifetime by means of drain breakdown voltage measurements. The circles represent experimental data.

If lateral bipolar transistor structures (= MOSFETs with body contacts) are not available, an estimation of the effective recombination lifetime can be obtained from a graph representing the drain breakdown voltage of n-channel SOI MOSFETs as a function of the effective gate length.[62] Indeed, experimental values of BV_{DS}/BV_{CBO}, can be plotted as a function of channel length on an abacus such as that which is presented in Figure 5.7.3 of Section 5.7. BV_{DS} is the measured drain breakdown voltage, and BV_{CBO} is the intrinsic breakdown voltage of the drain junction, which is approximately equal to the BV_{DS} of a very long channel device ($L_{eff} \cong L = 50...100$ μm). Figure 3.4.6 illustrates the method and gives the example of measurements carried out on devices with an effective recombination lifetime τ_{rec} of approximately 70 nsec.

3.5. Silicon-oxide interfaces

3.5.1. Capacitance measurements

Classical C-V techniques can be employed to measure the charges in the oxide layers and at the Si-SiO$_2$ interfaces of SOI devices, but the interpretation of the data is rather difficult. Indeed, the direct measurement of the capacitance of the whole SOI structure yields a complicated C-V curve (Figure 3.5.1) in which one can find contributions from the accumulation, depletion and inversion states at the front and back interfaces of the silicon film as well as at the top of the silicon substrate. Although analytical models of the SOI capacitor are available [63], it is usually easier to simulate the metal-insulator-semiconductor-insulator-semiconductor (MISIS) structure numerically and to compare the simulation results with the measurements in order to obtain an estimation of the charges in the oxide and at the Si/SiO$_2$ interfaces.[64]

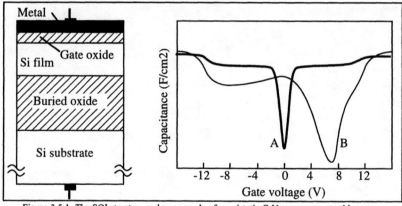

Figure 3.5.1: The SOI structure and an example of quasistatic C-V curves measured between the top metal electrode and the substrate. A: curve with one minimum; B: curve with two minima.

It is possible to measure separately the different capacitance components from MOS capacitors fabricated using a conventional SOI-CMOS process. Interdigitated capacitors are often used to minimize the parasitic resistance of the lightly doped channel region. Figure 3.5.2 presents the schematic configuration of such a capacitor. One assumes that the silicon film underneath the gate is not fully depleted. C_1 is the gate capacitance, C_2 is the capacitance across the buried oxide underneath the gate region, C_3 is the capacitance across the buried oxide underneath the P$^+$ diffusion, and C_4 is the capacitance between the metal patterns (line + pad) and the substrate (Figure 3.5.2). The capacitance C_4 is actually composed of two capacitances, C_{4F} and C_{4G} corresponding to the metal lines needed to contact the P$^+$ diffusion

93

(film contact) and the gate, respectively. The equivalent circuit representing the structure can be reduced to three capacitors, C_A, C_B and C_C, where

$$C_A = C_1$$
$$C_B = C_2 + C_3 + C_{4F}$$
$$C_C = C_{4G} \qquad (3.5.1)$$

Access to the capacitors is obtained through three terminals: the gate pad, the film contact pad, and the silicon substrate (Figure 3.5.2).[65]

If capacitance measurement is carried out between two of these terminals and leaving the third one floating, (*e.g.* measuring C_1 between the gate and film electrodes), altered C-V curves are obtained due to the presence of the two other capacitors. Better results are obtained by measuring C-V curves between one contact and the two others sorted to one another. For instance, $C_{G/FS}$ is measured between the gate terminal and the substrate and film connected together.

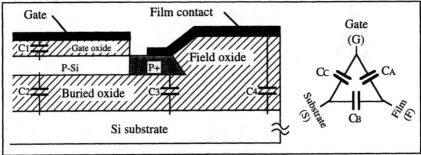

Figure 3.5.2: Schematic configuration of an SOI capacitor fabricated using a standard CMOS process and equivalent capacitor network.

From the equivalent circuit of Figure 3.5.2, we have for the three possible measurement configurations:

$$C_{G/FS} = C_C(V_{GS}) + C_A(V_{GF})$$
$$C_{F/SG} = C_A(-V_{GF}) - C_B(V_{FS})$$
$$C_{S/GF} = C_B(-V_{FS}) + C_C(-V_{GS}) \qquad (3.5.2)$$

Combining these equations, one obtains:

$$C_A(V_{GF}) + C_A(-V_{GF}) = C_{G/FS}(V_{G/FS}) + C_{F/SG}(V_{F/SG}) - C_{S/GF}(-V_{S/GF}) \qquad (3.5.3)$$

as well as two similar expressions for C_B and C_C, obtained by circular permutations. Unfortunately, C_A, C_B and C_C cannot be derived independently from the above relationships because they are asymmetrical with respect to the applied bias voltage (i.e., $C_A(V_{GF}) \neq C_A(-V_{GF})$). Nevertheless, $C_C = C_{4G}$ can be measured independently using a metal field capacitor, such as an unconnected metal pad, of capacitance C_T. We then obtain: $C_C(V_{GS})=C_{4G}=C_T(V_{GS})\dfrac{A_G}{A_T}$ and $C_{4F}=C_T\dfrac{A_F}{A_T}$, where A_G, A_F, and A_T are the areas of the capacitors C_{4G}, C_{4F}, and C_T, respectively. Now, C_A and C_B can easily be found using (3.5.2), and one obtains from (3.5.1): $C_1=C_A$ and $C_2+C_3 = C_B-C_{4F}$. Separation of C2 and C3 can be obtained by measuring two test structures with different area ratios for the capacitors C_2 and C_3. Finally, C-V analysis can be performed on the C_1 and C_2 curves.

3.5.2. Charge pumping

The charge pumping technique [66,67] is very efficient for characterizing Si/SiO₂ interfaces. It can be used with small area devices and can yield the distribution of interface states in the band gap. In the case of SOI devices, the front and back interfaces can be characterized independently from one another. The principle of the charge pumping technique in an bulk device is the following. Source and drain are connected together and slightly reverse biased, with respect to the substrate. A periodical triangular or trapezoidal signal, ΔV_G, with frequency f is applied to the gate. ΔV_G is sufficiently large to switch the silicon surface underneath the gate from accumulation to strong inversion (Figure 3.5.3). When the device goes into inversion, minority carriers are provided by the source and the drain to form the inversion channel. Some of these carriers are trapped by interface states. When the gate voltage is switched to produce an accumulation layer in the device, the inversion carriers (electrons in an n-channel device) disappear swiftly towards source and drain, and the minority carriers (electrons) which were trapped in the surface states now recombine with majority carriers from the substrate. This hole current constitutes the charge-pumping current I_{cp}.

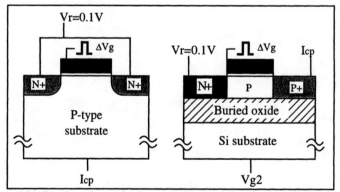

Figure 3.5.3: Experimental set-up for charge-pumping measurement in a bulk MOSFET (left) and an SOI gated PIN diode (right). I_{cp} is the charge-pumping current.

If the gate is pulsed at a frequency f, the charge-pumping current is given by:

$$I_{cp} = q^2\ \overline{N_{it}}\ A\ \Delta\Phi_s\ f \qquad (3.5.4)$$

95

where $\overline{N_{it}}$ is the average interface trap density, $A = W.L$ is the channel area, and $q\Delta\Phi_s$ is the energy range scanned within the bandgap. This basic expression can be extended to more sophisticated pumping techniques using gate offset bias and trapezoidal gate pulsed with different rise and fall times, in which case the energy distribution of the surface states across the bandgap can be obtained.

In the case of an SOI MOSFET, the charge-pumping current can be measured through a substrate contact. Better results are, however, obtained by carrying the measurement on a PIN (or $P^+P^-N^+$) gated diode (Figure 3.5.3).[68] In inversion, minority carriers are provided by the N^+ cathode, and the charge-pumping current is measured at the P^+ anode. Measurement of the front-interface trap density is obtained by pulsing the front gate, while back-interface trap density can be measured by pulsing the silicon substrate (back gate). Figure 3.5.4 presents the charge-pumping current measured on a SIMOX PIN gated diode as a function of the pulse frequency. Front and back interfaces were measured separately, by pulsing either the front gate or the back gate. The charge pumping current is larger when the back interface is probed, indicating a larger density of traps, N_{it} , than at the front interface. One can also observe that the back interface states encompass two components: a high density of slow states having a cutoff frequency of $\cong 1$-10 kHz and a lower density of fast states, indicated by the dotted line in Figure 3.5.4. Using devices with different channel widths,the charge-pumping technique can be used to measure the trap density at the interface between the silicon islands and the LOCOS isolation.[69]

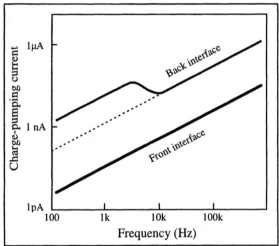

Figure 3.5.4: Charge-pumping current measured on a SIMOX PIN diode, as a function of frequency.

It is worth mentioning that the surface photovoltage technique (see Section 3.4.1) can be used to measure the interface trap density at the silicon film / buried oxide interface. Being non-desructive, it can be employed to map the interface trap density on SOI wafers before device processing.[70]

3.5.3. Ψ-MOSFET

The acronym Ψ-MOSFET (Psi-MOSFET) stands for "pseudo-MOSFET" or "point-contact MOSFET". It allows one to perform transistor-like measurements on SOI wafers without the need for actual device processing.[71,72] This method can, therefore, be used as a quick-turnaround tool for SOI wafer characterization. It is based on the fact that it is possible to obtain well-behaved transistor-like characteristics from an SOI wafer using a 2-point probe tool as source and drain contacts, and using that the substrate as a gate (Figure 3.5.5). Depending on the polarity of the gate voltage, either accumulation or inversion can be induced at the bottom of the silicon film, such that n-channel and p-channel-like characteristics can be obtained from a same wafer. Increasing the probe pressure reduces the series resistance of the source and drain contacts. For example, an increase in probe pressure from 10 to 30 grams improves the measured transconductance by 35%. The transconductance then saturates for higher pressures.

The drain current of a Ψ-MOSFET flows in the inversion or accumulation channel at the bottom of the SOI layer as well as through the neutral region of the film. The simplest case is, therefore, that of a **fully depleted** film where the current flowing through the silicon film itself is negligible. The measured current follows standard MOSFET theory:

$$I_D = f_g \, C_{ox2} \frac{\mu_0}{1+\theta(V_{G2}-V_{T2/FB2})} (V_{G2} - V_{T2/FB2}) \, V_D \qquad (3.5.5)$$

where f_g is a geometrical factor replacing the classical W/L term and which accounts for the non-parallel distribution of the current lines.

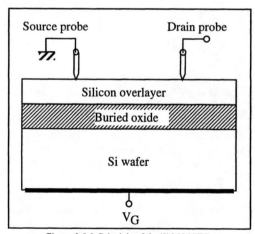

Figure 3.5.5: Principle of the Ψ-MOSFET.

In most practical configurations, the value of f_g is equal to $\cong 0.75$. $V_{T2/FB2}$ is either the threshold voltage in the case of inversion channel, or the flat-band voltage in the case of an accumulation channel. θ is a mobility attenuation factor which depends on the source and drain resistance: $\theta = \theta_0 + f_g\mu_0 C_{ox2}R_{SD}$. In a "regular", processed back-channel MOSFET the value of θ can be very small $(0.01\ V^{-1})$. In the Ψ-MOSFET the resistance of the metal-semiconductor contacts are significant, and θ reaches values higher than $0.05\ V^{-1}$. C_{ox} is the buried oxide capacitance.

In the case of a **partially depleted** film (P-type, for instance) one can distinguish three regimes of operation:

◊   For a sufficiently positive back gate bias the back interface is inverted $(V_{G2}>V_{T2})$ and the back depletion depth is maximum. The measured drain current is the sum of the current given by Equation (3.5.5) and the current in the quasi-neutral ("bulk") region of the silicon film.

◊   For $V_{FB2}<V_{G2}<V_{T2}$ current flows only in the quasi-neutral ("bulk") region of the silicon film, the thickness of which is equal to $t_{si}-x_d$. According to Equation (5.8.10) the extension of the depletion zone is given by:

$$x_d = \frac{-\varepsilon_{si}}{C_{ox2}} + \sqrt{\varepsilon_{si}^2/C_{ox2}^2 + \frac{2\varepsilon_{si}}{qN_a}(V_{G2} - V_{FB2})}$$

It is, however, convenient to assume that the bulk current decreases linearly with the gate voltage and to write:

$$I_D = q\, f_g\, \mu_0\, N_a\, V_D\, (t_{si}-x_d) = f_g\, \mu_0\, C_{ox2}\, V_D\, (V_0-V_{G2})$$

where $V_0=V_{FB2} + \dfrac{qN_a t_{si}}{C_{ox2}}$. V_0 is the gate voltage that *would* induce full depletion (Figure 3.5.6).

◊   For a negative gate bias $(V_{G2}<V_{FB2})$ the current flows both in the accumulation channel (Equation 3.5.5) and through the neutral ("bulk") silicon film $(x_d=0)$.

Different parameters of the SOI wafer can be extracted from Ψ-MOSFET measurements. From $I_D(V_{G2})$ and $\dfrac{I_D}{\sqrt{g_m}}(V_{G2})$ characteristics the mobilities, the back threshold voltage and the back flat-band voltage can be extracted.[73,74,75] Variations of fixed oxide charges (from wafer to wafer, from batch to batch or after exposure to ionizing radiations) can be monitored through the variation of V_{FB}. The back interface trap density can be determined from the subthreshold swing:

$$S = \frac{kT}{q}\ln(10)\left(1 + \frac{C_d}{C_{ox2}} + \frac{qN_{it}}{C_{ox2}}\right)$$

It is also worthwhile mentioning that the Ψ-MOSFET can be used in the pulsed mode to measure the carrier generation lifetime in the SOI layer, using a Zerbst-like technique similar to that described by Equations (3.4.4) to (3.4.9).[76,77]

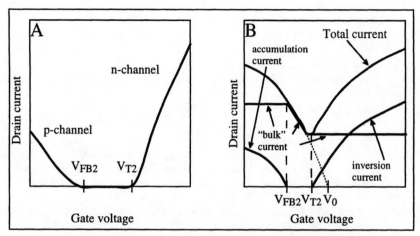

<u>Figure 3.5.6</u>: I-V_G characteristics of a Ψ-MOSFET. A: fully depleted film; B: partially depleted film.[78]

References

1   S. Cristoloveanu and S.S. Li, Electrical Characterization of Silicon-On-Insulator Materials and Devices, Kluwer Academic Publishers, 1995

2   S.N. Bunker, P. Sioshansi, M.M. Sanfacon, S.P. Tobin, Appl. Phys. Lett. Vol. 50, p. 1900, 1987

3   Z. Knittl, Optics of Thin Films, Wiley, New York, p. 37, 1976

4   F. Van de Wiele, Solid-State Imaging, NATO Advanced Study Institutes Series, Noordhoff, Leyden, p. 29, 1976

5   J.P. Colinge and F. Van de Wiele, J. Appl. Phys, Vol. 52, p. 4769, 1981

6   D.E. Aspenes, Properties of Silicon, Published by INSPEC (IEE), p. 59, 1988

7   T.I. Kamins and J.P. Colinge, Electronics Letters, Vol. 22, p. 1236, 1986

8   J. Vanhellemont, J.P. Colinge, A. De Veirman, J. Van Landuyt, W. Skorupa, M. Voelskow, and H. Bartsch, Proceedings of the 4th International Symposium on Silicon-on-Insulator Technology and Devices, Ed. by D. Schmidt, the Electrochemical Society, Vol. 90-6, p. 187, 1990

9   Y.S. Chang and S.S. Li, in "silicon-on-Insulator Technology and Devices", Ed. by S. Cristoloveanu, Electrochemical Society Proceedings Vol. 94-11, p. 154, 1994

10   H.J. Hovel, in "Semiconductors and Semimetals", Vol. 11 "Solar Cells", Ed. by R.K. Wilardson and A.C. Beer, Academic Press, p. 203, 1975

11   P.L. Swart and B.M. Lacquet, Journal of Electronic Materials, Vol. 19, No. 12, p. 1383, 1990

12   P.L. Swart and B.M. Lacquet, Proceedings IEEE SOS/SOI Conference, p. 153, 1989

13   P.L. Swart and B.M. Lacquet, Journal of Electronic Materials, Vol. 19, No. 8, p. 809, 1990

14   P.L. Swart and B.M. Lacquet, Journal of Applied Physics, Vol. 70, No. 2, p. 1069, 1991

15   P.L. Swart and B.M. Lacquet, Nuclear Instruments and Methods in Physics Research, Vol. 84, No. 2, p. 281, 1994

16   R.M.A. Azzam and N.M. Bashara, Ellipsometry and Polarized Light, Elsevier Science Publishers, North-Holland Personal Edition, chapter 1, 1987

17   D.A.G. Bruggeman, Annalen der Physik, Vol. 5, p. 636, 1935

18   J. Whitfield and S. Thomas, IEEE Electron Device Letters, Vol. 7, p. 347, 1986

19   D.C. Joy, D.E. Newbury, and D.L. Davidson, J. Appl. Phys., Vol. 53, p. R81, 1982

20   K.A. Bezjian, H.I. Smith, J.M. Carter, and M.W. Geis, J. Electrochem. Soc, Vol. 129, p. 1848, 1982

21   W.K. Chu, J.W. Mayer, and M.A. Nicolet, Backscattering Spectrometry, Academic Press, N.Y., 1978

22   M.T. Duffy, J.F. Corboy, G.W. Cullen, R.T. Smith, R.A. Soltis, G. Harbeke, J.R. Sandercock, and M. Blumenfeld, J. Crystal Growth, Vol. 58, p. 10, 1982

23   G. Harbeke and L. Jastrzebski, J. Electrochem. Society, Vol. 137, p. 696, 1990

24   W.C. Dash, J. Appl. Phys, Vol. 27, p. 1993, 1956

25   D.G. Schimmel, J. Electrochem. Soc, vol 126, p. 479, 1979

26   F. Secco d'Aragona, J. Electrochem. Soc., Vol. 119, p. 948, 1972

27   E. Sirtl and A. Adler, Zeitung für Metallkunde, Vol. 52, p. 529, 1961

28   M. Wright Jenkins, J. Electrochem. Soc., Vol. 124, p. 757, 1977

29   T.R. Guilinger, M.J. Kelly, J.W. Medernach, S.S. Tsao, J.O. Steveson, and H.D.T. Jones, Proceedings IEEE SOS/SOI Technology Conference, p. 93, 1989

30   M.J. Kelly, T.R. Guilinger, J.W. Medernach, S.S. Tsao, H.D.T. Jones, and J.O. Steveson, Proceedings of the fourth international Symposium on Silicon-on-Insulator Technology and Devices, ed. by D.N. Schmidt, Vol. 90-6, The Electrochemical Society, p. 120, 1990

31   T.R. Guilinger, M.J. Kelly, J.W. Medernach, S.S. Tsao, J.O. Steveson, and H.D.T. Jones, Proceedings IEEE SOS/SOI Technology Conference, p. 93, 1989

32   M.J. Kelly, T.R. Guilinger, J.W. Medernach, S.S. Tsao, H.D.T. Jones, and J.O. Steveson, Proceedings of the fourth international Symposium on Silicon-on-Insulator Technology and Devices, ed. by D.N. Schmidt, Vol. 90-6, The Electrochemical Society, p. 120, 1990

33   K.K. Ng, G.K. Celler, E.J. Povilonis, R.C. Frye, H.J. Leamy, and S.M. Sze, IEEE Electron Device Letters, Vol. 2, p. 316, 1981

34   J.P. Colinge, H. Morel, and J.P. Chante, IEEE Trans. on Electron Devices, Vol. 30, p. 197, 1983

35   T. Nishimura, K. Sugahara, S. Kusunoki, and Y. Akasaka, Ext. Abstracts of the 17th Conference of on Solid-State Devices and Materials, Tokyo, p. 1147, 1985

36   T.I. Kamins, Electronics Letters, Vol. 23, p. 175, 1987

37   I. De Wolf, J. Vanhellemont, H.E. Maes, A. Romano-Rodriguez, and H. Norström, Proceedings of the fifth International Symposium on Silicon-on-Insulator Technology and devices, Ed. by: K. Izumi, S. Cristoloveanu, P.L.F. Hemment, and G.W. Cullen, The Electrochemical Society Proceedings volume 92-13, p. 307, 1992

38   E. Martin, A. Pérez-Rodríguez, J. Jimenez, and J.R. Morante, in "silicon-on-Insulator Technology and Devices", Ed. by S. Cristoloveanu, Electrochemical Society Proceedings Vol. 94-11, p. 185, 1994

39   J. Macía, T. Jawhari, A. Pérez-Rodríguez, and J.R. Morante, in "silicon-on-Insulator Technology and Devices", Ed. by S. Cristoloveanu, Electrochemical Society Proceedings Vol. 94-11, p. 148, 1994

40   L. Jastrzebski, J.T. McGinn, P. Zanzucchi, and B. Cords, J. Electrochem. Soc., Vol. 137, p. 306, 1990

41   S. Cristoloveanu, J. Pumfrey, E. Scheid, P.L.F. Hemment, and R.P. Arrowsmith, Electronics Letters, Vol. 21, p. 802, 1985

42   M.B. Shabani, T. Yoshimi, H. Abe, T. Nakai, and B. Cords, in "Silicon-On-Insulator Technology and Devices VII", Ed. by. P.L.F. Hemment, S. Cristoloveanu, K. Izumi, T. Houston, and S. Wilson, Electrochemical Society Proceedings Vol. 96-3,p. 162, 1996

43   T.S. Moss, Optical Properties of Semiconductors, Butterworths, London, Chapter 4, 1959

44   A.M. Goodman, J. Appl. Phys, Vol. 53, p. 7561, 1982

45   L. Jastrzebski, G. Cullen, and R. Soydan, J. Electrochem. Society, Vol. 137, p. 303, 1990

46   M.A. Guerra, Proceedings of the 4th International Symposium on Silicon-on-Insulator Technology and Devices, Ed. by D. Schmidt, the Electrochemical Society, Vol. 90-6, p. 21, 1990

47   K. Nauka, M. Cao, and F. Assaderaghi, Proceedings of the IEEE International SOI Conference, p. 52, 1995

48   J.L. Freeouf and S.T. Liu, Proceedings of the IEEE International SOI Conference, p. 74, 1995

49   J.L. Freeouf, N. Braslau, and M. Wittmer, Applied Physics Letters, Vol. 63, p. 189, 1993

50   H.S. Chen, F.T. Brady, S.S. Li, and W.A. Krull, IEEE Electron Device Letters, Vol. 10, p. 496, 1989

51   H.S. Chen and S.S. Li, Proceedings of the 4th International Symposium on Silicon-on-Insulator Technology and Devices, Ed. by D. Schmidt, the Electrochemical Society, Vol. 90-6, p. 328, 1990

52   D.P. Vu and J.C. Pfister, Appl. Phys. Letters, Vol. 47, p. 950, 1985

53   T. Elewa, H. Haddara, and S. Cristoloveanu, in "Solid-State Devices", Ed. By. G. Soncini and P.U. Calzolari, Elsevier Science Publishers (North-Holland), p. 599, 1988

54   H.S. Chen, F.T. Brady, S.S. Li, and W.A. Krull, IEEE Electron Device Letters, Vol. 10, p. 496, 1989

55   *ibidem*

56   H.S. Chen and S.S. Li, Proceedings of the 4th International Symposium on Silicon-on-Insulator Technology and Devices, Ed. by D. Schmidt, the Electrochemical Society, Vol. 90-6, p. 328, 1990

57   D.P. Vu and J.C. Pfister, Appl. Phys. Letters, Vol. 47, p. 950, 1985

58   T. Elewa, H. Haddara, and S. Cristoloveanu, in "Solid-State Devices", Ed. By. G. Soncini and P.U. Calzolari, Elsevier Science Publishers (North-Holland), p. 599, 1988

59   M. Zerbst, Z. Angew. Phys., Vol. 22, p. 30, 1966

60   P.K. McLarty, T. Elewa, B. Mazhari, M. Mukherjee, T. Ouisse, S. Cristoloveanu, D.E. Ioannou, and D.P. Vu, Proceedings IEEE SOS/SOI Technology Conference, p. 54, 1989

61   T. Elewa, Ph.D. Thesis, ENSERG-LPCS, Grenoble (France), p. 90, July 1990

62   M. Haond and J.P. Colinge, Electronics Letters, Vol. 25, p.1640 , 1989

63   D. Flandre and F. Van De Wiele, IEEE Electron Device Letters, Vol. 9, p. 296, 1988

64   M. Gaitan and P. Roitman, Proceedings IEEE SOS/SOI Technology Conference, p. 48, 1989

65   J.H. Lee and S. Cristoloveanu, IEEE Electron Device Letters, Vol. 7, p. 537, 1986

66   J.S. Brugler and P.G.A. Jespers, IEEE Trans. Electron Devices, Vol. 16, p. 297, 1969

67   G. Groeseneken, H.E. Maes, N. Beltran, and R.F. Dekeersmaecker, IEEE Trans. Electron Devices, Vol. 31, p. 42, 1984

68   T. Elewa, H. Haddara, S. Cristoloveanu and M. Bruel, J. de Physique, Vol. 49, No 9-C4, p. C4-137, 1988

69   Y. Li and T.P. Ma, International Symposium on VLSI Technology, Systems, and Applications, Proceedings of Technical Papers, p. 123, 1997

70   K. Nauka, Microelectronic Engineering, Vol. 36, No. 1-4, p. 351, 1997

71   S. Cristoloveanu and S. Williams, IEEE Electron Device Letters, Vol. 31, p. 102, 1992

72   S. Cristoloveanu and S.S. Li, <u>Electrical Characterization of Silicon-On-Insulator Materials and Devices</u>, Kluwer Academic Publishers, p. 104, 1995

73   T. Ouisse, P. Morfouli, O. Faynot, H. Seghir, J. Margail, and S. Cristoloveanu, Proceedings of the IEEE International SOI Conference, p. 30, 1992

74   S. Cristoloveanu, A. Ionescu, C. Maleville, D. Munteanu, M. Gri, B. Aspar, M. Bruel, and A.J. Auberton-Hervé, in "Silicon-On-Insulator Technology and Devices VII", Ed. by. P.L.F. Hemment, S. Cristoloveanu, K. Izumi, T. Houston, and S. Wilson, Electrochemical Society Proceedings Vol. 96-3, p. 142, 1996

75   S. Wiliams, S. Cristoloveanu, and G. Campisi, Materials Science Engineering, Vol. B12, p. 191, 1992

76   A.M. Ionescu, S. Cristoloveanu, S.R. Wilson, A. Rusu, A. Chovet, and H. Seghir, Nuclear Instr. and Methods in Phys. Res., Vol. 112, p. 228, 1996

77   A.M. Ionescu, S. Cristoloveanu, D. Munteanu, T. Elewa, and M. Gri, Solid-State Electronics, Vol. 39, No. 12, p. 1753, 1996

78   *ibidem*

CHAPTER 4 - SOI CMOS Technology

Complementary MOS (CMOS) is, by far, the technology of choice for the realization of integrated circuits on SOI substrates. This Chapter will compare CMOS processing on bulk silicon and on SOI wafers. Processing of both thin and thicker SOI films will be discussed. We will assume that circuit processing is carried out on commercially available substrates, such as SIMOX wafers. It is worthwhile keeping in mind that, unlike in the case of SOS, SOI wafers contain only silicon and silicon dioxide, and that the appearance of SOI wafers is very similar to that of bulk silicon wafers. As a consequence, SOI circuit processing can be carried out in standard bulk silicon processing lines. Mixed batches (containing both bulk and SOI substrates) can be processed as well.

4.1. Comparison between bulk and SOI processing

Processing techniques for the fabrication of CMOS circuits in bulk silicon and in SOI are very similar. Figure 4.1.1 presents cross-sections of CMOS inverters made in bulk (p-well technology), in "thick-film" SOI (200-500 nm-thick films), and in thin-film SOI.

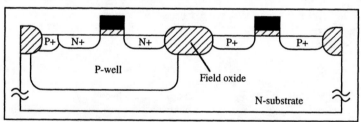

Figure 4.1.1.A: Cross-section of a bulk CMOS inverter.

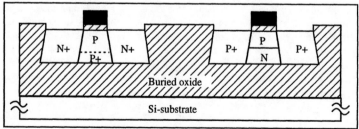

Figure 4.1.1.B: Cross-section of a "thick-film" SOI CMOS inverter.

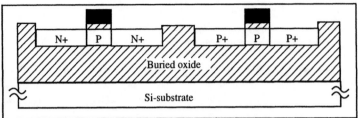

Figure 4.1.1.C: Cross-section of a thin-film SOI CMOS inverter.

The cross sections of Figure 4.1.1 are, of course, quite schematic. For instance, bulk CMOS can be much more complicated than in Figure 4.1.1.A and can make use of an epitaxial substrate, twin wells, or retrograde wells. From the cross sections, it is obvious that SOI processing, and more specifically thin-film SOI processing, is simpler than bulk processing. For instance, there is no need to create diffused wells in SOI. The anti-punchthrough implant used in bulk CMOS is kept unchanged in the case of a p-channel SOI device, and is replaced by a back-channel leakage suppression implant for the n-channel device (Figure 4.1.1.B).

In thin-film, fully-depleted SOI devices, these deep implants are unnecessary, and the entire impurity profile in the channel area is determined by a single shallow implant. In the frequent case where N^+ polysilicon is used as gate material, p-type impurities (boron) are used to control the threshold voltage in both the n-channel and the p-channel devices (Figure 4.1.1.C). Table 4.1.1 compares simplified CMOS process flows for bulk, "thick-film", partially-depleted (PD) SOI, and thin-film, fully-depleted (FD) SOI. It can be observed that SOI processing is simpler (less steps), and that at least one mask step is saved (two mask steps can be saved if the p-channel anti-punchthrough implant is a blanket implant, later compensated in the n-channel devices by a "heavy" back channel-stop implant). The simplification of the process is even more dramatic if thin-film (100 nm or less) devices are fabricated. The simplicity of processing can be further simplified, such that full CMOS front end process can be realized with only 5 mask steps.[1]

There might be a concern regarding the quality of the gate oxide of SOI devices. Early reports on the properties of gate oxides indicated higher leakage and charge trapping characteristics in oxides grown on SIMOX than on bulk silicon.[2] This was due to the poor quality of the material, carbon and metal contamination. In 1990, the material quality had improved, such that it was found that the breakdown field of thin gate oxides (12.5 nm) grown on SIMOX was 10-15% lower than that of similar oxides grown on bulk silicon (10 MV/cm *vs.* 11.5 MV/cm), and that the spread of the distribution around the mean value was quite similar to what is measured in bulk.[3] This problem seems to have disappeared in modern SIMOX material. For example, one has found an oxide breakdown field of 11.5 MV/cm in 250Å SIMOX gate oxides and cumulative oxide breakdown plots in SIMOX have been found to be similar to those measured on bulk silicon.[4,5] It is also worth mentioning that he quality of oxides grown on SOI wafers can be increased by using a pre-process wafer polishing step.[6]

Bulk CMOS	PD SOI CMOS	FD SOI CMOS
Oxidation	Oxidation	Oxidation
Well litho*		
Well doping & drive-in		
Nitride deposition	Nitride deposition	Nitride deposition
Active area litho*	Active area litho*	Active area litho*
Nitride etch	Nitride etch	Nitride etch
Field implant litho*	Field implant litho*	Field implant litho*#
Field implant	Field implant	Field implant#
Field oxide growth	Field oxide growth	Field oxide growth
Nitride strip	Nitride strip	Nitride strip
P-channel litho*	P-channel litho*	
Anti-punchthru implant	Anti-punchthru implant	
Gate oxide growth	Gate oxide growth	Gate oxide growth
P-ch Vth implant	P-ch Vth implant	P-ch Vth implant
N-channel Vth litho*	N-channel Vth litho*	N-channel Vth litho*
Anti punchthru implant	Back channel implant	
N-ch Vth implant	N-ch Vth implant	N-ch Vth implant
Poly deposition & doping	Poly deposition & doping	Poly deposition & doping
Gate litho* & etch	Gate litho* & etch	Gate litho* & etch
P+ S&D litho*	P+ S&D litho*	P+ S&D litho*
P+ S&D implant	P+ S&D implant	P+ S&D implant
N+ S&D litho*	N+ S&D litho*	N+ S&D litho*
N+ S&D implant	N+ S&D implant	N+ S&D implant
S&D reoxidation	S&D reoxidation	S&D reoxidation
Dielectric deposition	Dielectric deposition	Dielectric deposition
Contact hole litho*	Contact hole litho*	Contact hole litho*
Contact hole opening	Contact hole opening	Contact hole opening
Metallization	Metallization	Metallization
Metal litho*	Metal litho*	Metal litho*
Metal patterning	Metal patterning	Metal patterning
Sintering	Sintering	Sintering

Table 4.1.1: Comparison of bulk, "thick-film" (partially depleted - PD) SOI and thin-film, fully-depleted (FD) CMOS process flows. N^+ polysilicon is used as gate material. (The "*" symbol indicates a lithography step, the "#" indicates that the step may be optional.)

We will now describe into more detail some particularities of SOI processing: isolation techniques, doping profiles, and source and drain resistance issues.

4.2. Isolation techniques

There are many different ways of isolating the active silicon islands from one another. All these are simpler than the isolation schemes used in bulk silicon technology, owing to the presence of a buried insulator underneath the silicon film which provides an intrinsic vertical isolation. In a sense, one could say that a part of the complexity of the device isolation process used in bulk technology has been transferred to the wafer manufacturing stage (the fabrication of silicon-on-insulator material). Three of the main isolation techniques will be described next.

107

The LOCOS (local isolation of silicon) technique is undoubtedly the most popular isolation scheme used in bulk CMOS. It is a well-known, reliable and well-controlled process presenting little yield hazards. It consists in the following steps: a thin "pad" oxide is grown, and a silicon nitride layer is deposited. Using a mask step, the nitride is patterned to define the active silicon areas. Boron is then implanted around those islands which will contain n-channel devices (channel-stop implant), and the field oxide is thermally grown. No oxide is grown where the silicon is protected by nitride. A nitride and pad oxide stripping step completes the process. The exact same process can be used in SOI for lateral device isolation. During the field oxide growth step, the silicon film which is not covered with nitride is completely consumed, and the thermally-grown field oxide reaches through to the buried oxide (Figure 4.2.1).

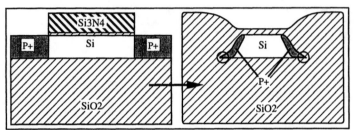

Figure 4.2.1: LOCOS isolation. The circles at the edges of the silicon island (left) indicate where source-to-drain edge leakage may occur.

The lateral encroachment of the field oxide, called "bird's beak", is proportional to the thickness of the grown oxide. This encroachment has typical values of a fraction of a micrometer, and hampers the use of the LOCOS technique for submicron bulk processing, where 0.5...0.8 mm-thick field oxides are grown. The local stresses in the silicon is also related to the thickness of the field oxide. In an SOI process the thickness of the oxide which must be grown to fully isolate the silicon islands is 2.5 - 3 times the thickness of the silicon film. If a 100 nm-thick SOI film is used, one has to grow a field oxide of only 250-300 nm. This implies less stress in the silicon, and, most importantly, a significantly smaller bird's beak. As a consequence, it seems that the LOCOS isolation can be used to realize devices with smaller geometries in SOI than in bulk.

In bulk devices, the P^+ implant (channel-stop implant) which is carried out prior to growing the field oxide prevents surface inversion of the silicon underneath the field oxide. Such an inversion would lead to leakage between n-channel devices. In SOI devices, the same implant is used for preventing source-to-drain leakage from taking place along the edges of the silicon island. Indeed, an inversion layer can form at the tip of the bottom corners of the islands (Figure 4.2.1). This inversion layer degrades the subthreshold slope of the device and is the source of leakage current when the device is turned off.

The edge leakage current is caused by the presence of a parasitic edge transistor in parallel with the main transistor. It has the same gate length as the main device, but has a much smaller width. If the P^+ impurity concentration at the edges of the silicon island is high enough, the threshold voltage of the edge device is higher than that of the main device, and the edge current is overshadowed by the main device current for all values of gate voltage, such that the presence of the edge transistor has no detrimental effect on the overall device characteristics. If, on the other hand, the P^+ impurity concentration at the edges of the silicon island is too low, the threshold voltage of the edge device is lower than that of the main device, and the edge transistor current dominates the current characteristics of the overall device at low gate voltage values (Figure 4.2.2). This results in a degradation of the subthreshold characteristics and an increase of the off-state leakage current.

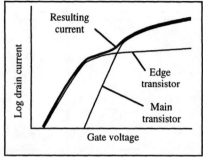

Figure 4.2.2: Subthreshold characteristics of a transistor with edge leakage.

One problem of the LOCOS isolation is that most of the boron implanted in the silicon segregates into the oxide. As a result, the doping at the tip of the bottom corners of the islands is quite low and edge leakage appears. One solution consists it performing a recessed LOCOS (half the silicon thickness is etched in the field regions before the oxide is grown. This procedure allow one to reduce the oxidation time, and, therefore, the boron segregation, and to eliminate the tip of the bottom corners of the islands (Figure 4.2.3). Sacrificial oxidation can be used as an alternative to silicon etching, prior to field oxide growth.[7]

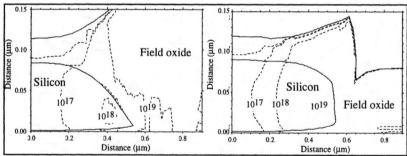

Figure 4.2.3: LOCOS isolation (left) and recessed LOCOS isolation (right). The boron concentrations are expressed in atoms/cm³. The dotted lines represent doping isoconcentration contours.

Figure 4.2.4 presents the $I_D(V_G)$ characteristics of n-channel transistors made using either a LOCOS or a recessed LOCOS isolation. Although the effect of edge conduction can hardly be seen when the back-gate voltage is equal to 0 volt, edge leakage rapidly occurs when a negative back-gate bias is applied, which gives rise to a kink in the subthreshold $I_D(V_G)$ characteristics. This kink increases the values of both the subthreshold slope and the OFF leakage current. When the recessed LOCOS process is used no kink can be seen in the subthreshold $I_D(V_G)$ characteristics, even at large negative back-gate bias where ideal subthreshold characteristics are observed until back accumulation is induced, at the largest negative gate biases. Another way of minimizing the segregation of boron into the oxides consists in performing the edge implant after LOCOS growth (Local Implantation post Field oxidation, LIF)[8] or even after gate patterning. [9,10]

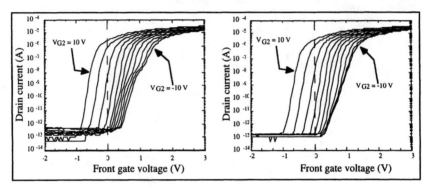

Figure 4.2.4: $I_D(V_G)$ characteristics of n-channel transistors with LOCOS isolation (left) and recessed LOCOS isolation (right). VD=50 mV. The back-gate voltage ranges from -10 to 10 volts in 2-volt steps. W/L=20μm/5μm.[11]

The mesa isolation technique is another way of isolating silicon islands from one another. This technique is attractive because of its simplicity. It simply consists into patterning the silicon into islands -or "mesas"- using a mask step and a silicon etch step. Passivation of the island edges is performed at the gate oxidation step, where the gate dielectric is grown not only on top of the silicon islands, but on their edges as well (Figure 4.2.5). This technique has been extensively used in SOS processing, where a KOH solution was used to etch the silicon and produce sloped island edges.

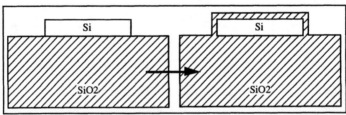

Figure 4.2.5: Mesa isolation

Several problems are associated with the mesa isolation. It is well known that the oxidation of silicon corners produces SiO_2 layers with non-uniform thickness. Indeed, the thickness of the oxide grown on the corners of an island can be 30 to 50% thinner than that grown on its top surface.[12] This thinning of the oxide depends on the oxidation temperature (the effect is more pronounced if oxidation is carried out below or close to the SiO_2 viscous flow temperature (965°C) [13]). Oxidation is also known to sharpen silicon corners [14], such that corners sharper than an angle of 45 degrees are produced. This effect is enhanced if more than a single oxidation step is performed (*i.e.*: if a sacrificial oxide is grown and stripped prior to gate oxidation). The thinning of the oxide and the sharpening of the corners both contribute to a reduction of the gate oxide breakdown voltage observed when mesa isolation is used. Sidewall leakage may also be observed, as in the case where LOCOS isolation is used. In a mesa process, the gate oxide and the gate material covers both the top of the silicon island and its edges. Therefore, there exist lateral (edge) transistors in parallel to the main (top) device. Furthermore, due to charge sharing between the main and the edge devices, the threshold voltage is reduced at the corner of the island [15]. This can produce a kink in the subthreshold characteristics as well as leakage currents similar to those described in Figure 4.2.2. Both the oxide breakdown and the leakage currents can be improved by using P+ sidewall doping and mesa edge rounding techniques [16,17].

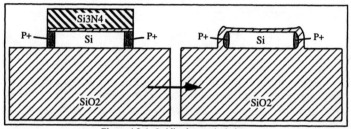

Figure 4.2.6: Oxidized mesa isolation.

A third isolation technique, called the oxidized mesa technique, results from the combination of mesa and LOCOS processes. As in the LOCOS formation, a nitride layer is patterned to define the active silicon areas. Boron is implanted around those islands which will contain n-channel devices. Some of the boron is then driven laterally in the silicon located underneath the nitride, after which the silicon is etched away to form mesas. The sidewalls of the silicon islands are then oxidized (Figure 4.2.6), and the nitride is stripped. The oxidized mesa process has several advantages. The corners of the silicon islands are rounded during the lateral oxidation step. This increases the breakdown voltage of gate oxide by over 30%, compared to the mesa isolation process.[18,19] In addition, there are no regions where the silicon is extremely thin, as it is the case when LOCOS isolation is used. This contributes to improve the subthreshold characteristics and reduce leakage currents.

4.3. Doping profile

The optimization of the doping profile in SOI MOSFETs serves two main purposes: the adjustment of the front threshold voltage and the elimination of back-channel leakage. Different gate materials can be used. The doping profile in a device realized with a P+ polysilicon gate will, of course be totally different from that of a device having an N+ polysilicon as gate material.[20,21,22] We will only consider here the case of N+ poly gate, which is, by far, the most common.

In "thick-film", partially depleted devices, the silicon film is thick enough ($t_{si} \geq$ ...200... nm) for the front implant controlling the front threshold voltage and the back implant controlling the back threshold voltage to be carried out independently. The doping profile of the (enhancement-mode) n-channel transistors presents a double hump (Figure 4.3.1.A). Indeed, a superficial boron implant is used to adjust the threshold voltage to the desired value,

111

and a higher energy implant is carried out to give the back interface a high enough threshold voltage ($\geq$ ...10... V) to avoid back-channel leakage problems.

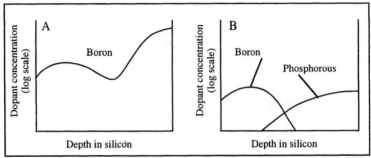

Figure 4.3.1: Doping profiles in "thick-film" SOI MOSFETs. A: n-channel device. B: p-channel device.

The doping profile of a "thick-film" p-channel transistor is shown in Figure 4.3.1.B. It is similar to the profile found in a buried-channel bulk device. A deep n-type (usually phosphorous) implant is used to avoid leakage between source and drain and to control the drain punchthrough voltage, while a shallow p-type implant is used to adjust the front threshold voltage.

In the case of thin-film ($t_{si} \leq$ ...100... nm) MOS devices, there is no room to create anything else than an almost flat doping profile (this "flat" profile is only the center portion of the Gaussian-like profile produced by the implantation), and front and back threshold voltages cannot be adjusted independently. The doping profiles of thin-film n-channel and p-channel devices are presented in Figure 4.3.2. For a front gate oxide thickness of ...20.. nm, the p-type impurity concentration is ...10^{17} cm^{-3}... in the n-channel transistor and ...4×10^{16} cm^{-3}... in the p-channel device. The latter operates as an accumulation-mode device, the channel region of which is fully depleted of holes (and, therefore, non-conducting) when the device is turned off. An accumulation channel forms at the top Si-SiO$_2$ interface when a negative gate voltage is applied. The n-channel transistor is a standard enhancement-mode device.

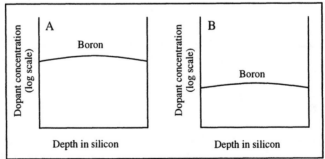

Figure 4.3.2: Doping profiles in thin-film SOI MOSFETs. A: n-channel device. B: p-channel device.

The front threshold voltage of fully depleted SOI MOSFETs is a function of the silicon film thickness. Indeed, the total depletion charge, Q_{depl}, is equal to $-qN_a t_{si}$. Using Equations 5.3.10 and 5.3.12 the dependence of the threshold voltage on the film thickness can be established. Typically the dV_{th}/dt_{si} variation is on the order of 10 mV per nanometer. The front threshold voltage varies also with the amount of charges in the BOX and the interface state density at the Si-SiO_2 back interface. This is the reason why some people use partially depleted MOSFETs instead of fully depleted devices. The following observations, however, give additional credit to fully depleted devices:

◊ Using the concept of "constant dose" doping instead of that of "constant concentration" doping, it is possible to significantly reduce the dV_{th}/dt_{si} variation. This approach proposes to optimize the channel implantation energy in such a way that the dependence of the implanted dose (cm^{-2}) on the film thickness is minimized.[23,24]

◊ In a fully depleted device the variation of drain current with film thickness (dI_D/dt_{si}) is significantly lower than the dV_{th}/dt_{si} variation. Indeed, when the film thickness is increased the body factor decreases, which partially compensates for the current decrease caused by the increase of threshold voltage.[25]

◊ The variation of threshold voltage with temperature (dV_{th}/dT) is 2 to 3 times lower in fully depleted (FD) MOSFETs than in partially depleted (PD) devices (see section 7.2.2). Because devices in a circuit operate at different temperatures due to the self-heating effect as shown in Section 5.7.4, threshold voltage variation can become unacceptable for PD devices. Although FD devices suffer from a greater spread of threshold voltage caused by silicon film thickness fluctuations, the overall threshold voltage variation, including the self-heating effect, may turn out to be smaller when using fully depleted devices rather than partially depleted MOSFETs, as illustrated in Figure 4.3.3.

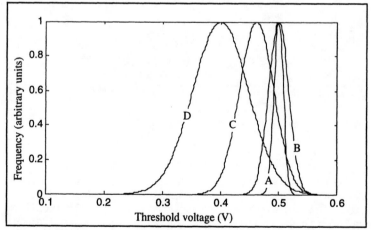

Figure 4.3.3: Example of statistical threshold voltage distribution in A: PDSOI without heating effect; B: FD SOI without heating effect; C: FDSOI with heating effect; D: PD SOI with heating effect The maximum temperature variation is 100°C.

The optimization of the doping concentration in thin-film, fully-depleted devices is a matter of balance between two effects. Firstly, the doping concentration must be low enough to ensure full depletion, and, secondly, it must be high enough to provide the device with a suitable (*i.e.*: large enough) threshold voltage. Finding such a balance is usually not a problem in p-channel devices, but it requires some attention for n-channel transistors. Simulations show that both doping concentration and silicon film thickness have to be optimized in order to produce useful values of threshold voltage.[26] Figure 4.3.4 illustrates the problem. The subthreshold slope of SOI n-channel MOSFETs can be used as a signature of the operation in the fully-depleted mode. If the slope is, say, smaller than 70 mV/decade, the device is fully depleted. If it is larger than, say, 80 mV/decade, it operates in the partially-depleted mode. The simulation of Figure 4.3.3 was carried out for a gate oxide thickness of 15 nm, a front-gate oxide charge density of 5×10^{10} cm^{-2} and a buried oxide charge density of 10^{11} cm^{-2}. When the SOI film is relatively thick (200 nm), the device always operates in the partially-depleted mode, unless the doping concentration is reduced to such low values that the threshold voltage is too low for useful applications (<0.1V). If the silicon film thickness is reduced to 100 nm, fully-depleted operation is maintained even if the channel doping concentration is increased to produce threshold voltages up to 300-400 mV. Further reduction of the silicon film thickness (70 nm) allows one to reach a threshold voltage of 0.7 V while still operating in the fully-depleted mode (Figure 4.3.4).

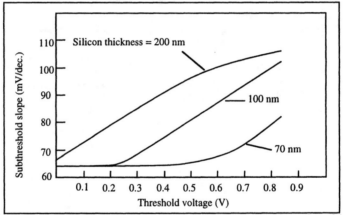

<u>Figure 4.3.4</u>: Subthreshold slope as a function of threshold voltage for various values of the SOI film thickness (n-channel device).

4.4. Source and drain resistance problems

In thin-film SOI technology, the source and drain sheet resistance can reach high values which can jeopardize the speed performances of the circuits (the sheet resistance is roughly inversely proportional to the film thickness). Therefore, it becomes imperative to form a silicide on the sources and the drains to reduce their sheet resistance. Titanium silicide ($TiSi_2$) is the most widely used silicide in SOI technology, although the use of cobalt and nickel silicide has been reported as well.[27,28] In bulk processes, a titanium thickness of ...60

nm... is usually deposited on the silicon in order to form the silicide. In SOI devices, thinner titanium films have to be used to form the silicide.

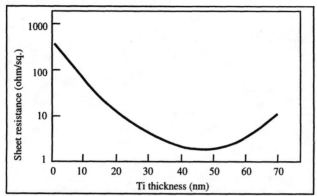

<u>Figure 4.4.1</u>: Source and drain sheet resistance as a function of the thickness of the deposited titanium thickness (100 nm-thick silicon film).

Figure 4.4.1 presents the source and drain sheet resistance (over N^+ and P^+ junctions) of thin-film SOI MOSFETs as a function of the deposited titanium thickness. The silicon film thickness ranges from 65 to 125 nm. The source and drain sheet resistance can be reduced from ...300 Ω/square... to 2 Ω/square by sputtering 45 nm of titanium and forming the silicide in a two-step annealing process. The use of a thicker titanium layer led to shorts between the gate and the source and drain in the reported experiment [29,30].

More recent results have confirmed that there exist a process window for the thickness of the deposited titanium. For 100 nm-thick devices, the optimal titanium thickness is between 35 to 45 nm. Thinner metal yields a non-continuous silicide layer with a high resistivity, and thicker metal layers tend to consume all the silicon of the source and drain regions. Since silicon is the diffusing species in titanium silicide, silicon from underneath the gate oxides diffuses into the silicide, once the entire silicon thickness has been consumed, which leads to the formation of voids underneath the gate edges and to non-functional devices.[31]

In addition to the formation of voids in the silicon film, there is another reason to keep the silicide from reaching the buried oxide. Indeed, the series resistance of a silicided SOI junction abruptly increases when the thickness of the silicide approaches that of the silicon film. This effect is due to a reduction of the contact area once the silicide consumes the silicon layer because the horizontal portion directly underneath the silicide is no longer available for contact.[32] The effective area through which current can flow is then drastically reduced and the resistance increases (Figure 4.4.2). The optimum silicide thickness appears to be approximately 80% of the total silicon film thickness.

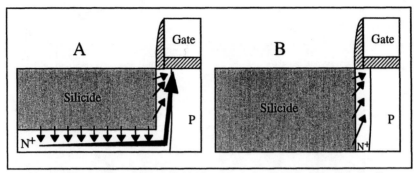

Figure 4.4.2: Current flow path in a silicided junction. A: the silicide is thinner than the silicon film; B: the film is fully silicided.

Another way of fabricating thin SOI devices without compromising the source and drain (S&D) resistance consists in using different silicon film thicknesses for the channel and the S&D regions. Indeed, the use of a thinner silicon film in the channel region will provide the desired fully depleted, thin-film SOI MOSFET features, and the used of thicker silicon for the sources and drains will decrease the S&D resistance. This can be achieved in two ways. In a first approach, a thin silicon film is used, and selective epitaxial growth is used to increase the thickness of the S&D regions (elevated source and drain technique).[33,34] In a second technique a thicker silicon film is used and the channel areas are thinned using a LOCOS technique (recessed channel technique, presented in Figure 4.4.3).[35,36,37,38] It is even possible to self-align the recessed region to the polysilicon gate.[39]

Finally, the S&D resistance can be reduced as well by using selective metal deposition (*e.g.* tungsten) on the source and drain.[40]

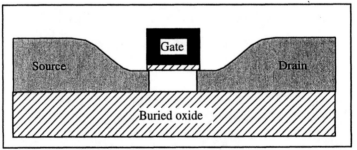

Figure 4.4.3: Thin-film SOI MOSFET with recessed channel.

116

4.5. SOI MOSFET design

There exist different types of designs of SOI transistors. The densest and most common layout is presented in Figure 4.5.1.A. It consists of a rectangular active area, a gate, and contact holes. In the case of an n-channel device (which will be illustrated throughout this Section) the pattern of the active area is surrounded by another mask pattern (dark field if positive resist is employed) which is used for the field implant, the N-channel V_{th} adjust and back-channel stop implants, as well as for the N$^+$ source and drain implant steps. Similarly, the p-channel transistors are enclosed in a pattern defining the P$^+$ source and drain implant. The latter pattern can also be used to create a P$^+$ body contact in n-channel devices.

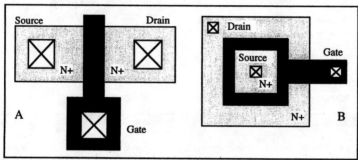

Figure 4.5.1: Layout of SOI MOSFETs. A: "normal" device. B: edgeless device.

When the application in which a circuit is used may lead to edge leakage problems (such as in devices submitted to ionizing radiations which can generate huge amounts of oxide charges in the oxide at the edges of the silicon islands), "edgeless" device designs can be utilized (Figure 4.5.1.B). In such a device, the silicon island (active area) presents no edge underneath the gate between the source and the drain. It is, however, worth noting that edgeless devices occupy much more silicon real estate than conventional devices, and are not used where integration density is a prime concern.

Some applications require devices having body contacts. Indeed, contacting the silicon underneath the gate effectively suppresses the kink effect as well as parasitic lateral bipolar effects. Several schemes exist to provide the transistor body with a contact. The conventional contact is presented in Figure 4.5.2.A. It consists into a P$^+$ diffusion which is in contact with the P-type silicon underneath the gate. Such a device can also be used as a lateral bipolar transistor, the P$^+$ diffusion being the base contact, and the source and drain being used as emitter and collector, respectively. In transistors with large gate width, the presence of a single body contact at one end of the channel region may not be sufficient to suppress kink or bipolar effects. These effects can indeed take place underneath the gate, "far" from the body contact, the efficiency of which is reduced by the high resistance of the weakly doped channel region. The H-gate MOSFET design helps solving this problem, since body contacts are present at both ends of the channel (Figure 4.5.2.B). Furthermore, the H-gate device offers no direct edge leakage path between source and drain (the edges run only from N^+ to P^+ diffusions) [41].

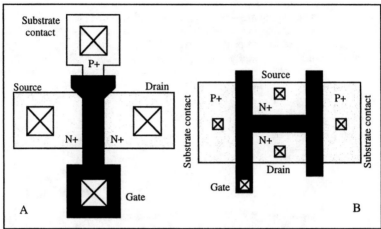

Figure 4.5.2: Transistors with body contact. A: "normal" contact. B: H-gate device.

A third type of body contact, more compact than the previous ones, is shown in Figure 4.5.3. The P+ body ties are created on the side of the N^+ source diffusion. As in the case of the H-gate device, there is no direct edge leakage path between source and drain (the edges of the active area under the gate run only from N^+ to P^+ diffusions). If the device is very wide, additional P^+ regions can be formed in the source (such that a P^+-N^+-P^+...N^+-P^+ structure is produced). This device has the drawback of being asymmetrical (source and drain cannot be swapped), and the effective channel width, W_{eff}, is smaller than the width of the active area [42].

It is worth mentioning that body contacts are used in "thick-film", partially-depleted devices only. In thin-film devices, the full depletion gives the silicon below the gate an almost infinite resistivity which renders body contacts totally ineffective.

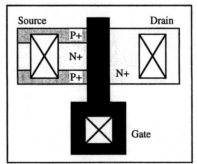

Figure 4.5.3: N-channel transistor with body tie at the source.

118

4.6. SOI-bulk CMOS design comparison

Generally speaking, SOI CMOS technology offers a higher integration density than bulk CMOS. This becomes evident from comparison between the layout of a bulk CMOS inverter and that of an SOI CMOS inverter (Figures 4.6.1 and 4.6.2).

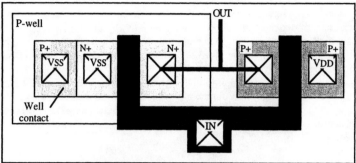

Figure 4.6.1: Layout of a bulk CMOS inverter.

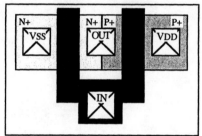

Figure 4.6.2: Layout of an SOI CMOS inverter.

This higher density results mainly from the absence of wells in SOI. A second cause of density increase is the possibility offered by SOI of having a direct contact between P^+ and N^+ junctions (such as the drains of the n-channel and the p-channel devices of Figure 4.6.2). The number of contact holes per gate is also lower in SOI than in bulk. This reduces a source of fabrication yield hazard, compared to bulk.

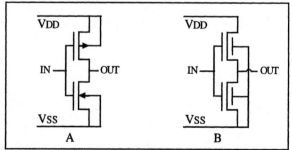

Figure 4.6.3: Back-gate (body) bias configuration in bulk (A) and SOI (B) CMOS inverters.

119

One of the major differences between SOI and bulk design is the difference of body effect and of body/back gate bias conditions. The body effect induced by the back gate (dV_{th1}/dV_{G2}) is negligible in partially depleted devices. The expressions for the body effect in thin-film devices (the dependence of threshold voltage on back-gate bias) can be derived from Sections 5.3.2 and 5.8 for n-channel and p-channel devices, respectively. Furthermore, the back-gate bias configuration of SOI MOSFETs is different from the substrate bias used in bulk. Let us take the example of a simple CMOS inverter (Figure 4.6.3). In bulk CMOS, the body of the n-channel device is connected to ground (V_{SS}), while the body of the p-channel transistor is connected to V_{DD} (usually +5 V). Hence, the potential of the body is the same as that of the source in both types of devices ($V_{sub} = 0$). In the SOI inverter, the back gate (the underlying silicon wafer) is common to both n- and p-type devices. It is usually grounded. Hence, the back-gate voltage is 0 V for the n-channel device, but it is equal to $-V_{DD}$ for the p-channel transistor, the source voltage being always used as a reference. As a consequence, SOI p-channel transistors have usually to be designed for operating with a back-gate bias, V_{G2}, which is equal to $-V_{DD}$.

References

1   H. Onishi, K. Imai, Y. Matsubara, Y. Yamada, T. Tamura, T. Sakai, and T. Horiuchi, Symposium on VLSI Technology, Digest of Technical Papers, p. 33, 1997

2   C.T. Lee and A. Burns, IEEE Electron Device Letters, Vol. 9, p. 235, 1988

3   P.H. Woerlee, C. Juffermans, H. Lifka, W. Manders, F. M. Oude Lansink, G.M. Paulzen, P. Sheridan, and A. Walker, Technical Digest of IEDM, p. 583, 1990

4   P.K. Karulkar, IEEE Electron Device Letters, Vol. 14, p. 80, 1993

5   I.K. Kim, W.T. Kang, J.H. Lee, S. Yu, S.C. Lee, K. Yeom, Y.G. Kim, D.H. Lee, G. Cha, B.H. Lee, S.I. Lee, K.C. Park, T.E. Shim, and C.G. Hwang, Technical Digest of IEDM, p. 605, 1996

6   W.M. Huang, Z.A. Ma, M. Racanelli, D. Hughes, S. Ajuria, G. Huffman, T.P. Ong, P.K. Ko, C. Hu, and B.Y. Hwang, Technical Digest of IEDM, p. 735, 1993

7   T. Ohno, Y. Kado, M. Harada, and T. Tschuiya, IEEE Transactions on Electron Devices, Vol. 42, No. 8, p. 1481, 1995

8   H.S. Kim, S.B. Lee, D.U. Choi, J.H. Shim, K.C. Lee, K.P. Lee, K.N. Kim, and J.W. Park, Technical Digest of the Symposium on VLSI Technology, p. 143, 1995

9   M. Racanelli, W.M. Huang, H.C. Shin, J. Foerstner, J. Ford, H. Park, S. Cheng, T. Wetteroth, S. Hong, H. Shin, and S.R. Wilson, in "Silicon-on-Insulator Technology and Devices VII", Ed. by. P.L.F. Hemment, S. Cristoloveanu, K. Izumi, T. Houston, and S. Wilson, Proceedings of the Electrochemical Society, Vol. 96-3, p. 422, 1996

10   M. Racanelli, W.M. Huang, H.C. Shin, J. Foerstner, B.Y Hwang, S. Cheng, P.L. Fejes, H. Park, T. Wetteroth, S. Hong, H. Shin, and S.R. Wilson, Technical Digest of the IEDM, p. 885, 1995

11   J.P. Colinge, A. Crahay, D. De Ceuster, V. Dessart, and B. Gentinne, Electronics Letters, Vol. 32, No. 19, p. 1834, 1996

12   S.S. Tsao, D.M. Fletwood, V. Kaushik, A.K. Datye, L. Pfeiffer, and G.K. Celler, Proceedings IEEE SOS/SOI Technology Workshop, p. 33, 1987

13   R.K. Smeltzer and J.T. McGinn, Proceedings IEEE SOS/SOI Technology Workshop, p. 32, 1987

14   R.B. Marcus and T.T. Sheng, J. Electrochem. Soc., Vol. 129, p. 1278, 1982

15   M. Matloubian, R. Sundaresan, and H. Lu, Proceedings IEEE SOS/SOI Technology Workshop, p. 80, 1988

16   M. Haond and O. Le Néel, Proceedings IEEE SOS/SOI Technology Conference, p. 132, 1990

17   O. Le Néel, M.D. Bruni, J. Galvier, and M. Haond, in "ESSDERC 90", Adam Hilger Publisher, Ed. by. W. Eccleston and P.J. Rosser, p. 13, 1990

18   M. Haond, O. Le Néel, G. Mascarin, and J.P. Gonchond, Proceedings IEEE SOS/SOI Technology Conference, p. 68, 1989

19   M. Haond, O. Le Neel, G. Mascarin, and J.P. Gonchond, in ESSDERC'89, European Solid-State Device Research Conference, Berlin, Ed. by. A. Heuberger, H. Ryssel and P. Lang, Springer-Verlag, p. 893, 1989

20   T. Aoki, M. Tomizawa, and A. Yoshii, IEEE Trans. on Electron Devices, Vol. 36, p. 1725, 1989

21   T. Nishimura, Y. Yamaguchi, H. Miyatake, and Y. Akasaka, Proceedings IEEE SOS/SOI Technology Conference, p. 132, 1989

22   M. Haond, in ESSDERC'89, European Solid-State Device Research Conference, Berlin, Ed. by. A. Heuberger, H. Ryssel and P. Lang, Springer-Verlag, p. 881, 1989

23   M.J. Sherony, L.T. Su, J.E. Chung, and D.A. Antoniadis, IEEE Electron Device letters, Vol. 16, No. 3, p. 100, 1995

24   D.A. Antoniadis, Proceedings of the IEEE International SOI Conference, p. 1, 1995

25   P. Smeys, U. Magnusson, J.P. Colinge, Proceedings of ESSDERC'92, Microelectronic Engineering, Vol 19, p. 823, 1992

26   J.P. Colinge, Proceedings IEEE SOS/SOI Technology Conference, p. 13, 1989

27   L.T. Su, M.J. Sherony, H. Hu, J.E. Chung, and D.A. Antoniadis, IEEE Electron Device Letters, Vol. 15, No. 9, p. 363, 1994

28   F. Deng, R.A. Johnson, W.B. Dubbelday, G.A. Garcia, P.M. Asbeck, and S.S. Lau, Proceedings of the IEEE International SOI Conference, p. 78, 1996

29   T. Nishimura, Y. Yamaguchi, H. Miyatake, and Y. Akasaka, Proceedings IEEE SOS/SOI Technology Conference, p. 132, 1989

30   Y. Yamaguchi, T. Nishimura, Y. Akasaka, and K. Fujibayashi, IEEE Transactions on Electrons Devices, Vol. 39, No. 5, p.1179, 1992

31   J. Foerstner, J. Jones, M. Huang, B.Y. Hwang, M. Racanelli, J. Tsao, and N.D. Theodore, Proceedings of the IEEE International SOI Conference, p. 86, 1993

32   L.T. Su, M.J. Sherony, H. Hu, J.E. Chung and D.A. Antoniadis, Technical Digest of IEDM, p. 723, 1993

33   J.M. Hwang, R. Wise, E. Yee, T. Houston, and G.P. Pollack, Digest of Technical papers, Symposium on VLSI Technology, p. 33, 1994

34   M. Cao, T. Kamins, P. Vande Voorde, C. Diaz, and W. Greene, IEEE Electron Device Letters, Vol. 18, No. 6, p. 251, 1997

35   M. Chan, F. Assaderaghi, S.A. Parke, S.S. Yuen, C. Hu, and P.K. Ko, Proceedings of the IEEE International SOI Conference, p. 172, 1993

36   O. Faynot and B. Giffard, IEEE Electron Device Letters, Vol. 15, p. 175, 1994

37   A. Toriumi, J. Koga, H. Satake, and A. Ohata, Technical Digest of IEDM, p. 847, 1995

38   C. Raynaud, J.L. Pelloie, O. Faynot, B. Dunne, and J. Hartmann, Proceedings of the IEEE International SOI Conference, p. 12, 1995

39   J.H. Lee, H.C. Shin, J.S. Lyu, B.W. Kim, and Y.J. Park, Proceedings of the IEEE International SOI Conference, p. 122, 1996

40   D. Hisamoto, K. Nakamura, M. Saito, N. Kobayashi, S. Kimura, R. Nagai, T. Nishida, and E. Takeda, Technical Digest of IEDM, p. 829, 1992

41   N.K. Annamalai and M.C. Biwer, IEEE Trans. Nuclear Science, Vol. 35, p. 1372, 1988

42   Y. Omura and K. Izumi, IEEE Transactions on Electron Devices, Vol. 35, p. 1391, 1988

CHAPTER 5 - The SOI MOSFET

Although most types of devices can be fabricated in SOI films, the preferred application field for Silicon-on-Insulator technology is undeniably CMOS. Other types of devices (bipolar devices, novel devices, etc.) will be reviewed in Chapter 6. SOI MOSFETs exhibit interesting properties which make them particularly attractive for applications such as rad-hard circuits, deep-submicron devices and high-temperature electronics. The properties of the SOI MOSFET operating in a harsh environment will be described in Chapter 7.

5.1. Capacitances

5.1.1. Source and drain capacitance

Contrarily to bulk CMOS where both junction and field oxide isolations are used, CMOS SOI devices are dielectrically isolated from one another. This rules out latch-up between devices, as indicated in Chapter I. Similarly, there is no leakage path between devices, while surface leakage problems and field transistor action may occur in bulk technologies. Full dielectric isolation can be interesting for monolithic integration of both high-voltage devices and low-voltage CMOS on a single chip.

In bulk MOS devices, the parasitic drain (or source)-to-substrate (or well) capacitance consists of two components: the capacitance between the drain and the substrate, and the capacitance between the drain and the channel-stop implant under the field oxide (Figure 5.1.1.A). As devices are shrunk to smaller geometries, higher substrate doping concentrations are used, and hence the junction capacitance increases as well. In SOI devices (with reach-through junctions), the junction capacitance has only one component: the capacitance of the MOS structure comprising the junction (gate electrode of the MOS structure under consideration), the buried oxide (gate oxide of the MOS structure), and the underlying silicon substrate (substrate of the MOS structure). This parasitic capacitance can only be smaller than the capacitance of the buried oxide (Figure 5.1.1.B), which is typically lower than the junction capacitance of a bulk MOSFET. This reduction of parasitic capacitances contributed to the excellent speed performances observed in CMOS/SOI circuits. In addition, the buried oxide thickness does not have to be scaled down, as device geometries are reduced. This reinforces the capacitance advantage of SOI over bulk, as technologies evolve towards submicron dimensions.

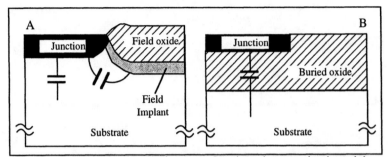

Figure 5.1.1: Parasitic junction capacitances. A: Capacitance between a junction and the substrate and between the junction and the field (channel-stop) implant in a bulk device. B: Capacitance between a junction and the substrate, across the buried oxide, in an SOI device.

The presence of a buried oxide underneath the devices reduces not only the junction capacitances, but it reduces some other capacitances as well (all the capacitances between the silicon substrates and another terminal). Table 5.1.1. presents typical capacitances of a bulk and an SOI 1-micrometer processes.[1] The capacitances are given in fF/μm^2. The reduction of capacitance is, of course, most noticeable between the junctions and the substrate, but one can also observe that even the metal 1-to-substrate capacitance can be reduced by 40% by using SOI substrates rather than bulk silicon wafers.

Capacitor type	SOI (SIMOX)	Bulk	Gain (SOI vs. bulk)
Gate	1.3	1.3	1
Junction-to-substrate	0.05	0.2 ... 0.35	4 ... 7
Polysilicon-to-substrate	0.04	0.1	2.5
Metal 1-to-substrate	0.027	0.05	1.85
Metal 2-to-substrate	0.018	0.021	1.16

Table 5.1.1: Parasitic capacitances (fF/μm^2) found in typical bulk and SOI 1-μm CMOS processes.

The drain capacitance can be calculated as a function of the supply voltage. In the case of a bulk device, it is given by the classical expression for the capacitance of a PN junction:

$$C = \sqrt{\frac{q\varepsilon_s}{2} \frac{N_a N_d}{(N_a+N_d)}} \frac{1}{\sqrt{\Phi_o - V_d}}$$

where V_d is the voltage across the junction (*i.e.* the drain-substrate voltage), Φ_o is the built-in junction potential, and N_d and N_a are the doping concentrations in the N- and P-type regions, respectively. In the case of an SOI MOSFET, the drain capacitance is easily obtained from the MOS capacitor theory:

$$C = \frac{C_{BOX}}{\sqrt{1 + \frac{2C_{BOX}^2 V_d}{qN_a\varepsilon_{si}}}}$$

where C_{BOX} is the capacitance of the buried oxide, V_d is the drain-substrate voltage, and N_a is the substrate doping concentration. Typical values of capacitances are presented in Figure 5.1.2.

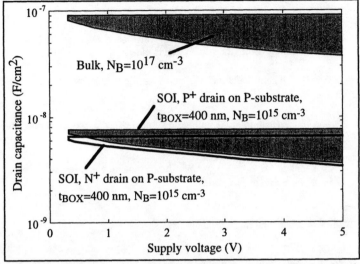

Figure 5.1.2: Parasitic junction capacitance per unit area as a function of supply voltage in bulk CMOS with constant substrate doping concentration (10^{17} cm^{-3}) and in standard SIMOX with a substrate doping concentration of 10^{17} cm^{-3}.[2]

5.1.2. Gate capacitance

Anomalous intrinsic gate capacitance characteristics (when compared to bulk) may be obtained, in some modes of operation of SOI MOSFETs, as a result of the possibility of multiple conduction paths (*e.g.*: a back channel), source (or drain) coupling with the gate through the floating body, combinations of front and back oxide capacitances, as well as to floating substrate effects and impact-ionization related phenomena [3,4]:

◊   The subthreshold front gate capacitance for low V_D values and for inverted film back interface conditions includes the back depletion region or the fully depleted film capacitance and exceeds by far the conventional (bulk) extrinsic value.

◊   The saturation intrinsic source-to-gate capacitance is equal to 0.72, instead of 0.66 times the total gate oxide capacitance (WLC_{ox1}) when the device is operated in partial depletion, as a result of a unique capacitive coupling through the source junction, floating body, and pinch-off depleted region.

◊   Impact ionization causes a steady increase in the source-to-gate capacitance with drain voltage in partial as well as in full depletion device operation.

◊   The activation of the parasitic bipolar transistor induces a decrease in the drain-to-gate capacitance, reducing it down to unusual negative values, for very high V_D values.

In accumulation-mode devices, the gate capacitance exhibits the following features [5]:

◊   A two-step dependence of the gate capacitance on the gate voltage which tends to disappear as a positive back-gate bias is applied. This effect is due to the presence of a "body" conduction within the silicon film thickness.

◊   A significant value of the gate capacitance below threshold, despite the negligiblecurrent

◊   A lower (than in bulk) saturation value for C_{GS} in moderate accumulation and below threshold.

◊   A kink in the $C_{GD}(V_{G1})$ curves in the transition region between the triode and the saturation regimes of operation.

5.2. Distinction between thick- and thin-film devices

All SOI MOSFETs are not alike. Their physics is highly dependent on the thickness of the silicon film on to which they are made. Three types of devices can be distinguished, depending on both the silicon film thickness and the channel doping concentration: the thick-film and the thin-film devices, as well as the "medium thickness" device, which can exhibit either a thin- or a thick-film behavior, depending on the back-gate bias. Figure 5.2.1 presents the band diagrams of a bulk, a thick-film SOI, and a thin-film SOI n-channel device at threshold.

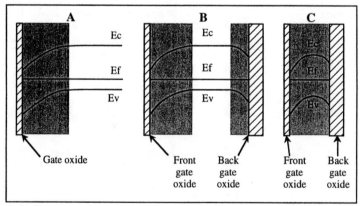

Figure 5.2.1: Band diagram in a bulk (A), a thick-film SOI (B), and a thin-film SOI device. All devices are represented at threshold (front gate voltage = threshold voltage). The shaded areas represent the depleted zones. SOI devices are represented for a condition of weak inversion (below threshold) at the back interface.

◊   In a *bulk* device (Figure 5.2.1.A), the depletion zone extends from the Si-SiO$_2$ interface down to the maximum depletion width, x$_{dmax}$, which is classically given by:

$$x_{dmax} = \sqrt{\frac{4\varepsilon_{si}.\Phi_F}{q\,N_a}}\,,\ \Phi_F \text{ being the Fermi potential, which is equal to } \frac{kT}{q}\ln\!\left(\frac{N_a}{n_i}\right).$$

◊   In a *thick-film SOI* device (Figure 5.2.1.B), the silicon film thickness is larger than twice the value of x$_{dmax}$. In such a case, there is no interaction between the depletion zones arising from the front and the back interfaces, and there exists a piece of neutral silicon beneath the front depletion zone. Such a device is called **partially depleted (PD)**. If this neutral piece of silicon, called "body", is connected to ground by a "body contact", the characteristics of the device will exactly be those of a bulk device. If, however, the body is left electrically floating, the device will basically behave as a bulk device, but with the notable exception of two parasitic effects: the first one is called the "kink effect" (see Section 5.6), and the second one is the presence of a parasitic open-base NPN bipolar transistor between source and drain (see Section 5.8).

◊   In a *thin-film SOI* device (Figure 5.2.1.C), the silicon film thickness is smaller than x$_{dmax}$. In this case, the silicon film is fully depleted at threshold, irrespective of the bias applied to the back gate (with the exception of the possible presence of thin accumulation or inversion layers at the back interface, if a large negative or positive bias is applied to the back gate, respectively). Such a device is called **fully depleted (FD)**. Fully depleted SOI devices are virtually free of the kink effect, if their back interface is not in accumulation. Among all types of SOI devices, fully depleted devices with depleted back interface exhibit the most attractive properties, such as low electric fields, high transconductance, excellent short-channel behavior, and a quasi-ideal subthreshold slope. Thin-film SOI MOSFETs are often referred to as fully-depleted devices.

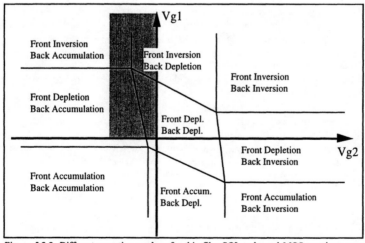

Figure 5.2.2: Different operation modes of a thin-film SOI n-channel MOS transistor as a function of front-gate bias (V$_{G1}$) and back-gate bias (V$_{G2}$) - (linear regime, low drain voltage). The shaded area represents the normal mode of operation.

127

Because either of the front and back interface can be in accumulation, depletion or inversion, one can number 9 modes of operation in a thin-film SOI transistor as a function of V_{G1} and V_{G2} (Figure 5.2.2).[6] Most of these operation modes are not of practical use, however. The useful operation modes are indicated by the shaded area of Figure 5.2.2.

To be more general, it should be mentioned that the presence of accumulation, depletion or inversion layers is also a function of the drain voltage, and that the back interface can, for instance, be accumulated near the source and depleted near the drain. Such a mode of operation will be analyzed later on. It is also worthwhile noting that most of the "attractive properties" of fully depleted devices, with depleted back interface (high transconductance, etc.), can be disabled by the presence of an accumulation layer at the back interface.

The previous remarks are valid for enhancement-mode MOSFETs, which constitute the most popular class of SOI devices (at least in the n-channel case). It is worth noting, however, that another type of devices can be realized. These are accumulation-mode (or "deep-depletion") devices. The operation mode of a MOSFET (enhancement or accumulation) depends on fabrication parameters, the most important being the type of gate material used (Table 5.2.1).[7] We will, however, consider that thin-film SOI devices are enhancement-mode devices, unless otherwise specified.

	N+ poly gate	P+ poly gate
n-channel device	Inversion	Accumulation
p-channel device	Accumulation	Inversion

Table 5.2.1: Different operation modes of a thin-film SOI MOSFET as a function of gate material.

◊ *Medium-thickness* SOI devices are an intermediate case between thick- and thin-film devices, and are obtained in those cases where $x_{dmax} < t_{si} < 2x_{dmax}$, t_{si} being the silicon film thickness. If the back-gate bias is such that the front and back depletion zones do not touch each other, or if the back interface is neutral or accumulated, the transistor will behave as a thick-film device. If, on the other hand, the presence of a back-gate bias induces an overlap between the front and back depletion zones, the device will be fully depleted, and it will behave as a thin-film device.

The merits of the different types of SOI MOSFETs are reported in Table 5.2.2, where some electrical properties of the devices are compared. Bulk silicon devices are taken as a reference. One can see that thin-film, fully depleted devices without accumulation at the back interface offer the most attractive properties for ULSI applications. The popularity of partially depleted devices is due to the independence of their threshold voltage on silicon film thickness and on charges in the buried oxide.

The following section of this chapter will describe the properties and the characteristics of thick- and thin-film MOSFETs. Medium-thickness devices behave in a rather complex way and can be treated as either thick- or thin-film devices, as a function of back-gate, front-gate or drain bias conditions. Indeed, in such a device the back-interface charge condition, for instance, can vary between accumulation, neutrality and depletion along the device length from source to drain.

	Bulk	Thick-film SOI	Thin-film SOI Back accum.	Thin-film SOI Fully depleted
Mobility	0	0	0 / -	+
Transconductance	0	0	0 / -	+
Short-channel effect	0	0	+	0 / +
S&D capacitance	0	+	+	+
Hot carriers	0	0 / +	0 / -	+
Subthreshold slope	0	0	0 / -	+
Vth sensitivity on tsi	0	0	-	-
Kink	0	-	-	0
Parasitic bipolar	0	-	-	0 / -
Total-dose hardness	0	0 / +	0 / -	-
SEU hardness	0	+	+	+
Soft-error hardness	0	+	+	+

Table 5.2.2: Comparison of some of the electrical properties of thick-film, thin-film with accumulated back interface, and thin-film fully depleted SOI devices. The bulk device is given as a reference. 0,+ and - mean "similar to bulk", "better than bulk" and "worse than bulk", respectively.

5.3. I-V Characteristics

5.3.1. Threshold voltage

The threshold voltage of an enhancement-mode **bulk** n-channel MOSFET is classically given by:[8]

$$V_{th} = V_{FB} + 2\Phi_F + \frac{q N_a x_{dmax}}{C_{ox}} \qquad (5.3.1)$$

where V_{FB} is the flatband voltage, equal to $\Phi_{MS} - \frac{Q_{ox}}{C_{ox}}$ (we will neglect the presence of fast surface states, N_{it}, in the present analysis), Φ_F is the Fermi potential, equal to $\frac{kT}{q} \ln\left(\frac{N_a}{n_i}\right)$, and x_{dmax} is the maximum depletion width, equal to $\sqrt{\frac{4\varepsilon_{si} \Phi_F}{q N_a}}$.

In a **thick-film** SOI device ($t_{si} > 2 x_{dmax}$), there can be no interaction between the front and back depletion zones. In that case, the threshold voltage is the same as in a bulk transistor and is given by Equation (5.3.1).

The threshold voltage of a **thin-film**, fully depleted, enhancement-mode n-channel SOI device [9] (Figure 5.3.1) can be obtained by solving the Poisson equation, using the depletion approximation: $\frac{d^2\Phi}{dx^2} = \frac{q N_a}{\varepsilon_{si}}$, which when integrated twice yields the potential as a function of depth in the silicon film, x:

$$\Phi(x) = \frac{q N_a}{2 \, \varepsilon_{si}} x^2 + (\frac{\Phi_{s2} - \Phi_{s1}}{t_{si}} - \frac{q N_a t_{si}}{2 \, \varepsilon_{si}}) x + \Phi_{s1} \qquad (5.3.2)$$

where Φ_{s1} and Φ_{s2} are the potentials at the front and back silicon/oxide interfaces, respectively (Figure 5.3.2). The doping concentration, N_a, is assumed to be constant.

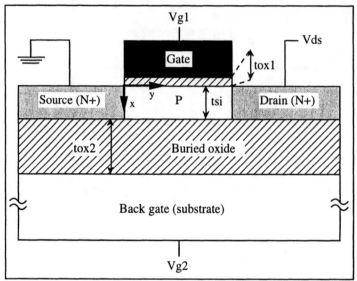

Figure 5.3.1: Cross-section of a thin-film, n-channel SOI MOSFET illustrating some of the notations used in this section.

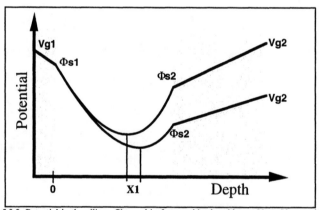

Figure 5.3.2: Potential in the silicon film and in front and back oxides, at $V_{G1} = V_{th1}$ and for two back-gate bias conditions. x_1 is the point of minimum potential. The depletion zone between x=0 and x=x_1 is controlled by the front gate, and the depletion zone between x=x_1 and the back Si-SiO_2 interface is controlled by the back gate. The shaded areas represent the gate (left) and buried (right) oxides.

The electric field in the silicon film is given by:

$$E(x) = \frac{-q\,N_a}{\varepsilon_{si}}x - (\frac{\Phi_{s2} - \Phi_{s1}}{t_{si}} - \frac{q\,N_a\,t_{si}}{2\,\varepsilon_{si}}) \qquad (5.3.3)$$

The front surface electric field, E_{s1} (at x=0), can be calculated from (5.3.3) and is given by:

$$E_{s1} = (\frac{\Phi_{s1} - \Phi_{s2}}{t_{si}} + \frac{q\,N_a\,t_{si}}{2\,\varepsilon_{si}}) \qquad (5.3.4)$$

Applying Gauss' theorem at the front interface, one obtains the potential drop across the gate oxide, Φ_{ox1}:

$$\Phi_{ox1} = \frac{\varepsilon_{si}\,E_{s1} - Q_{ox1} - Q_{inv1}}{C_{ox1}} \qquad (5.3.5)$$

where Q_{ox1} is the fixed charge density at the front Si-SiO$_2$ interface, Q_{inv1} is the front channel inversion charge ($Q_{inv1} < 0$), and C_{ox1} is the front gate oxide capacitance. Similarly, applying Gauss' theorem at the back interface and using (5.3.4) yields the potential drop across the buried oxide, Φ_{ox2}:

$$\Phi_{ox2} = -\frac{\varepsilon_{si}\,E_{s1} - q\,N_a\,t_{si} + Q_{ox2} + Q_{s2}}{C_{ox2}} \qquad (5.3.6)$$

where Q_{s2} is the charge in a possible back inversion ($Q_{s2} < 0$) or accumulation ($Q_{s2} > 0$) layer.

The front and back gate voltages, V_{G1} and V_{G2}, are given by:

$$V_{G1} = \Phi_{s1} + \Phi_{ox1} + \Phi_{MS1} \quad \text{and} \quad V_{G2} = \Phi_{s2} + \Phi_{ox2} + \Phi_{MS2} \qquad (5.3.7)$$

where Φ_{MS1}, Φ_{MS2} are the front and back work function differences, respectively.

By combining (5.3.4), (5.3.5) and (5.3.7), we obtain the relationship between the front gate voltage and the surface potentials:

$$V_{G1} = \Phi_{MS1} - \frac{Q_{ox1}}{C_{ox1}} + \left(1 + \frac{C_{si}}{C_{ox1}}\right)\Phi_{s1} - \frac{C_{si}}{C_{ox1}}\Phi_{s2} - \frac{\frac{1}{2}Q_{depl} + Q_{inv1}}{C_{ox1}} \qquad (5.3.8)$$

where $C_{si} = \varepsilon_{si}/t_{si}$ and Q_{depl} is the total depletion charge in the silicon film, which is equal to $-q\,N_a\,t_{si}$.

131

Similarly, one finds the relationship between the back gate voltage and the surface potentials:

$$V_{G2} = \Phi_{MS2} - \frac{Q_{ox2}}{C_{ox2}} - \frac{C_{si}}{C_{ox2}} \Phi_{s1} + \left(1 + \frac{C_{si}}{C_{ox2}}\right) \Phi_{s2} - \frac{\frac{1}{2} Q_{depl} + Q_{s2}}{C_{ox2}} \tag{5.3.9}$$

Equations (5.3.8) and (5.3.9) are the key relations which describe the charge coupling between the front and back gates in a fully depleted SOI MOSFET. Combining them yields the dependence of the (front) threshold voltage on back-gate bias and device parameters.

We will now detail the expression of the threshold voltage of the thin-film SOI MOSFET as a function of the different possible steady-state charge conditions at the back interface.

If the back surface is **accumulated**, Φ_{s2} is pinned to approximately 0V. The threshold voltage $V_{th1,acc2}$ is, therefore, obtained from Equation (5.3.8), where $V_{th1,acc2} = V_{G1}$ is calculated at $\Phi_{s2} = 0$, $Q_{inv1}=0$, and $\Phi_{s1} = 2\Phi_F$. The result is:

$$V_{th1,acc2} = \Phi_{MS1} - \frac{Q_{ox1}}{C_{ox1}} + \left(1 + \frac{C_{si}}{C_{ox1}}\right) 2\Phi_F - \frac{Q_{depl}}{2C_{ox1}} \tag{5.3.10}$$

If the back surface is **inverted**, Φ_{s2} is pinned to approximately $2\Phi_F$. The front threshold voltage $V_{th1,inv2}$ is, therefore, obtained from Equation (5.3.8), where $V_{th1,inv2} = V_{G1}$ is calculated at $\Phi_{s2} = 2\Phi_F$, $Q_{inv1}=0$, and $\Phi_{s1} = 2\Phi_F$. The result is:

$$V_{th1,inv2} = \Phi_{MS1} - \frac{Q_{ox1}}{C_{ox1}} + 2\Phi_F - \frac{Q_{depl}}{2C_{ox1}} \tag{5.3.11}$$

It is worth noting that in this case, the device is still ON even if $V_{G1} < V_{th1,inv2}$ since the back interface is inverted, and that the device is, therefore, useless for any practical circuit application.

If the back surface is depleted, Φ_{s2} depends on the back-gate voltage, V_{G2}, and its value can range between 0 and $2\Phi_F$. The value of back-gate voltage for which the back interface reaches accumulation (the front interface being at threshold), $V_{G2,acc}$, is given by Equation (5.3.9) where $\Phi_{s1} = 2\Phi_F$, $\Phi_{s2} = 0$, and $Q_{s2}=0$. Similarly, the value of back-gate voltage for which the back interface reaches inversion, $V_{G2,inv}$, is given by the same equation where $\Phi_{s1} = 2\Phi_F$, $\Phi_{s2} = 2\Phi_F$, and $Q_{s2}=0$. When $V_{G2,acc} < V_{G2} < V_{G2,inv}$ the front threshold voltage is obtained by combining Equations (5.3.8) and (5.3.9) with $\Phi_{s1} = 2\Phi_F$ and $Q_{inv1} = Q_{s2} =0$. The result is:

$$V_{th1,depl2} = V_{th1,acc2} - \frac{C_{si} C_{ox2}}{C_{ox1} (C_{si} + C_{ox2})} (V_{G2} - V_{G2,acc}) \tag{5.3.12}$$

The above relationships are valid if the thickness of the inversion or accumulation layers are small with respect to the silicon film thickness. This may no longer be the case in ultra-thin-film devices, in which case the width of the accumulation/inversion zones must be subtracted from the silicon film thickness to obtain an effective silicon thickness (= the effective width of the depleted layer) to be used in place of t_{si} in the above relationships, in a first approximation. In very thin films (< 10 nm) complex interaction can take place between the front inversion channel and a bottom accumulation layer. This interaction has been reported to give rise to a decrease of carrier mobility (due to an intense electric field) and to tunneling between the inversion and the accumulation layer.[10]

The influence of the back-gate bias on the $I_D(V_{GS})$ characteristics of a thin-film SOI n-channel MOSFET with low V_{DS} is shown in Figure 5.3.3. In the left portion of the graph (region A), the back interface is inverted. As a consequence, current flows in the device even if the front-gate voltage is negative. Φ_{s2} is pinned to $2\Phi_F$, and the front threshold voltage is fixed at a constant value. The apparent shift of the curves to the left with increased back bias is actually a shift of the curves upwards due to the increase of the back channel current. In the right portion of the graph (region C), the back interface is accumulated, the front threshold voltage is constant, and no shift of the curves to the right can be obtained by further negative increase of the back bias. In region B, the back interface is depleted, and the front threshold voltage depends linearly on the back bias. The anomalous slope found in region *a* is explained by the following effect. As the back and front interfaces are depleted, and as the back interface is close to inversion, any increase of V_{G1} will push the point of minimum potential deeper into the silicon film, and therefore reduce the back threshold voltage. This can lead to creation of an inversion channel at the back interface. Thus, one observes a situation where a variation of the bias in the front gate creates and modulates an inversion channel at the back interface [11], which nicely illustrates the importance of the interaction between front and back gate in thin SOI devices.

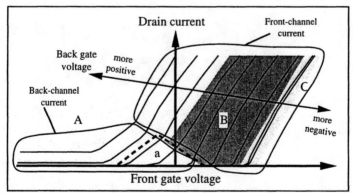

Figure 5.3.3: Linear $I_D(V_{G1})$ characteristics of a thin-film SOI n-channel MOSFET for different VG2 values. Different back interface conditions are outlined by the shaded areas: inversion (A), depletion (B), accumulation (C), and depletion/inversion regime depending on the front gate voltage (a).

133

5.3.2. Body effect

In a bulk device, the body effect is defined as the dependence of the threshold voltage on the substrate bias. In an SOI transistor, it is similarly defined as the dependence of the threshold voltage on the back-gate bias.

In a **bulk** n-channel transistor, the threshold voltage can be written as [12]:

$$V_{th} = \Phi_{MS} + 2\Phi_F - \frac{Q_{ox}}{C_{ox}} + \frac{Q_b}{C_{ox}} \quad \text{with} \quad Q_b = \sqrt{2\varepsilon_{si}\, q\, N_a\, (2\Phi_F\text{-}V_B)}$$

which can be rewritten:

$$V_{th} = \Phi_{MS} + 2\Phi_F - \frac{Q_{ox}}{C_{ox}} + \frac{\sqrt{2\varepsilon_{si}\, q\, N_a\, (2\Phi_F\text{-}V_B)}}{C_{ox}}$$

and, by defining $\gamma = \dfrac{\sqrt{2\varepsilon_{si}\, q\, N_a}}{C_{ox}}$

one obtains: $V_{th} = \Phi_{MS} + 2\Phi_F - \dfrac{Q_{ox}}{C_{ox}} + \gamma\sqrt{2\Phi_F} + \gamma\left(\sqrt{2\Phi_F\text{-}V_B} - \sqrt{2\Phi_F}\right)$

The last term depicts the dependence of threshold voltage on substrate bias (body effect). When a negative bias is applied to the substrate (with respect to the source), the threshold voltage increases as a square-root function of the substrate bias. If the threshold voltage with zero substrate bias is referred to as V_{th0}, one can write:

$$V_{th}(V_B) = V_{th0} + \gamma\left(\sqrt{2\Phi_F\text{-}V_B} - \sqrt{2\Phi_F}\right)$$

γ is called the body-effect parameter (unit: $V^{1/2}$).

The "body effect" (or, more accurately, the back-gate effect) can be neglected (γ=0) in **thick-film** SOI devices, because there is no coupling between front and back gate.

In a **thin-film**, fully depleted SOI device, the "body effect" (or, more accurately, the back-gate effect) can be obtained from Equation (5.3.12):

$$\frac{dV_{th1}}{dV_{G2}} = - \frac{C_{si}\, C_{ox2}}{C_{ox1}\, (C_{si} + C_{ox2})} = \frac{-\varepsilon_{si}\, C_{ox2}}{C_{ox1}\, (t_{si}\, C_{ox2} + \varepsilon_{si})} = \gamma \qquad (5.3.13)$$

The symbol γ is chosen by analogy with the case of a bulk device. It should be noted that γ is dimensionless in the case of thin-film SOI transistors, and that the threshold voltage dependence on back-gate bias is linear. In most cases, the following approximation can be made: $\gamma \cong -\dfrac{t_{ox1}}{t_{ox2}}$ Expression (5.3.13) is valid when the film is fully depleted only. In a first order approximation, one can consider that Φ_{S2} is pinned at $2\Phi_F$ when the back interface is inverted, and that further increase of the back-gate bias will no longer modify the front threshold voltage ($\gamma \cong 0$). Similarly, when a large negative back-gate bias is applied, the back interface is accumulated, Φ_{S2} is pinned at 0 V, and further negative increase of the back-gate bias will not modify the front threshold voltage ($\gamma \cong 0$). Variation of the front gate threshold voltage with back-gate bias, based on these assumptions, is represented on Figure 5.3.4. In a real case, however, the back surface potential can exceed $2\Phi_F$ (back inversion) or become smaller than 0V (back accumulation), these excursions being limited to a few $\dfrac{kT}{q}$. As a result, the front threshold voltage slightly further increases (decreases) when the back-gate voltage is increased beyond the threshold of back accumulation (inversion).

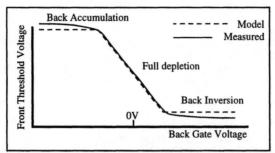

Figure 5.3.4: Variation of the front-gate threshold voltage with back-gate bias.

It is worth noting that relationship (5.3.13) is independent of the doping concentration, N_a. If C_{ox1} and C_{ox2} are known, (5.3.13) can be used to determine the thickness of the silicon film [13] The dependence of the front-gate threshold voltage on back-gate bias decreases with increasing t_{ox2}. When t_{ox2} is very thick ($C_{ox2} \cong 0$), the front threshold voltage is virtually independent of the back-gate bias. In real SOI devices, the back-gate material is not a metal, as considered in this first-order model, but a silicon substrate, the surface of which can become inverted, depleted or accumulated as a function of the back-gate bias conditions. This variation of the substrate surface potential has some influence on the device threshold voltage, but this influence is small, and can be neglected as long as the thickness of the buried oxide layer is large compared to that of the front gate oxide.

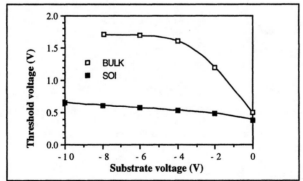

Figure 5.3.5: Dependence of threshold voltage on back bias (body effect) in bulk and fully depleted SOI MOSFETs.[14]

Although direct comparison between the "body-effect" parameters γ of bulk and thin-film SOI devices cannot be made, because of their different units ($V^{1/2}$ for γ_{bulk}, while γ_{SOI}

has no dimensions), it is clear that the back-gate effect of SOI devices is much smaller than the body effect of bulk MOSFETs. Figure 5.3.5 presents the experimental variation of threshold voltage in a bulk and a thin-film, fully depleted, SOI n-channel transistor. When a -8V back (substrate) bias is applied to the bulk device, a 1.2V increase of threshold voltage is observed. Under the same bias conditions, the threshold voltage of the SOI device increases only by 0.2V.

The reduced body effect is an important feature of SOI devices. Indeed, the body effect reduces the current drive capability of transistors whose source is not directly connected to ground, such as transfer gates, nMOS load devices, and differential input pairs. A higher gain can, therefore, be expected from SOI gates than from their bulk counterparts.

5.3.3. Short-channel effects

There are numerous effects caused by the reduction of channel length in MOSFETs.[15] In this Section we will more specifically deal with the so-called "short-channel effect" which results in a roll-off of the threshold voltage in short-channel devices. It is due to the loss of control by the gate of a part of the depletion zone below it. In other words, the depletion charge controlled by the gate is no longer equal to $Q_{depl} = q\,N_a\,x_{dmax}$ (bulk MOSFET case), but to a fraction of it, which we will call Q_{d1}. This reduction of the depletion charge, due to the encroachment from the source and drain, becomes significant in short-channel devices, and brings about a lowering of threshold voltage obtained by substituting Q_{d1} to Q_{depl} in Equation (5.3.1). In a bulk MOSFET, Q_{d1} can be geometrically represented by the area of a trapezoid (Figure 5.3.6). In a long-channel device, the lengths of the upper and the lower base of the trapezoid are almost equal to L, the channel length. In a short-channel device, the upper base length is still equal to L, but the lower base is significantly shorter (it can even disappear, as in Figure 5.3.6). The value of Q_{d1} can be approximated by: [16]

$$Q_{d1} = Q_{depl}\left(1 - \frac{r_j}{L}\left(\sqrt{1 + \frac{2x_{dmax}}{r_j}} - 1\right)\right)$$

where r_j is the source and drain junction depth.

In a thin-film SOI device, the depletion charge controlled by the gate is given by: $Q_{d1} = Q_{depl}\left(1 - \frac{d}{L}\right)$ where d is a distance defined by Figure 5.3.6 and $Q_{depl} = q\,N_a\,t_{si}$. Calculating the value of d is rather complex and necessitates iterative calculation of the back-surface potential. A description of the calculation method is given in [17].

A much cruder, empirical but reasonably usable expression of Q_{d1} can be obtained for fully-depleted MOSFETs using the following approximation, adapted from [18,19] (Figure 5.3.7). Q_{d1} is given by the trapezoid area: $Q_{d1} = Q_{depl}\left(1 - \alpha\,\frac{d_S + d_D}{2L}\right)$ where $Q_{depl} = q\,N_a\,x_1$, $\alpha = x_1/x_{dmax}$, d_S and d_D are the bases of the triangular charges controlled by source and drain in a virtual bulk device having the same doping concentration, N_a, as the SOI MOSFET under consideration. x_1 is the depth of minimum potential in the silicon film (see Figure 5.3.2), and is equal to $\dfrac{t_{si}}{2} + \dfrac{\varepsilon_{si}}{qN_a t_{si}}$ ($2\Phi_F$-Φ_{s2}). The values of d_S and d_D can be approximated by the depletion depths generated by the source and drain the junctions:

$$d_S = \sqrt{\frac{2\varepsilon_{si}}{qN_a}(E_g/2 + \Phi_F)} \quad \text{and} \quad d_D = \sqrt{\frac{2\varepsilon_{si}}{qN_a}(E_g/2 + \Phi_F + V_{DS})}$$

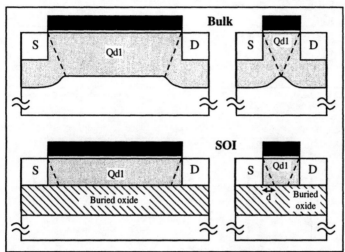

<u>Figure 5.3.6</u>: Distribution of depletion charges in long-channel (left) and short-channel (right) bulk and thin-film SOI MOSFETs. Q_{d1} is the depletion charge controlled by the gate.

This simplified model has been used to plot Figure 5.3.8 by replacing $qN_a x_{dmax}$ by Q_{d1} in Equation (5.3.1). One can see that threshold voltage roll-off starts to occur at significantly smaller gate lengths in thin SOI transistors than in bulk devices (this can also be seen in Figure 5.3.6, since Q_{d1} retains a reasonable trapezoid shape in the short-channel SOI device, while it shows a triangular shape ($Q_{d1} \cong \dfrac{Q_{depl}}{2}$) in the bulk case).

The short-channel effect is smaller in thin-film devices with accumulation at the back side than in thin-film fully depleted devices, but both of them show less short-channel effect than the bulk devices [20,21]. It seems that an optimum control of the space charge in the silicon film by the gate (which would further minimize the short-channel effect) could be obtained by using double-gate devices (one gate below the active silicon film and one above it).[22] Making such devices is a real technological challenge, but some practical solutions have already been proposed.[23]

137

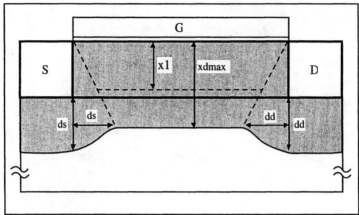

Figure 5.3.7: Cross section of an SOI MOSFET and the equivalent bulk device, used for approximating the short-channel effect.

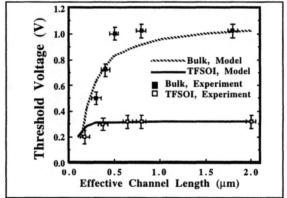

Figure 5.3.8: Threshold voltage as a function of gate length in bulk and thin-film (100 nm) SOI n-channel MOSFETs [24].

Another short-channel effect, called drain-induced conductivity enhancement (DICE) is also due to charge sharing between the gate and the junctions [25]. DICE is caused by the reduction of the depletion charge controlled by the gate due to size increase of the drain junction-related depletion zone, which itself increases with V_{DS}. In order to model the DICE effect, an analysis involving the solution of the two-dimensional Laplace equation (in a fully depleted device) must be carried out. It can be shown that DICE is lower in thin-film SOI devices than in bulk devices [26].

5.3.4. I-V characteristics

The expression of the current characteristics $I_D(V_{G1}, V_{G2}, V_{DS})$ of a **thick-film** SOI MOS transistor is identical to that of a bulk MOSFET, with the expression of the kink effect and the presence of parasitic bipolar effects, which will be discussed later in this Chapter.

Derivation of the current characteristics of a **thin-film**, fully depleted, SOI device can be done using assumptions of the classical gradual-channel approximation: constant mobility as a function of y, uniform doping of the silicon film in the channel region, and negligible diffusion current. Once again, we will consider the case of an n-channel device [27]. A more complete analysis, including short-channel effects can be found in Ref. [28]. Using Ohm's law in an elemental section of the inversion channel, one can write:

$$I_D = - W \, \mu_n \, Q_{inv1}(y) \frac{d\Phi_{s1}(y)}{dy} \qquad (5.3.14)$$

where W is the width of the channel, and μ_n is the mobility of the electrons in the inversion layer. Integration of the previous expression from source (y=0) to drain (y=L) yields:

$$I_D = - \frac{W}{L} \, \mu_n \int_{2\Phi_F}^{2\Phi_F+V_{DS}} Q_{inv1}(y) \, d\Phi_{s1}(y) \qquad (5.3.15)$$

Assuming full depletion in the silicon film and assuming that the thickness of the inversion and accumulation layers at the interfaces is equal to zero, one obtains the inversion charge density in the front channel, $Q_{inv1}(y)$, from Equation (5.3.8):

$$-Q_{inv1}(y) = C_{ox1} \left(V_{G1} - \Phi_{MS1} + \frac{Q_{ox1}}{C_{ox1}} - \left(1 + \frac{C_{si}}{C_{ox1}}\right)\Phi_{s1}(y) + \frac{C_{si}}{C_{ox1}} \Phi_{s2}(y) + \frac{Q_{depl}}{2C_{ox1}}\right) \qquad (5.3.16)$$

where the back surface potential, $\Phi_{s2}(y)$, can be extracted from (5.3.9):

$$\Phi_{s2}(y) = \frac{C_{ox2}}{C_{ox2}+C_{si}} \left(V_{G2} - \Phi_{MS2} + \frac{Q_{ox2}}{C_{ox2}} + \frac{C_{si}}{C_{ox2}}\Phi_{s1}(y) + \frac{Q_{depl}}{2C_{ox2}} - \frac{Q_{s2}(y)}{C_{ox2}} \right) \qquad (5.3.17)$$

Different cases can now be distinguished for the back interface: the film can be fully depleted from source to drain (DS+DD = "depleted source + depleted drain"), accumulated from source to drain (AS+AD), accumulated near the source and depleted near the drain (AS+DD), inverted from source to drain (IS+ID), and inverted near the source and depleted near the drain (IS+DD). Because back-channel inversion is generally undesirable, we will describe only the (DS+DD), (AS+AD) and (AS+DD) cases.

*   The back interface is accumulated from source to drain (AS+AD) under the condition that $V_{G2} < V_{G2,acc}(L)$

with
$$V_{G2,acc}(L) = V_{G2,acc} - \frac{C_{si}}{C_{ox2}} V_{DS}$$
(5.3.18.a)

where
$$V_{G2,acc} = \Phi_{MS2} - \frac{Q_{ox2}}{C_{ox2}} - \frac{C_{si}}{C_{ox2}} 2\Phi_F - \frac{Q_{depl}}{2C_{ox2}}$$
(5.3.18.b)

which is obtained from (5.3.9), assuming $\Phi_{s1} = 2\Phi_F$, $\Phi_{s2} = 0$, and $Q_{s2}=0$. Equation (5.3.18.a) is also obtained from (5.3.9) with $\Phi_{s1}(L) = 2\Phi_F + V_{DS}$. Note that $V_{G2,acc}(L) < V_{G2,acc}(0)$.

From (5.3.15) and (5.3.16) one obtains:

$$I_{D,acc2} = \frac{W}{L} \mu_n C_{ox1} \left((V_{G1}-V_{th1,acc2}) V_{DS} - \left(1 + \frac{C_{si}}{C_{ox1}}\right) \frac{V_{DS}^2}{2} \right)$$
(5.3.19)

where $V_{th1,acc2}$ is given by (5.3.10).

The drain saturation voltage is obtained from (5.3.19) under the condition that $dI_D/dV_{DS} = 0$:

$$V_{Dsat,acc2} = \frac{V_{G1}-V_{th1.acc2}}{1 + \frac{C_{si}}{C_{ox1}}}$$
(5.3.20)

and the saturation current is given by:

$$I_{Dsat,acc2} = \frac{1}{2} \frac{W}{L} \frac{\mu_n C_{ox1}}{1 + C_{si}/C_{ox1}} (V_{G1}-V_{th1,acc2})^2$$
(5.3.21)

*   The back interface is depleted from source to drain (DS+DD) under the condition that $V_{G2,acc} < V_{G2} < V_{G2,inv}$

where
$$V_{G2,inv} = \Phi_{MS2} - \frac{Q_{ox2}}{C_{ox2}} + 2\Phi_F - \frac{Q_{depl}}{2C_{ox2}}$$
(5.3.22)

which is obtained from (5.3.9), assuming $\Phi_{s1} = \Phi_{s2} = 2\Phi_F$, and $Q_{s2}=0$.

For the back interface, depleted from source to drain, we have $Q_{s2}(y)=0$. Using (5.3.15), (5.3.16) and (5.3.17), one obtains:

$$I_{D,depl2} = \frac{W}{L} \mu_n C_{ox1} \left((V_{G1}-V_{th1,depl2}) V_{DS} - \left(1 + \frac{C_{si} C_{ox2}}{C_{ox1} (C_{si}+C_{ox2})}\right) \frac{V_{DS}^2}{2} \right)$$
(5.3.23)

where the front threshold voltage under depleted back interface conditions, $V_{th1,depl2}$, is given by (5.3.12).

The drain saturation voltage is obtained from (5.3.23) under the condition that $dI_D/dV_{DS} = 0$:

$$V_{Dsat,depl2} = \frac{V_{G1}-V_{th1.depl2}}{1+\dfrac{C_{si}\,C_{ox2}}{C_{ox1}\,(C_{si}+C_{ox2})}} \tag{5.3.24}$$

and the saturation current is given by:

$$I_{Dsat,depl2} = \frac{1}{2}\frac{W}{L}\frac{\mu_n\,C_{ox1}}{1+\dfrac{C_{si}\,C_{ox2}}{C_{ox1}\,(C_{si}+C_{ox2})}}\,(V_{G1}-V_{th1,depl2})^2 \tag{5.3.25}$$

* If the back surface is accumulated near the source and depleted near the drain (AS+DD), the accumulation layer occurs at the back interface from $y=0$ to $y=y_t$, where y_t is the point where $\Phi_{s2}(y_t)=0$ and $Q_{s2}(y_t)=0$.

The drain current is then given by the following expression:

$$I_{D,AS+DD} = -\frac{W}{L}\mu_n\left(\int_{2\Phi_F}^{\Phi_{s1}(y_t)} Q_{inv1}(y)\,d\Phi_{s1}(y) + \int_{\Phi_{s1}(y_t)}^{2\Phi_F+V_{DS}} Q_{inv1}(y)\,d\Phi_{s1}(y)\right) \tag{5.3.26}$$

where $\Phi_{s1}(y_t)$ is given by combining (5.3.9), where $\Phi_{s2}(y_t)=0$ and $Q_{s2}(y_t)=0$, and (5.3.18.b):

$$\Phi_{s1}(y_t)= 2\Phi_F + \frac{C_{ox2}}{C_{si}}\,(V_{G2,acc} - V_{G2}) \tag{5.3.27}$$

Combining (5.3.26) and (5.3.27), one obtains:

$$\begin{aligned}I_{D,AS+DD} = \frac{W}{L}\mu_n\,C_{ox1}\Big\{ &(V_{G1}-V_{th1.acc2})V_{DS}\left(1+\frac{C_{si}\,C_{ox2}}{C_{ox1}\,(C_{si}+C_{ox2})}\right)\frac{V_{DS}^2}{2}\\ &-\frac{C_{si}\,C_{ox2}}{C_{ox1}\,(C_{si}+C_{ox2})}\,V_{DS}\,(V_{G2,acc} - V_{G2})\\ &+\frac{C_{si}\,C_{ox2}}{C_{ox1}\,(C_{si}+C_{ox2})}\frac{C_{ox2}}{C_{si}}\,(V_{G2,acc} - V_{G2})^2 \Big\}\end{aligned} \tag{5.3.28}$$

The drain saturation voltage is obtained from (5.3.28) under the condition that $dI_D/dV_{DS} = 0$:

$$V_{Dsat,AS+DD} = \frac{V_{G1}-V_{th1.acc2} - \dfrac{C_{si}\,C_{ox2}}{C_{ox1}\,(C_{si}+C_{ox2})}\,(V_{G2,acc}-V_{G2})}{1 - \dfrac{C_{si}\,C_{ox2}}{C_{ox1}\,(C_{si}+C_{ox2})}} \tag{5.3.29}$$

and the drain saturation current is given by:

$$I_{Dsat,AS+DD} = \frac{W}{L} \frac{\mu_n\,C_{ox1}}{1+\dfrac{C_{si}C_{ox2}}{C_{ox1}(C_{si}+C_{ox2})}} \left\{ (V_{G1}-V_{th1,acc2})^2 \right.$$

$$-\frac{2C_{si}C_{ox2}}{C_{ox1}(C_{si}+C_{ox2})}(V_{G1}-V_{th1,acc2})(V_{G2,acc}-V_{G2})$$

$$\left. +\frac{C_{ox2}^2\,(C_{si}+C_{ox1})}{C_{ox1}^2\,(C_{si}+C_{ox2})}(V_{G2,acc}-V_{G2})^2 \right\} \qquad (5.3.30)$$

The saturation current in a thin-film SOI MOSFET (fully depleted with depleted back interface or fully depleted with accumulated back interface) is given by Equations (5.3.21) and (5.3.25) and can be written in the general form:

$$I_{Dsat} \cong \frac{W\,\mu_n\,C_{ox1}}{2\,L\,(1+\alpha)}[V_{G1}-V_{th}]^2 \qquad (5.3.31)$$

where $\alpha = C_{si}/C_{ox1}$ in a fully depleted device with accumulation at the back interface, and $\alpha = \dfrac{C_{si}\,C_{ox2}}{C_{ox1}\,(C_{si}+C_{ox2})}$ in a fully depleted transistor with depleted back interface. The AS+DD case basically follows similar rules, but it is more complicated and will not be described here. In a bulk transistor, expression (5.3.31) is also valid [29] if $\alpha = C_D/C_{ox}$ with C_D being equal to $\dfrac{\varepsilon_{si}}{x_{dmax}}$. Indeed, the Spice Level3 bulk MOSFET model [30] assumes $\alpha = \dfrac{\gamma}{2\sqrt{2\Phi f - V_{BS}}}$ with $\gamma = \dfrac{\sqrt{2\,\varepsilon_{si}\,q\,N_a}}{C_{ox}}$, which yields $\alpha = \dfrac{\varepsilon_{si}}{x_{dmax}\,C_{ox}}$ after straightforward manipulation. The values of the threshold voltages used in (5.3.31) can be extracted from the relationships (5.3.10) and (5.3.12) for the SOI devices, and V_{th} has its usual meaning in the case of a bulk transistor. It is worth noting that the values of α are in the following sequence:

$$\alpha_{\text{fully depleted SOI}} < \alpha_{\text{bulk}} < \alpha_{\text{back accum SOI}}$$

so that the drain saturation current is highest in the fully depleted device, lower in the bulk device, and even lower in the device with back accumulation. It is also interesting to note that α represents the ratio C_b/C_{ox1} of two capacitors. Indeed, C_{ox1} is the gate oxide capacitance, and C_b is the capacitance between the inversion channel and the back-gate electrode. The value of C_b is either that of the depletion capacitance $C_D=\varepsilon_{si}/x_{dmax}$ in a bulk device, the silicon film capacitance $C_{si}=\varepsilon_{si}/t_{si}$ in a fully depleted device with back interface accumulation, or the value given by the series association of C_{si} and C_{ox2} ($C_b = \dfrac{C_{si}C_{ox2}}{(C_{si}+C_{ox2})}$) in a fully depleted device.

This high saturation current in thin-film, fully depleted SOI MOSFETs brings about an increase of current drive (compared to bulk devices) which largely contributes to the excellent speed performances of thin-film fully depleted (FD) SOI CMOS circuits. Figure 5.3.9 compares the drain saturation current in a bulk and a thin-film FD SOI device, as a function of $V_{G(1)}$-V_{th}. The larger current drive of the SOI device is evident [31] (it is actually 20-30% higher that the drive of the bulk device). The superior current drive capability of FD SOI devices gets somewhat degraded as short gate lengths are considered, because of velocity saturation effects. It has been shown, however, that FD SOI devices still present a 25% current drive improvement over bulk devices for a gate length of 0.2 μm.[32]

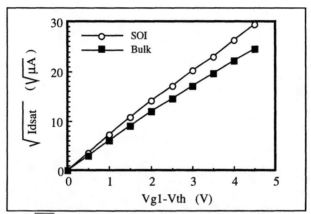

Figure 5.3.9: $\sqrt{I_{Dsat}}$ as a function of V_{G1}-V_{th} in a bulk and a thin-film, fully depleted SOI device with same technological parameters.

It is worth noting that the SOI MOSFET, which was considered so far as a four-terminal device (source, drain, front gate and back gate), is actually a five-terminal device, the fifth terminal being the body. In many applications, the body is left unconnected, but its presence has considerable influence on the dynamic characteristics of the device. This influence is more pronounced in thick-film devices than in thin, fully depleted, devices. A charge-based model of the five-terminal SOI MOSFET can be found in Ref. [33]. Detailed analysis of the short-channel effects is described in Ref. [34].

5.3.5. C∞-continuous model

In some cases, such as for the simulation of analog circuits, it is necessary to have a model which is continuous in all regimes of operation (weak or strong inversion), and whose derivatives are also continuous (C∞-continuous model). The EKV model and the ENSERG model belong to that class of models.[35,36,37]

We will describe here a C∞-continuous model which has been developed for fully depleted SOI MOSFETs.[38] We will keep the notations used in Figure 5.3.1. According to charge-sheet models [39] the relationship between the inversion charge density at the front interface, Q_n, and the front surface potential, Ψ_{s1}, is linear and can be written as:

$$Q_n = -C_{ox1}\left(V_{G1} - V_{FB1} + \frac{Q_b}{2C_{ox1}} - \alpha\left(V_{G2} - V_{FB2} + \frac{Q_b}{2C_{ox2}}\right) - n\Psi_{s1}\right) \qquad (5.3.32)$$

where $Q_b = -qN_at_{si}$, $\alpha = \dfrac{C_{si}\,C_{ox2}}{C_{ox1}\,(C_{si} + C_{ox2})}$ and $n = 1 + \alpha$.

If there is no inversion at the back interface the drain current is written as the sum of the drift and diffusion terms of Q_n:

$$I_D = -W\,\mu_n\,Q_n\,\frac{dV_{ch}}{dy} = -W\,\mu_n\left(Q_n\,\frac{d\Psi_{s1}}{dy} - u_T\,\frac{dQ_n}{dy}\right) \qquad (5.3.33)$$

where μ_n is the electron mobility, V_{ch} is the channel potential and $u_T = kT/q$. Using Equations (5.3.32) and (5.3.33) and integrating along the channel one obtains:

$$\frac{d\Psi_{s1}}{dy} = \frac{1}{nC_{ox1}}\frac{dQ_n}{dy} \Rightarrow I_D\int_0^L dy = W\,\mu_n\int_S^D\left(u_T - \frac{Q_n}{nC_{ox1}}\right)dQ_n$$

$$\Downarrow$$

$$I_D = \mu_n\,\frac{W}{L}\left(u_T(Q_{nd} - Q_{ns}) - \frac{Q_{nd}^2 - Q_{ns}^2}{2\,n\,C_{ox1}}\right) \qquad (5.3.34)$$

where Q_{ns} and Q_{nd} are the inversion charge densities at the source and drain edges, calculated for V_{ch} being equal to V_S and V_D, respectively. Following the EKV model [40] one can develop an explicit model using an approximate expression for Q_n that tends to the desired limits in weak and strong inversion:

$$Q_n = -C_{ox}nu_TS_{NT}\,\ln\left[1 + \frac{-Q_0/C_{ox}}{nu_TS_{NT}}\exp\left(\frac{V_G - V_{Ti} - nV_{ch}}{nu_T}\right) + \exp\left(\frac{V_G - V_T - nV_{ch}}{S_{NT}nu_T}\right)\right] \qquad (5.3.35)$$

In the latter expression the second term of the logarithm dominates in weak inversion and the third term dominates in strong inversion. The parameter S_{NT} (<1) controls the transition from weak to strong inversion. Q_0 is the inversion charge density at $V_{G1} = V_{Ti}$, V_{Ti} being the threshold voltage which corresponds to a surface potential of $2\Phi_F + V_{ch}$. (*i.e.*: $V_{Ti} = V_{T0i} - \alpha\,V_{G2}$ where V_{T0i} is the threshold voltage then $V_{G2} = 0$). This threshold value appears in the weak inversion term of (5.3.35), and one has to introduce a different threshold voltage value in the strong inversion term, $V_T = V_{T0} - \alpha\,V_{G2}$, to take into account the fact that in strong inversion the surface potential is always larger than $2\Phi_F + V_{ch}$ by a few kT/q. The use of two threshold voltages (a higher threshold in strong than in weak inversion) may seem awkward, but it is physically more correct than other approaches which, for instance, overestimate field mobility degradation in order to fit the model with experimental data. V_{T0} and V_{T0i} can be considered as fitting parameters obtained from parameter extraction, or as threshold voltage parameters which can be derived from the device physical characteristics in the following manner:

$$V_{T0i} = V_{FB1} - \frac{Q_b}{2C_{ox1}} - \alpha \left(\frac{Q_b}{2C_{ox2}} - V_{FB2} \right) + 2 n \Phi_F \qquad (5.3.36a)$$

and

$$V_{T0} = V_{FB1} - \frac{Q_{Db}}{2C_{ox1}} - \alpha \left(\frac{Q_b}{2C_{ox2}} - V_{FB2} \right) + n \Phi_B \qquad (5.3.36b)$$

where Φ_B is larger than $2\Phi_F$ by a few kT/q.

Equation (5.3.34) can also be obtained under an equivalent way by using in (5.3.33) the expression of dV_{ch} in terms of Q_n, which can be derived from the unified charge-control model [41] adapted to the SOI MOSFET:

$$V_{G1} - V_T - n V_{ch} = n u_T \ln \left(\frac{Q_n}{Q_0} \right) - \frac{Q_n - Q_0}{C_{ox1}} \qquad (5.3.37)$$

The latter equation allows one to extract simple but accurate analytical expressions for the conductances. For instance, by differentiating (5.3.37) with respect to V_{G1}, one finds:

$$\frac{dQ_n}{dV_{G1}} = \frac{Q_n}{nu_T - Q_n/C_{ox1}} . \qquad (5.3.38)$$

By differentiating (5.3.34) with respect to the front gate voltage and by replacing dQ_n/dV_{G1} by the right-hand term of (5.3.38) the front-gate transconductance can be obtained:

$$g_{m1} = \frac{dI_D}{dV_{G1}} = \frac{\mu_n W}{nL} (Q_{nd} - Q_{ns}) \qquad (5.3.39a)$$

Similarly, by differentiating (5.3.34) with respect to the drain voltage, the back-gate voltage and the source voltage, one obtains the drain conductance, the back-gate transconductance, and the source conductance, respectively:

$$g_d = \frac{dI_D}{dV_D} = -\mu_n \frac{W}{L} Q_{nd} \qquad (5.3.39b)$$

$$g_{m2} = \frac{dI_D}{dV_{G2}} = \alpha \frac{\mu_n W}{nL} (Q_{nd} - Q_{ns}) \qquad (5.3.39c)$$

$$g_s = \frac{dI_D}{dV_s} = \mu_n \frac{W}{L} Q_{ns} \qquad (5.3.39d)$$

The above expressions become explicit functions of the applied voltages, by calculating Q_{ns} and Q_{nd} using Equation (5.3.35) with $V_{ch}=V_S$ and $V_{ch}=V_D$, respectively. More complete expressions, including mobility degradation, DIBL, DICE and short-channel effects, can be found in Ref. [42].

The charge densities are obtained by integrating the charge densities along the channel. Using (5.3.32) and (5.3.33), one finds a relationship between dQ_n and dy which can be integrated to yield the total electron charge in the channel:

$$dy = -\frac{W\mu_n}{I_D}\left(Q_n \, d\Psi_{s1} - u_T \, dQ_n\right) \quad \text{and} \quad d\Psi_{s1} = \frac{dQ_n}{nC_{ox1}}$$

$$\Downarrow$$

$$dy = -\frac{W\mu_n}{I_D}\left(\frac{Q_n}{nC_{ox1}} - u_T\right)dQ_n$$

$$\Downarrow$$

$$W\int_0^L Q_n dy = \frac{W^2\mu_n}{I_D}\int_0^L\left(\frac{Q_n^2}{nC_{ox1}} - u_T Q_n\right)dQ_n$$

Using (5.3.34) one obtains:

$$Q_N = \frac{WL\left[u_T \dfrac{Q_{nd}+Q_{ns}}{2} - \dfrac{Q_{ns}^2 + Q_{ns}Q_{nd} + Q_{nd}^2}{3nC_{ox1}}\right]}{u_T - \dfrac{Q_{ns} + Q_{nd}}{2nC_{ox1}}} \tag{5.3.40}$$

The drain and source charges are given by:

$$Q_D = W\int_0^L \frac{y}{L} Q_n dy$$

$$= WL\left\{\frac{C_{ox1}u_T n}{2} + \frac{n^2 C_{ox1}^2 \mu_n W}{3 L I_D}\left(u_T - \frac{Q_{nd}}{nC_{ox1}}\right)^3 - \frac{n^3 C_{ox1}^3 \mu_n^2 W^2}{15 L^2 I_D^2}\left[\left(u_T - \frac{Q_{ns}}{nC_{ox1}}\right)^5 - \left(u_T - \frac{Q_{nd}}{nC_{ox1}}\right)^5\right]\right\} \tag{5.3.41a}$$

and

$$Q_S = Q_N - Q_D \tag{5.3.41b}$$

The front and back gate charges, Q_{G1} and Q_{G2}, are obtained by integrating the front and back charge densities ($Q_{g1} = C_{ox1}[V_{G1}-V_{FB1}-\Psi_{s1}]$ and $Q_{g2} = C_{ox2}[V_{G2}-V_{FB2}-\Psi_{s2}]$) along the channel, respectively:

$$Q_{G1}=WLC_{ox1}\left[\frac{\alpha}{n}V_{g1} - \frac{Q_b}{2nC_{ox1}} - \frac{\alpha}{n}\left(V_{g2}+\frac{Q_b}{2C_{ox2}}\right)\right] - \frac{Q_N}{n} - WLQ_{ox1} \tag{5.3.41c}$$

$$Q_{G2}=WLC_{ox2}\left[V_{g2}\left(1-\frac{C_{ox2}}{C_{ox2}+C_{si}}\right) - \frac{\alpha C_{si}}{n(C_{ox2}+C_{si})} - \frac{Q_b}{2(C_{ox2}+C_{si})}\left\{1+\frac{C_{si}}{n}\left(\frac{1}{C_{ox1}} + \frac{\alpha}{C_{ox2}}\right)\right\} - V_{g1}\frac{C_{si}}{n(C_{si}+C_{ox2})}\right]$$
$$- Q_N\frac{C_{si}C_{ox2}}{nC_{ox1}(C_{si}+C_{ox2})} - WLQ_{ox2} \tag{5.3.41d}$$

where $V_{g1}=V_{G1}-V_{FB1}$ and $V_{g2}=V_{G2}-V_{FB2}$. Q_{ox1} and Q_{ox2} are the fixed charge densities in the front and back oxides, respectively. The total depletion charge in the silicon film is equal to $-Q_{G1}-Q_{G2}-Q_{ox1}-Q_{ox2}=WLQ_b$.

The intrinsic capacitances of the transistor are given by:

$$C_{ij} = \chi_{ij} \frac{dQ_i}{dV_j}$$ (5.3.42)

where $\chi_{ij} = 1$ if $i=j$ and $\chi_{ij} = -1$ if $i \neq j$. These expressions are valid up to a frequency limit of $\mu_n(V_{G1}-V_{FB1}-nV_S)/(3nL^2)$, which is over 1 GHz for a 1 μm-long device with a front gate oxide thickness of 30 nm.

5.4. Transconductance and mobility

The current drive capability of an SOI MOSFET depends basically on three factors: the saturation current, the transconductance and the mobility of the carriers in the inversion layer. The saturation current has been dealt with in Section 5.3, and it has been shown that I_{Dsat} is larger in a thin-film, fully depleted, SOI MOSFET with depleted back interface than in the corresponding bulk device. On the other hand, I_{Dsat} is lower in a thin-film, fully depleted SOI MOSFET with accumulated back interface than in a bulk device. The transconductance and the channel mobility are described next.

5.4.1 Transconductance

The transconductance of a MOSFET, g_m, is a measure of the effectiveness of the control of the drain current by the gate voltage. In a bulk n-channel MOSFET in saturation, it is given by:

$$g_m = dI_{Dsat}/dV_G \quad \text{(for } V_{DS} > V_{Dsat}\text{)}$$

$$= \frac{W}{L} \mu_n C_{ox} (V_G-V_{th}) + 4\Phi_F \left(\frac{C_D}{C_{ox}}\right)^2 \left(1 - \sqrt{1 + \left(\frac{C_{ox}}{C_{depl}}\right)^2 \frac{V_G-V_{FB}}{2\Phi_F}}\right)$$

with $C_D = \varepsilon_{si}/x_{dmax}$ [43]

which can be approximated by (from 5.3.31):

$$I_{Dsat} \cong \frac{W \mu_n C_{ox1}}{2 L (1 + \alpha)} (V_{G1}-V_{th})^2 \quad \text{with } \alpha = \frac{\varepsilon_{si}}{x_{dmax} C_{ox}}$$

In a thin-film, fully depleted, SOI MOSFET, the transconductance can be obtained from (5.3.21) and (5.3.25):

$$g_m = dI_{Dsat}/dV_{G1} = \frac{W \mu_n C_{ox1}}{L (1 + \alpha)} (V_{G1}-V_{th})$$ (5.4.1)

147

for $V_{DS} > V_{Dsat}$ and with $\alpha = C_{si}/C_{ox1}$ in a fully depleted device with accumulation at the back interface, and $\alpha = \dfrac{C_{si} \, C_{ox2}}{C_{ox1} \, (C_{si}+C_{ox2})}$ in a fully depleted transistor with depleted back interface [44].

As in the case of the analysis of the saturation current (Section 5.3), α represents the ratio C_b/C_{ox1} of two capacitors. Indeed, C_{ox1} is the gate oxide capacitance, and C_b is the capacitance between the inversion channel and the back-gate electrode. The value of C_b is either that of the depletion capacitance $C_D = \varepsilon_{si}/x_{dmax}$ in a bulk device, the silicon film capacitance $C_{si} = \varepsilon_{si}/t_{si}$ in a fully depleted device with back interface accumulation, or the value given by the series association of C_{si} and C_{ox2} ($C_b = \dfrac{C_{si}C_{ox2}}{(C_{si}+C_{ox2})}$) in a fully depleted device. It is once again worth noting that the values of α are typically in the following sequence:

$$\alpha_{\text{fully depleted SOI}} < \alpha_{\text{bulk}} < \alpha_{\text{back accum SOI}}$$

such that the transconductance is highest in the fully depleted device, lower in the bulk device, and even lower in the device with back accumulation.

5.4.2. g_m/I_D

The maximum voltage gain of a MOSFET is obtained when the value of g_m/I_D is largest, since the gain of a transistor is given by $\dfrac{\Delta V_{out}}{\Delta V_{in}} = \dfrac{\Delta I_d}{g_D} \dfrac{1}{\Delta V_{in}} = \dfrac{\Delta V_{in}}{g_D} \dfrac{g_m}{\Delta V_{in}} \dfrac{1}{\Delta V_{in}} = \dfrac{g_m}{g_D} = \dfrac{g_m}{I_D}$ V_A, [45] where g_D is the output drain conductance, and V_A is the Early voltage. The Early voltage V_A of a FD SOI transistor is basically identical to that of a bulk device.

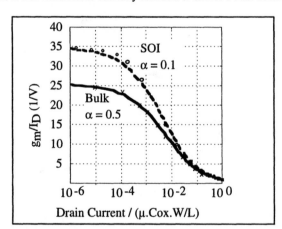

Figure 5.4.1: g_m/I_D ratios in saturation (V_D = 2.5 V) for bulk (α=0.5) and SOI (α=0.1) MOSFETs

The largest value of g_m/I_D appears in the weak inversion regime for MOS transistors.[46] The value of g_m/I_D can be rewritten:

$$\frac{g_m}{I_D} = \frac{dI_D}{I_D \, dV_G} = \frac{\ln(10)}{S} = \frac{q}{(1+\alpha)kT} = \frac{q}{nkT}$$

(5.4.2)

where S is the subthreshold swing and n is the body factor (see Section 5.5). In strong inversion, g_m/I_D becomes [47]:

$$\frac{g_m}{I_D} = \sqrt{\frac{2 \, \mu \, C_{ox} \, W/L}{(1+\alpha)I_D}} = \sqrt{\frac{2 \, \mu \, C_{ox} \, W/L}{n \, I_D}}$$

(5.4.3)

Because of the lower value of α in fully depleted SOI MOSFETs, values of g_m/I_D significantly higher than in bulk devices can be obtained (Figure 5.4.1).

5.4.3. Mobility

The mobility of the electrons within the inversion layer of an n-channel MOSFET has so far been considered to be constant. It is actually a function of the vertical electric field below the gate oxide. It can be approximated by:

$$\mu_n(y) = \mu_{max} \, [E_c/E_{eff}(y)]^c, \quad \text{for } E_{eff}(y) > E_c$$

(5.4.4)

where μ_{max}, E_c, and c are fitting parameters depending on the gate oxidation process and the device properties [48,49], and

$$E_{eff}(y) = E_{s1}(y) - \frac{Q_{inv1}(y)}{2\varepsilon_{si}}$$

(5.4.5)

The vertical electric field below the gate oxide is given by (5.3.4):

$$E_{s1}(y) = (\frac{\Phi_{s1}(y) - \Phi_{s2}(y)}{t_{si}} + \frac{q \, N_a \, t_{si}}{2 \, \varepsilon_{si}})$$

The inversion charge in the channel, $Q_{inv1}(y)$, can be obtained from (5.3.16):

$$-Q_{inv1}(y) = C_{ox1} \left(V_{G1} - \Phi_{MS1} + \frac{Q_{ox1}}{C_{ox1}} - \left(1 + \frac{C_{si}}{C_{ox1}}\right)\Phi_{s1}(y) + \frac{C_{si}}{C_{ox1}} \, \Phi_{s2}(y) + \frac{Q_{depl}}{2C_{ox1}} \right)$$

and the back surface potential, $\Phi_{s2}(y)$, is given by (5.3.17):

$$\Phi_{s2}(y) = \frac{C_{ox2}}{C_{ox2} + C_{si}} \left(V_{G2} - \Phi_{MS2} + \frac{Q_{ox2}}{C_{ox2}} + \frac{C_{si}}{C_{ox2}}\Phi_{s1}(y) + \frac{Q_{depl}}{2C_{ox2}} - \frac{Q_{s2}(y)}{C_{ox2}} \right)$$

149

The expression of the surface electric field can be simplified by considering the case of a thin, fully depleted, device operating with a low drain voltage ($V_{DS} \cong 0$), so that the front and back surface potentials are independent of y. If the back interface is depleted and close to inversion, $\Phi_{s1}-\Phi_{s2} \cong 0$, and the front surface electric field is equal to $\dfrac{q\,N_a\,t_{si}}{2\,\varepsilon_{si}}$. In this case, the surface electric field is lower than the surface electric field in the corresponding bulk device, which is equal to $\dfrac{qN_a x_{dmax}}{\varepsilon_{si}}$, since $t_{si} < x_{dmax}$. If the film is fully depleted, but if the film is not close to inversion, $E_{s1} \cong \dfrac{q\,N_a\,x_1}{\varepsilon_{si}}$ is a good approximation (x_1 is the point of minimum potential in the film - Figures 5.3.2 and 5.4.2), and the surface electric field is still lower than in a bulk devices, since $x_1 < t_{si} < x_{dmax}$.

Figure 5.4.2 presents the electric field as a function of depth, x, in a bulk and a thin-film, fully depleted, SOI device. The slope of the $E(x)$ lines are the same if both devices have the same doping concentration, N_a. The (front) surface electric field, $E_{s1}=E(x=0)$, is lower in the SOI device than in the bulk device. It is worth noting, however, that if the back of the fully depleted SOI device is accumulated ($\Phi_{s1}-\Phi_{s2} \cong 2\Phi_F$), the surface electric field can be larger than in the corresponding bulk device.

The increase of surface mobility in thin, fully depleted, SOI devices has been reported by several authors [50,51]. The maximum of mobility is obtained right above threshold, but it is rapidly overshadowed by the decrease of mobility caused by the increase of Q_{inv1} when V_{G1} is significantly larger than V_{th1} (see Equation (5.4.5)).

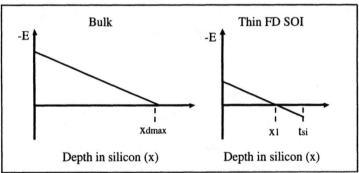

Figure 5.4.2: Electric field distribution in a bulk and a thin-film, fully depleted (FD) SOI device with depleted back interface.

Beyond Equation (5.4.5), the reduction of the surface electric field means a broadening of the inversion channel, which leads to reduced scattering of the carriers in the inversion channel by the Si/SiO$_2$ interface. In an extreme case, the whole thickness of the silicon film can become inverted. In that case, the inversion carriers located near the center of the film (*i.e.* at some distance from the interfaces) exhibit bulk-like mobility (in contrast to surface mobility) [52]. The complete description of thin-film devices with volume inversion implies quantum-mechanical considerations which lie beyond the scope of the present analysis.

5.5. Subthreshold slope

The inverse subthreshold slope (or, in short, the subthreshold slope, or subthreshold swing) is defined as the inverse of the slope of the $I_D(V_G)$ curve in the subthreshold regime, presented on a semilogarithmic plot (Figure 5.5.1).

$$S = \frac{d\,V_G}{d\,(\log I_D)} \tag{5.5.1}$$

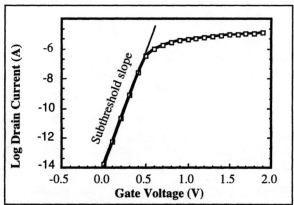

Figure 5.5.1: Semilogarithmic plot of $I_D(V_G)$ of an nMOS device.

The subthreshold current of an MOS transistor is a minority carrier diffusion current [53,54]. With analogy to bipolar transistor theory (current in the base), we find:

$$I_D = -q\,A\,D_n\,\frac{dn}{dy} = q\,A\,D_n\,\frac{n(0)-n(L)}{L} \tag{5.5.2}$$

with A being the cross sectional area of the channel, D_n is the diffusion coefficient of electrons, and n(0) and n(L) the electron (minority carrier) concentrations at the source and drain side. We have: $n(0)=n_{po}\,e^{\beta\Phi_s}$ and $n(L)=n_{po}\,e^{\beta(\Phi_s-V_D)}$ where Φ_s is the surface potential $\beta=\frac{q}{kT}$, and $n_{po}=\frac{n_i^2}{N_a}$. The inversion channel can be approximated by a box carrier profile with a uniform carrier concentration n(0) and a thickness d defined as the distance from the Si-SiO$_2$ interface at which the potential is lowered by $\frac{kT}{q}$ [55].

We have:
$$d = \frac{\frac{kT}{q}}{E_s} \quad \text{with } E_s = -\frac{d\Phi_s}{dx}.$$

Using Einstein's relationship $D_n = \frac{kT}{q}\,\mu_n$ and A=W.d, where W is the channel width, we find:

151

$$I_D = \left(\mu_n \frac{W}{L} q \left(\frac{kT}{q} \right)^2 \frac{n_i^2}{N_a} (1 - e^{-\beta V_D}) \right) \frac{e^{\beta \Phi_s}}{-\frac{d\Phi_s}{dx}} \tag{5.5.3}$$

The inverse subthreshold slope S is given by $\dfrac{dV_G}{d \log(I_D)}$

and is equal to $S = \dfrac{\ln(10)}{\dfrac{d \ln I_D}{d V_G}}$ $\tag{5.5.4}$

From Equation (5.5.3) we derive:

$$\frac{d \ln I_D}{dV_G} = \frac{1}{I_D} \frac{dI_D}{d\Phi_s} \frac{d\Phi_s}{dV_G} = C \frac{d\Phi_s}{dV_G} \tag{5.5.5}$$

with
$$C = \frac{q}{kT} - \frac{\dfrac{d}{d\Phi_s}\left(-\dfrac{d\Phi_s}{dx} \right)}{-\dfrac{d\Phi_s}{dx}} \tag{5.5.6}$$

The second term of C is a correction term which accounts for the reduction of current increase due to the increase of surface electric field with increasing surface current (a large surface electric field will decrease the channel thickness).

In the case of a **bulk** or a **thick** SOI device, we have:

$$E_s = -\frac{d\Phi_s}{dx} = \sqrt{\frac{2 q N_a \Phi_s}{\varepsilon_{si}}} = \frac{q N_a}{C_D}$$

with C_D being the depletion capacitance. We also have: $\dfrac{d}{d\Phi_s}\left(-\dfrac{d\Phi_s}{dx} \right) = \dfrac{C_D}{\varepsilon_{si}}$. The correction term is then given by:

$$\frac{\dfrac{d}{d\Phi_s}\left(-\dfrac{d\Phi_s}{dx} \right)}{-\dfrac{d\Phi_s}{dx}} = \frac{C_D^2}{q N_a \varepsilon_{si}} = \frac{1}{2\Phi_s} \tag{5.5.7}$$

Using the relationship between gate voltage and charges in the silicon and at the interfaces:

$$V_G = \Phi_{MS} + \Phi_s + \frac{-Q_D - Q_{ox} + C_{it} \Phi_s}{C_{ox}}$$

we find:
$$\frac{d\Phi_s}{dV_G} = \frac{C_{ox}}{C_{ox} + C_D + C_{it}} \tag{5.5.8}$$

where $C_D = dQ_D/d\Phi_S$, $Q_D = q N_a x_{dmax}$, and $C_{it} = q N_{it}$.

Usually, the correction factor is small compared to $\frac{q}{kT}$ (about 4% of it, considering $\Phi_s = \frac{2}{3}$ Φ_F), and will be neglected here. Using (5.5.4), (5.5.5) and (5.5.6), one finds:

$$S = \frac{kT}{q} \ln(10) \; (1 + \frac{C_D + C_{it}}{C_{ox}}) \tag{5.5.9}$$

or

$$S = \frac{kT}{q} \ln(10) \; (1 + \frac{C_D}{C_{ox}}) \tag{5.5.10}$$

if the interface traps are neglected.

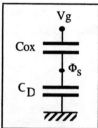

Figure 5.5.2: Equivalent capacitor network (bulk case).

The expression of the inverse subthreshold slope is the same for a thick-film (partially depleted) SOI device. In both cases, the right-side term of (5.5.10) can be represented by the capacitor network of Figure 5.5.2 in which we can observe that

$$C_{ox} \, d(V_G - \Phi_s) = I_G = C_D \, d\Phi_s \text{ and, hence, } \frac{d\,V_G}{d\,\Phi_s} = 1 + \frac{C_D}{C_{ox}}.$$

In the case of a thin-film, **fully depleted** SOI device (with depleted back interface), we have from Equations (5.3.8) and (5.3.9):

$$V_{G1} = \Phi_{MS1} - \frac{Q_{ox1}}{C_{ox1}} + \frac{q\,N_a\,t_{si}}{2\,C_{ox1}} + \Phi_{s1}(\frac{\varepsilon_{si}}{t_{si}\,C_{ox1}} + 1) + \Phi_{s2}(\frac{-\varepsilon_{si}}{t_{si}\,C_{ox1}}) \tag{5.5.11}$$

$$V_{G2} = \Phi_{MS2} - \frac{Q_{ox2}}{C_{ox2}} + \frac{q\,N_a\,t_{si}}{2\,C_{ox2}} + \Phi_{s1}(\frac{-\varepsilon_{si}}{t_{si}\,C_{ox2}}) + \Phi_{s2}(\frac{\varepsilon_{si}}{t_{si}\,C_{ox2}} + 1) \tag{5.5.12}$$

Writing $\varepsilon_{si}/t_{si} = C_{si}$, the capacitance of the depleted silicon film, and $q\,N_a\,t_{si} = Q_{depl}$, the depletion charge in the silicon film, we obtain:

$$V_{G1} = \Phi_{MS1} - \frac{Q_{ox1}}{C_{ox1}} + \frac{Q_{depl}}{2\,C_{ox1}} + \Phi_{s1}(\frac{C_{si}}{C_{ox1}} + 1) + \Phi_{s2}(\frac{-C_{si}}{C_{ox1}}) \tag{5.5.13}$$

$$V_{G2} = \Phi_{MS2} - \frac{Q_{ox2}}{C_{ox2}} + \frac{Q_{depl}}{2\,C_{ox2}} + \Phi_{s1}(\frac{-C_{si}}{C_{ox2}}) + \Phi_{s2}(\frac{C_{si}}{C_{ox2}} + 1) \tag{5.5.14}$$

Eliminating Φ_{s2} between these two equations, we can obtain Φ_{s1} as a function of V_{G1} and V_{G2}:

$$\Phi_{s1}\left((1+\frac{C_{si}}{C_{ox1}})\frac{C_{ox1}}{C_{si}}(1+\frac{C_{si}}{C_{ox2}})-\frac{C_{si}}{C_{ox2}}\right) =$$

$$V_{G2}-\Phi_{MS2}+\frac{Q_{ox2}}{C_{ox2}}-\left(\frac{Q_{depl}}{2\,C_{ox1}}+\Phi_{MS1}-\frac{Q_{ox1}}{C_{ox1}}-V_{G1}\right)\frac{C_{ox1}}{C_{si}}(1+\frac{C_{si}}{C_{ox2}})+\frac{Q_{depl}}{2\,C_{ox2}} \qquad (5.5.15)$$

Assuming that the back-channel conduction is negligible, we have to calculate the value of

$$\frac{d\ln I_D}{dV_{G1}}=\frac{1}{I_D}\frac{dI_D}{d\Phi_{s1}}\frac{d\Phi_{s1}}{dV_{G1}}$$

From Equation (5.5.15) we obtain:

$$\frac{d\Phi_{s1}}{dV_{G1}}=\frac{\frac{C_{ox1}}{C_{si}}(1+\frac{C_{si}}{C_{ox2}})}{(1+\frac{C_{si}}{C_{ox1}})\frac{C_{ox1}}{C_{si}}(1+\frac{C_{si}}{C_{ox2}})-\frac{C_{si}}{C_{ox2}}}=\frac{\frac{1}{C_{si}}+\frac{1}{C_{ox2}}}{\frac{1}{C_{ox1}}+\frac{1}{C_{si}}+\frac{1}{C_{ox2}}} \qquad (5.5.16)$$

This expression corresponds to the capacitor network of Figure 5.5.3.

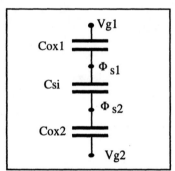

Figure 5.5.3: Equivalent capacitor network (thin-film, fully depleted SOI case).

It can be shown that the correction factor, C, is smaller in thin SOI devices than in bulk devices, and can be neglected.

Usually, $C_{ox2} \ll C_{ox1}$ and $C_{ox2} \ll C_{si}$. As a result, $\dfrac{d\Phi_{s1}}{dV_{G1}}$ tends to unity, and using (5.5.4), (5.5.5) and (5.5.6) one obtains:

$$S \cong \frac{kT}{q}\ln(10) \equiv S_0 \qquad (5.5.17)$$

The complete expression of the subthreshold slope, including the presence of interface states, is given by:

$$S = \frac{dV_{G1}}{d \log(I_D)} = \frac{kT}{q} \ln(10) \left[\left(1 + \frac{C_{it1}}{C_{ox1}} + \frac{C_{si}}{C_{ox1}} \right) - \frac{\frac{C_{si}}{C_{ox2}} \frac{C_{si}}{C_{ox1}}}{1 + \frac{C_{it2}}{C_{ox2}} + \frac{C_{si}}{C_{ox2}}} \right] \qquad (5.5.18)$$

where V_{G1}, $C_{it1}=qN_{it1}$, $C_{it2}=qN_{it2}$, C_{ox1} and C_{ox2} are the front gate voltage, the top interface traps capacitance, the bottom interface traps capacitance, the front gate oxide capacitance, and the bottom gate oxide capacitance, respectively. C_{si} is the silicon film capacitance, and is defined by $C_{si}=\varepsilon_{si}/t_{si}$, where t_{si} is the silicon film thickness. The expression between brackets in Equation (5.5.18) is usually close to unity. For instance, S/S_0 is equal to 1.07, when $t_{si}=90$ nm, $t_{ox1}=15$ nm, $t_{ox2} = 400$ nm, $N_{it1}=10^{10}$ cm^{-2}V^{-1}, and $N_{it2} =5x10^{10}$ cm^{-2}V^{-1}.

A mere comparison of Equations (5.5.10) and (5.5.17) shows that the inverse subthreshold slope of a thin-film, fully depleted, SOI MOSFET will be lower than that of a bulk or thick-film SOI device having the same parameters. The theoretical minimum value of S is 60 mV/decade at room temperature (*i.e.* a 60 mV increase of gate voltage results into a tenfold increase of subthreshold drain current). This also means that, in the subthreshold regime, any increase of gate bias ΔV_G will give rise to an increase of surface potential $\Delta\Phi_{s1}$ equal to ΔV_G (perfect coupling between V_G and Φ_{s1}).

A comparison between the subthreshold slopes in a thin-film SOI MOSFET and a thicker-film device is made in Figure 5.5.4, where the $I_D(V_G)$ characteristics of two n-channel devices (a thin-film device with $t_{si}=100$ nm and a device exhibiting thick-film behavior ($t_{si}=200$ nm) are presented [56]. The subthreshold slope of the thicker-film device is identical to that of a bulk device with same channel doping concentration. It is interesting to note that the thinner device has a slightly lower leakage current at $V_G=0V$ than the thicker device, although its threshold voltage is lower.

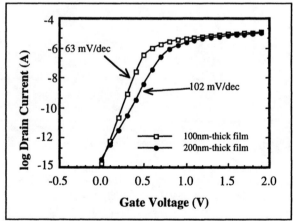

Figure 5.5.4: Simulated $I_D(V_G)$ characteristics of a 200 nm-thick, partially depleted, n-channel device and of a 100 nm-thick, fully depleted MOSFET.

155

In general, and neglecting the presence of interface states, one can write:

$$S = \frac{kT}{q} \ln (10) \, (1+ \alpha) \qquad\qquad (5.5.19)$$

where, as in the case of the analyses of the saturation current (Section 5.3) and the transconductance (Section 5.4), α represents the ratio C_b/C_{ox1} of two capacitors. Indeed, C_{ox1} is the gate oxide capacitance, and C_b is the capacitance between the inversion channel and the back-gate electrode. The value of C_b is either that of the depletion capacitance $C_D = \varepsilon_{si}/x_{dmax}$ in a bulk device, the silicon film capacitance $C_{si} = \varepsilon_{si}/t_{si}$ in a fully depleted device with back interface accumulation, or the value given by the series association of C_{si} and C_{ox2} ($C_b = \dfrac{C_{si}C_{ox2}}{(C_{si}+C_{ox2})}$) in a fully depleted device. It is once again worth noting that the values of α are typically in the following sequence:

$$\alpha_{\text{fully depleted SOI}} < \alpha_{\text{bulk}} < \alpha_{\text{back accum SOI}}$$

such that the inverse subthreshold slope has the lowest (*i.e.* best) value in the fully depleted device, it is larger in the bulk device, and even larger in the device with back accumulation.

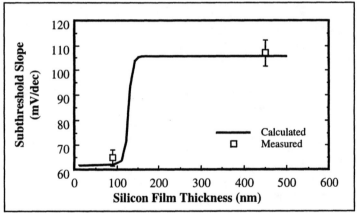

Figure 5.5.5: Simulated and measured subthreshold slope as a function of silicon film thickness. $N_a = 8 \times 10^{16}$ cm^{-3}, gate oxide thickness is 25 nm.[57]

The minimum theoretical value of 60 mV/dec at room temperature is never reached because of the presence of traps at the Si-SiO$_2$ interfaces and because the finite value of C_{ox2} (Figure 5.5.3), but values as low as 65 mV/dec can easily be obtained. The transition between the thin-film and the thick-film regimes is quite sharp, as illustrated in Figure 5.5.5, where the transition between partial and full depletion is quite clear.

The excellent value of the subthreshold slope in thin-film, fully-depleted SOI devices allows one to use smaller values of threshold voltage than in bulk (or thick SOI) devices without increasing the leakage current at $V_{G1}=0$. As a result, better speed performances can be obtained, especially at low supply voltages (2-3 volts). [58] As in bulk devices, an increase of the subthreshold slope is observed in short-channel devices, but thin SOI MOSFETs show less degradation than bulk transistors [59,60].

5.6. Impact ionization and high-field effects

Several parasitic phenomena related to impact ionization in the high electric field region near the drain appear in SOI MOSFETs. Some of them, such as the reduced drain breakdown voltage, are related to the parasitic NPN bipolar transistor found in the n-channel SOI MOSFET, and will be dealt with in Section 5.7. The present section will neglect body current multiplication by bipolar effect and focus on two phenomena. The first of these is the kink effect, and the second one is the hot-electron degradation. The kink effect is normally not found in bulk devices operating at room temperature when substrate contacts are provided, which is usually the case. Hot-electron degradation takes place in bulk MOSFETs having a short channel and constitutes a major reliability hazard in submicron devices. A comparison of hot-electron degradation phenomena between bulk and SOI devices will be made in Section 5.6.2.

5.6.1. Kink effect

The kink effect is characterized by the appearance of a "kink" in the output characteristics of an SOI MOSFET, as illustrated in Figure 5.6.1. The kink appears above a given drain voltage. It can be very strong in n-channel transistors, but is usually absent from p-channel devices. The kink effect is not observed in bulk devices at room temperature when substrate or well contacts are provided, but it can be observed in bulk MOSFETs operating at low temperatures [61] or in devices realized in floating (unconnected) p-wells. Finally, it has been observed from the very beginning of SOI technology, namely in SOS devices [62].

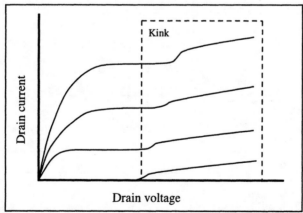

Figure 5.6.1: Illustration of the kink effect in the output characteristics of an n-channel SOI MOSFET.

The kink effect can be explained as follows. Let us consider a "thick"-film, **partially depleted** SOI n-channel transistor. When the drain voltage is high enough, the channel electrons can acquire sufficient energy in the high electric field zone near the drain to create electron-hole pairs, due to an impact ionization mechanism. The generated electrons rapidly move into the channel and the drain, while the holes (which are majority carriers in the p-type body) migrate towards the place of lowest potential, *i.e.*, the floating body (Figure 5.6.2.C).

The injection of holes into the floating body forward biases the source-body diode. The floating body reaches a positive potential which can be calculated by writing the following equation:

$$I_{holes,gen} = I_{so}\left(\exp\left(\frac{q\,V_{BS}}{n\,k\,T}\right) - 1\right)$$

where $I_{holes,\,gen}$ is the hole current generated near the drain (Figure 5.6.2.A), I_{so} is the saturation current of the source-body diode, V_{BS} is the potential of the floating body, and n is the ideality factor of the diode. The value of $I_{holes,gen}$ depends on different parameters, including the body potential. Its calculation requires iterative solving of a set of complex equations [63] and will not be described here.

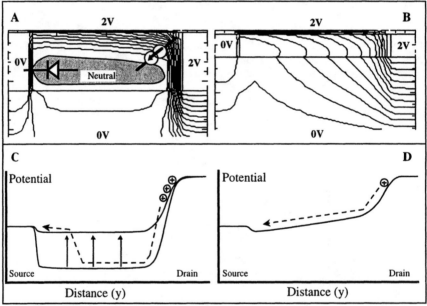

Figure 5.6.2:
A: Isopotential lines in a "thick"-film (30 nm, $N_a=8\times10^{16}$ cm^{-3}), partially depleted (PD) SOI n-channel MOSFET (one curve every 200 mV). The gray area is the neutral, floating body.
B: Isopotential lines in a thin-film (10 nm, $N_a=8\times10^{16}$ cm^{-3}), fully depleted (FD) device.
C: Potential in the neutral region from source to drain in the PD device before and after the onset of the kink effect (lower and upper curve, respectively).
D: Potential from source to drain in the FD device.

The increase of body potential (an increase of 700-800 mV) gives rise to a decrease of the threshold voltage. This is again a manifestation of the body effect (see Section 5.3.2). The threshold voltage shift can be calculated using the expression of the body effect in a *bulk* transistor where the substrate bias is replaced by the floating body potential. This decrease of threshold voltage induces an increase of the drain current as a function of drain voltage which can be observed in the output characteristics of the device (Figure 5.6.1), and which is called "kink effect". If the minority carrier lifetime in the silicon film is high enough, the kink effect can be reinforced by the NPN bipolar transistor structure present in the device (the "base" hole current is amplified by the bipolar gain, which gives rise to an increased net drain current, sometimes called "second kink").

Let us now consider the case of a thin-film, **fully depleted** SOI n-channel transistor (Figure 5.6.2.B). A mere comparison of Figures 5.6.2.A and 5.6.2.B shows that the electric field near the drain is lower in the fully depleted device than in the partially depleted one (the density of isopotential lines is lower). As a result, less electron-hole pair generation will take place in the fully depleted device [64]. As in the case of the partially depleted device, the generated electrons rapidly move into the channel and the drain, while the holes migrate towards the place of lowest potential, *i.e.*, near the source junction. However, contrarily to the case of the partially depleted transistor, the source-to-body diode is "already forward biased" (the body-source potential barrier is very small), due to the full depletion of the film (Figure 5.6.2.D), and the holes can readily recombine in the source without having to raise the body potential (there is no significant potential barrier between the body and the source). As a result, the body potential remains unchanged, the body effect is virtually equal to zero, and there is no threshold voltage decrease as a function of drain voltage. This explains why thin-film, fully depleted n-channel SOI MOSFETs are free of kink effect. If a negative back-gate bias is used, however, to induce an accumulation layer at the back interface, the device behaves as a partially depleted device, and the kink reappears.

For completeness, one can mention that the p-channel SOI transistors are usually free of kink effect, because the coefficient of pair generation by energetic holes is much lower than that of pair generation by energetic electrons. The kink effect is not observed in bulk devices if the majority carriers generated by impact ionization can escape into the substrate or to a well contact. If the well is left floating, or if the silicon is not conducting (*e.g.* at such low temperatures that the carriers are frozen out), the kink effect appears. Finally, the kink effect can be eliminated from partially depleted SOI MOSFETs if a substrate contact is provided for the removal of excess majority carriers from the device body (this technique is unfortunately not 100% effective because of the relatively high resistance of the body) (see also Section 5.7.1).

The presence of a floating substrate can also give rise to anomalous subthreshold slope effects [65]. Indeed, when the body floats, the (relatively weak) impact ionization which can occur near the drain in the subthreshold regime (in addition to the other carrier generation mechanisms) can result in holes being injected into the neutral body. The hole injection charges the body and produces a forward bias on the body-source junction which reduces the threshold voltage while the device operates in the subthreshold region. As a result, the subthreshold current "jumps" from a high-V_{th} characteristic to a low-V_{th} $I_D(V_{G1})$ curve, and a subthreshold slope lower than 60 mV/decade (at room temperature) can be observed.

5.6.2. Hot-electron degradation

Hot-carrier degradation is one of the most important factors affecting the reliability of modern MOS devices. During the last decade, the dimensions of the devices have been dramatically reduced, but the supply voltage of integrated circuits has remained fixed at a steady 5 volts. Submicron devices would need a lower supply voltage, such as 3.3 volts. Such a decrease of supply voltage, however, would reduce the speed performance of the circuits, and the pressure is high to keep compatibility with the 5V standard as long as possible.

The horizontal electric field in the transistor, which is roughly proportional to the ratio of the supply voltage to the gate length, increases as the device dimensions are reduced, and, in modern MOSFETs, the electric field near the drain can reach values high enough to affect the reliability of the device. Let us take the example of an n-channel device. When the transistor operates in the saturation mode, a high electric field can develop between the channel pinchoff and the drain junction. This electric field gives the electrons such a high energy that some of them can be injected into the gate oxide, thereby damaging the oxide-silicon interface [66]. At very high injection levels, gate current can even be measured [67]. The high-energy channel electrons can also create electron-hole pairs by impact ionization. In a bulk device, the generated holes escape into the substrate and constitute a substrate current. There exists a relationship between the substrate current and the gate current [68], and the lifetime of the device can be related to the magnitude of the hot-electron injection on the gate oxide. The relation between the device lifetime, τ, defined by (static) hot-carrier degradation of the gate oxide, can be correlated with the impact-ionization current according to the following relationship [69]:

$$\tau \propto \frac{W}{I_D} (M-1)^{-m} \tag{5.6.1}$$

where $m \cong 3$ [70], and M is the multiplication factor due to impact ionization, defined by $I_{body} = (M-1) I_{Dsat}$, where I_{body} is the hole current generated by impact ionization. In a fully depleted SOI MOSFET, the multiplication factor can be obtained by integrating the ionization coefficient over the high field region near the drain, and depends on both the drain and gate voltages [71]:

$$M-1 \cong \frac{A_i}{B_i} [V_{DS}-V_{Dsat}] \exp\left(-\frac{B_i\, l_c}{V_{DS}-V_{Dsat}}\right)$$

where l_c is a characteristic length defined by: $l_c = t_{si} \sqrt{\dfrac{C_{si}\, \beta}{2C_{ox1}\,(1+\alpha)}}$ with $\beta = 1$ if the back interface is accumulated, and $\beta = 1 + \dfrac{C_{si}}{C_{si}+C_{ox2}}$ if the back interface is depleted. A_i and B_i are impact ionization constants for electrons, assumed to be equal to 1.4×10^6 cm^{-1} and 2.6×10^6 V/cm, respectively. As in the case of the analysis of the saturation current, the transconductance and the subthreshold slope, α represents the ratio C_b/C_{ox1} of two capacitors. Indeed, C_{ox1} is the gate oxide capacitance, and C_b is the capacitance between the inversion channel and the back-gate electrode. The value of C_b is either that of the depletion capacitance $C_D=\varepsilon_{si}/x_{dmax}$ in a bulk device, the silicon film capacitance $C_{si}=\varepsilon_{si}/t_{si}$ in a fully depleted device with back interface accumulation, or the value given by the series association of C_{si} and C_{ox2} ($C_b = \dfrac{C_{si}C_{ox2}}{(C_{si}+C_{ox2})}$) in a fully depleted device. It is once again worth noting that the values of α are typically in the following sequence:

$$\alpha_{\text{fully depleted SOI}} < \alpha_{\text{bulk}} < \alpha_{\text{back accum SOI}}$$

such that the multiplication factor is lowest in the fully depleted device, higher in the bulk device, and the highest in the device with back accumulation.

Figure 5.6.3 presents the multiplication factor (M-1) in bulk and thin-film, fully depleted SOI MOSFETs of similar geometries. The multiplication factor, is related to the lifetime of the device through expression (5.6.1). Clearly, less carrier-electron degradation can be expected from the fully depleted SOI MOSFETs. Numerical modeling of the devices indeed shows that the electric field near the drain is minimized if fully-depleted SOI devices are used [72].

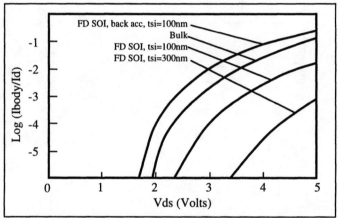

Figure 5.6.3: Multiplication factor (M-1)=I_{body}/I_D as function of drain voltage in bulk and fully-depleted SOI nMOSFETs. The SOI MOSFETs have a thickness of either 100 or 300 nm, with or without back accumulation. L=1µm, V_G-V_{th}=1.5 V.

It is worth noting that devices with back accumulation can present higher drain electric fields than bulk devices and can, therefore, be submitted to more hot-carrier degradation. Finally, the use of thicker SOI films permits one to further minimize hot-electron degradation, provided the device remains fully depleted. Confirmation of the reduced hot-electron degradation in fully depleted SOI devices has been experimentally obtained. Both threshold voltage shift and transconductance degradation as a function of stress time at high V_{DS} have been found to be significantly smaller in fully depleted devices than in bulk devices and SOI devices with back-interface accumulation.

It seems that there is no consensus on the hot-carrier degradation issues in SOI devices. Several papers have been published on that topic, leading to contradictory conclusions. Several publications support the idea that SOI devices are less prone to hot-carrier degradation than bulk devices, while some others draw the opposite conclusion. These discrepancies are partly due to the fact that it is almost impossible to make one-to-one comparisons between fully depleted SOI and bulk devices. Indeed, if the same channel doping concentration is used in both types of devices, the threshold voltages and saturation currents will be different. If the threshold voltages are similar, the doping profiles are different, etc. In addition, the drain breakdown characteristics are different as well. People seem to agree, however, on the following point: the peak drain electric field is lower in fully depleted SOI devices than in bulk devices, leading to less hot-carrier generation.[73,74,75,76]

However, because the degradation mechanism does not depend only on the hot-carrier generation rate, but also on many other different parameters (the quality of the buried oxide, the direction and magnitude of the vertical electric field, ...) the hot carriers degrade SOI devices in a different way than they do in bulk devices [77,78], mainly due to carrier injection in the buried oxide, although some authors report degradation of the front oxide only.[79]

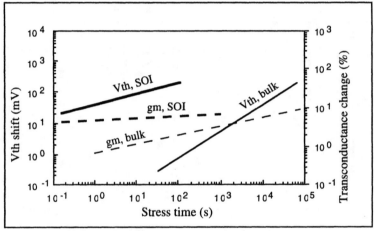

Figure 5.6.4: Threshold voltage shift and transconductance degradation in 0.5 μm thin-film SOI and bulk n-channel transistors as a function of stress time. V_{GS}=2 V, V_{DS}=6 V.[80]

Figure 5.6.4 presents the threshold voltage shift and the transconductance degradation in SOI thin-film (80 nm) and bulk transistors having a gate length of 0.5 μm. A much larger initial transconductance degradation, a more pronounced initial threshold voltage shift, and a weaker time dependence are observed in the SOI devices. The parameter which degrades the most in SOI is the threshold voltage. The degradation depends on both front- and back-gate bias.[81,82] Depending on the applied front and back gate bias either holes or electrons can be injected in the buried oxide. A t^n time dependence of the degradation is observed, with n=0.25 in the case of pure hole or pure electron injection, and n=0.5 when both holes and electrons are injected in the buried oxide.[83]

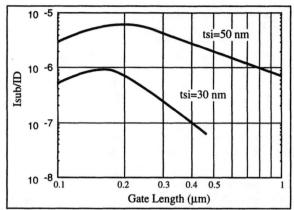

<u>Figure 5.6.5</u>: I_{sub}/I_D versus gate length in FD SOI MOSFETs for two silicon film thicknesses.[84]

It has also been observed that the substrate current generated by hot-carrier effects decreases in SOI devices once the channel length is shorter than 0.2 μm (Figure 5.6.5). This is caused by velocity saturation overshoot of the carriers in the channel. These can then reach the drain without loosing energy in an impact ionization process. This observation further supports the use of SOI for very short channel devices.

Finally, it appears that hot-carrier induced gate current characteristics of SOI devices are similar to those of bulk transistors.[85]

5.7. Floating-body and other parasitic effects

There exists a parasitic bipolar transistor in the MOS structure. If we consider an n-channel device, the N^+ source, the P-type body and the N^+ drain indeed form the emitter, the base, and the collector of an NPN bipolar transistor. In a bulk device, the base of the bipolar transistor is usually grounded by means of a substrate contact. In an SOI device, however, the body (= the base of the bipolar transistor) is usually left floating. This parasitic bipolar transistor is the origin of several undesirable effects in SOI devices: extremely high (almost infinite) subthreshold slope and reduction of the drain breakdown voltage.

5.7.1. Anomalous subthreshold slope

As we have seen at the end of Section 5.6.1, the generation of majority carriers (holes in the case of an n-channel transistor) by impact ionization near the drain can give rise to an increase of the body potential and decrease of the threshold voltage. Sometimes, a similar effect can occur at gate voltages lower than the threshold voltage . If the drain voltage is high enough, impact ionization can occur in the subthreshold region, even though the drain current is very small. This effect is observed in partially depleted devices and in fully depleted devices with back accumulation, and can be explained as follows. When the device is turned off, there is no impact ionization, and the body potential is equal to zero, since there is no base-to-source current. When the gate voltage is increased the weak inversion current can induce impact ionization in the high electric field region near the drain, holes are generated, the body potential increases, and the threshold voltage is reduced. Consequently, the whole $I_D(V_{GS})$ characteristic shifts to the left, and the current can increase with gate voltage with a slope larger than 60 mV/decade. In other words, inverse subthreshold slopes lower than the theoretical limit (in absence of multiplication effect) of 60 mV/decade can be observed.[86,87]

If the minority carrier lifetime in the silicon film is high enough, the parasitic bipolar transistor present in the NPN structure of the MOS device can amplify the base current (*i.e.* amplify the hole current generated by impact ionization near the drain). The base current is given by $I_{body} = (M-1) I_{Dsat} \equiv (M-1) I_{ch}$ (see Section 5.6) where I_{ch} is the channel current, and the resulting increase of drain current is given by $\Delta I_D = \beta I_{body} = \beta (M-1) I_{ch}$ (Figure 5.7.1). This increase of drain current constitutes a positive feedback loop on the current flowing through the device: the drain current suddenly increases, and an infinite subthreshold slope is observed (Figure 5.7.2).

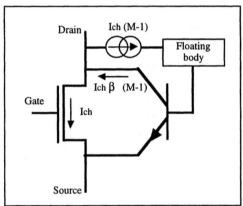

Figure 5.7.1: Parasitic bipolar transistor of the SOI MOSFET. I_{ch} is the channel current.

This phenomenon is known as the "single-transistor latchup".[88,89,90,91] It can be explained as follows (Figure 5.7.2). At low drain bias (curve a), a normal subthreshold slope is obtained, under forward as well as reverse gate voltage scan conditions. If the drain voltage is increased (curve b), the impact ionization near the drain raises the body potential (forward gate voltage scan case). This reduces the threshold voltage and leads to an increase of the drain current, which in turn increases impact ionization near the drain. When the gain of the positive feedback loop, $\beta (M-1)$, reaches unity, the current suddenly increases. The positive feedback is self-limiting: increased body bias also increases the drain saturation voltage which results in a lower electric field near the drain and a smaller impact ionization current. During a descending gate voltage scan (curve b), the impact ionization current under a large drain voltage keeps the body voltage high, which in turn keeps the threshold voltage of the device low, and a high drain current is observed until the positive feedback can no longer be maintained, so that $\beta (M-1) < 1$, and the drain current suddenly drops. It can be noticed that the gate voltage at which the current suddenly raises, during a forward scan, is higher than the

164

voltage at which the current suddenly falls, during a reverse scan. As a consequence, an hysteresis is observed in the $I_D(V_G)$ curve (curve b). If the drain bias is large enough, the positive feedback loop cannot be turned off once it has been triggered (β (M-1) remains larger or equal to 1), and the device cannot be turned off (curve c). It is worth noting that this parasitic phenomenon was not observed in devices made in early SOI material, where the recombination lifetime of minority carriers in the base was low and where the gain of the bipolar transistor, β, was small. The single-transistor latch effect is not observed if the body of the device is grounded, and it is reduced if the device is fully depleted.

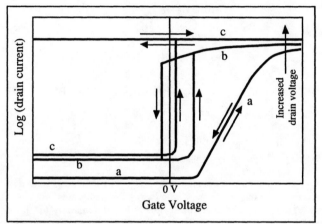

Figure 5.7.2: Illustration of the single-transistor latch. "Normal" subthreshold slope at low drain voltage (a), infinite subthreshold slope and hysteresis (b), and device "latch-up" (c).

5.7.2. Reduced drain breakdown voltage

It has been demonstrated that the peak electrical field at the drain junction of SOI devices is lower than what is commonly found in bulk devices if the junction is reaching through the silicon film to the buried oxide [92]. From this result, a higher junction breakdown voltage could be expected in SOI transistors than in bulk devices. Unfortunately, the SOI MOSFET includes a parasitic bipolar transistor with floating base. From bipolar transistor theory [93], the collector (drain) breakdown voltage with open base, BV_{CEO}, is smaller than when the base is grounded (BV_{CB0}). Both breakdown voltages relate as follows:

$$BV_{CEO} = \frac{BV_{CBO}}{\sqrt[n]{\beta}}$$

(5.7.1)

where β is the gain of the bipolar device, and n ranges typically between 3 and 6. The above relationship is quite a simplification of the breakdown mechanisms occurring in SOI MOSFETs, since both β and M-1 (the multiplication factor) depend on the drain voltage in a highly nonlinear fashion [94]. From bipolar transistor theory, one can also write:

165

$$\beta \cong 2 \, (L_n/L_B)^2 - 1 \qquad (5.7.2)$$

where L_B is the base width, which can be assumed, in first approximation, to be equal to the effective channel length, L, and L_n is the electron diffusion length: $L_n^2 = D_n \, \tau_n$, where D_n and τ_n are the diffusion coefficient and the lifetime of the minority carriers in the base, respectively. From (5.7.1) and (5.7.2), and taking BV_{CEO} as the drain breakdown voltage with floating body equal to BV_{DS}, one obtains:

$$BV_{DS} = \frac{BV_{CBO}}{\sqrt[n]{\dfrac{2D_n \, \tau_n}{L^2} - 1}} \qquad (5.7.3)$$

where L is the gate length.

Using Einstein's relationship $D_n = \dfrac{kT}{q} \, \mu_n$, an abacus relating BV_{DS}/BV_{CBO} to channel length can be plotted, with the minority carrier lifetime in the SOI material as parameter (Figure 5.7.3) [95]. It should, however, be noted that Equations (5.7.2) and (5.7.3), and, hence, Figure 5.7.3 are valid only if $L_n > L$, *i.e.* if β is significantly larger than 1, (*i.e.* when short channels and relatively good lifetime SOI material are used.)

This reduction of the drain breakdown voltage was not observed in devices made on early (and defective) SOI material, where the minority carrier lifetime was low, and its control constitutes one of the major challenges for today's SOI research activity [96], especially when submicron devices are considered. Possible solutions of the problem include the use of lightly-doped sources and drains (LDS and LDD)[97], the use of lifetime killers, controlled introduction of defects in the silicon film, growth of silicides on source and drain [98,99], the use of SiGe sources and drains [100,101,102], and the use of body contacts.

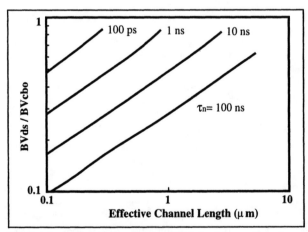

Figure 5.7.3: Relationship between the reduction of drain breakdown voltage, the gate length and the minority carrier lifetime (τ_n) in SOI MOSFETs.

Finally, it should be noted that the above model stems from the presence of a neutral base region, and is, therefore, valid only for partially depleted devices. The understanding of parasitic bipolar effects in fully depleted devices necessitates the development of a model for a bipolar transistor with fully depleted base [103,104]. Numerical simulations show that, even though no kink is discernible in fully depleted SOI MOSFETs, these devices are subjected to significant body charging by impact ionization, which portends significant parasitic bipolar effects. The drain breakdown voltage is controlled by the common-emitter bipolar breakdown voltage BV_{CEO} which occurs under the condition that $\beta(M-1)=1$. Both β and M are highly nonlinear. M increases strongly with V_{DS} due to the increasing drain electric field, and β decreases strongly with V_{DS} due to the prevalent high-injection in the body (base of the bipolar device). Numerical analysis also suggests that the use of an LDS structure can reduce β through the reduction of the bipolar emitter efficiency and concludes that an optimized LDD/LDS fully depleted SOI MOSFET renders submicron SOI CMOS a viable and advantageous technology.

5.7.3. Other floating-body effects

The presence of a floating body gives rise to a series of problems of transient effects, hysteresis, etc. All these effects have their origin in the charging/discharging of the floating body by currents coming from the source or the drain and in the capacitive coupling between the gate and the floating body. One can list the following effects:

◊   Transient leakage current in pass transistors, which can affect SOI SRAM and DRAM circuitry.[105,106,107,108,109]

◊   Drain current transient or overshoot when a voltage step is applied to the gate (capacitive coupling between the gate (or the drain) and the floating body).[110,111]

◊   Variations of drain current depending on device history.[112,113,114,115]

◊   Irregular signal propagation (frequency-dependent delay times, pulse stretching).[116,117,118]

◊   Degradation of logic states in dynamic circuits.[119]

◊   Double snapback of the $I_D(V_D)$ characteristics (see Section 8.3.1).[120,121,122]

The classical remedy to floating-body problems is the use of a body contact (or body tie) in partially depleted devices. In fully depleted devices there are, usually, no floating body problems and, if there are, these are much weaker than in partially depleted transistors.

5.7.4. Self heating

SOI transistors are thermally insulated from the substrate by the buried insulator. As a result, removal of excess heat generated by Joule effect within the device is less efficient than in bulk, which yields to substantial elevation of device temperature.[123,124] The conduction paths for excess heat are multiple: heat diffuses vertically through the buried oxide and laterally through the silicon island into the contacts and the metallization.[125]

The negative resistance[126], which can be seen in the output characteristics of SOI MOSFETs, is due to a mobility reduction effect caused by device heating.[127] This effect is clearly visible on the output curves once sufficient power is dissipated in the device. Because of the relatively low thermal conductivity of the buried oxide, the devices heat up by 50 to 150°C and a mobility reduction is observed.[128]. This effect has been included in device and circuit simulators.[129]

One should not forget, however, that this effect takes place as power is dissipated into the device. This is the case when the device is measured in a quasi-dc mode with a curve tracer or an HP4145, but not in an operating CMOS circuit. Indeed, in an operating CMOS circuit, there is virtually no current flowing through the devices in the standby mode, and power is dissipated in the devices only during switching for brief periods of time (< 1 nanosecond).

It has been shown by a pulsed measurement technique[130] that the time constants involved in the self heating of SOI transistors are on the order of several tens of nanoseconds, and that no negative resistance effect is observed when the devices are measured in the pulse mode, because the time during which power is dissipated is much shorter than the thermal time constant of the devices. A similar experiment reports that the self-heating does not influence the output characteristics of transistors, if the measurement is carried out at a slew rate higher than 20 V/μs.[131] It seems thus that the negative resistance is not a problem for digital circuits (with the possible exception of output buffers). There might, however, be an influence of the duty cycle and the frequency at which the devices are switched on the overall local temperature, which could modify the mobility.

As far as analog circuits are concerned, some effects are observed as well. For instance, the output conductance (g_D) of a transistor becomes frequency dependent. At low frequencies the self-heating mechanism can follow the signal, and a reduction of g_D is observed. Above 100 kHz the transconductance increases. Three transconductance increases are indeed observed in SIMOX devices for time constants approximately equal to 1μs, 100 ns, and 3 ns. There is a factor of 2 difference between the high- and the low-frequency output conductance.[132] The heat can also propagate from one device to another, and thermal coupling effects can be observed in sensitive structures such as current mirrors.[133]

5.8. Accumulation-mode p-channel MOSFET

Thick-film p-channel MOSFETs (with an N[+] poly gate) are usually buried-channel devices, and their characteristics are similar to those of bulk buried-channel pMOSFETs.

Thin-film, enhancement-mode p-channel MOSFETs can also be fabricated [134], but these devices exhibit large values of threshold voltage when they have a thin (20 nm or less) gate oxide and N[+] polysilicon as gate material. In order to obtain useful values of threshold voltage (around -0.7 V) when a thin gate oxide and an N[+] polysilicon gate are used, the body of the transistor has to have a p-type doping (Figure 5.8.1). Such a device is an accumulation-mode device (also called deep-depletion device), the characteristics of which will be derived in this section.

5.8.1. I-V characteristics

When the device is turned OFF, the silicon film is fully depleted due to the presence of positive interface charges and to the negative value of the work function difference between the N[+] polysilicon gate and the P-type body of the device. When the device is turned ON, the film is no longer fully depleted and conduction occurs both in the body of the device and in a surface accumulation channel [135,136].

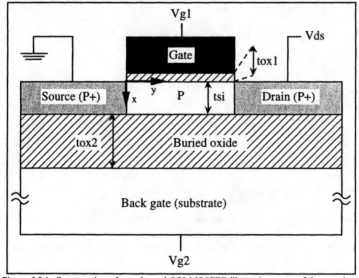

Figure 5.8.1: Cross-section of a p-channel SOI MOSFET illustrating some of the notations used in this Section.

When a zero bias is applied to the gate, the film is fully depleted due to both the presence of positive charges at the $Si\text{-}SiO_2$ interface and the use of an N[+]-polysilicon gate. The gate material-silicon work function difference is given by:

169

$$\Phi_{MS1} = -\frac{E_g}{2} - \frac{kT}{q} \ln \frac{N_a}{n_i} \tag{5.8.1}$$

where n_i is the intrinsic carrier concentration in silicon. When a negative gate voltage is applied, the overall hole concentration in the silicon film is increased. Threshold of accumulation is reached when $\Phi_{s1}=0$. The gate voltage needed to accomplish this is given by:

$$V_{th,acc} = \Phi_{MS1} - \frac{Q_{ox1}}{C_{ox1}} = V_{fb1} \tag{5.8.2}$$

Where V_{fb1} is the front flat-band voltage, Q_{ox1} is the charge density per unit area in the gate oxide, and C_{ox1} is the gate oxide capacitance per unit area. When the gate voltage, V_{G1}, is lower (i.e. larger in absolute value) than the accumulation threshold voltage, the accumulation charge per unit area in the channel is given by: $Q_{acc}(y) = -[V_{G1} - V_{fb1} - V(y)] C_{ox1}$, where $V(y)$ is the local potential along the channel ($y=0$ at the source junction and $y=L$ at the drain junction). Using the gradual channel approximation, one obtains the following expression for the current in the accumulation channel:

$$\int_0^{V_{DS}} dV = I_{acc} \int_0^L dR(y) \ , \ with \ dR(y) = \frac{dy}{W \mu_s Q_{acc}} \tag{5.8.3}$$

which yields after integration:

$$I_{acc} = \frac{W}{L} \mu_s C_{ox1} \left((V_{G1} - V_{fb1}) V_{DS} - \frac{V_{DS}^2}{2} \right) \tag{5.8.4}$$

in the linear regime ($V_{DS} > V_{G1} - V_{fb1}$), and

$$I_{acc} = \frac{W}{L} \mu_s \frac{C_{ox1}}{2} (V_{G1} - V_{fb1})^2 \tag{5.8.5}$$

above saturation ($V_{DS} < V_{G1} - V_{fb1}$).

μ_s is the hole surface mobility, which is equal to $\dfrac{\mu_{s0}}{1+\theta(V_{fb1}-V_{G1})}$, with μ_{s0} being the zero-field surface mobility, and θ being a field mobility reduction factor.

The other current flowing from source to drain is the body current, which appears if there is a portion of non-depleted silicon below the channel. As far as this current component is concerned, similarity is found between the operation of this MOS device and that of a JFET. Conduction between source and drain occurs in the neutral part of the body. The width of this conduction path is modulated by the vertical extension of the depletion zones related to the front and the back gate. The front gate depletion depth can be found by solving the Poisson equation using the depletion approximation:

$$\frac{d^2\Phi(x)}{dx^2} = \frac{q N_a}{\varepsilon_{si}} \tag{5.8.6}$$

with $\Phi(x)$ being the potential in the x-direction (in the depth of the semiconductor film), and N_a the acceptor doping concentration.
We find:

$$-E(x) = \frac{q\,N_a\,x}{\varepsilon_{si}} + A \quad \text{and} \quad \Phi(x) = \frac{q\,N_a\,x^2}{2\varepsilon_{si}} + Ax + B$$

(A and B are integration constants) with the following boundary conditions: $E(x_{depl})=0$ and $\Phi(x_{depl})=0$ where x_{depl} is the depletion depth. From the above expressions, we find the surface potential:

$$\Phi(0) = \frac{q\,N_a\,x_{depl}^2}{2\varepsilon_{si}}$$

Using Gauss' law, we finally obtain:

$$V_G - V_{FB} = \frac{q\,N_a\,x_{depl}^2}{2\varepsilon_{si}} + \frac{q\,N_a\,x_{depl}}{C_{ox}} \qquad (5.8.7)$$

The gate voltage for which the depletion depth equals the silicon thickness is:

$$V_G = V_{FB} + \frac{q\,N_a\,t_{si}^2}{2\varepsilon_{si}} + \frac{q\,N_a\,t_{si}}{C_{ox}} \equiv V_{FB} + V_{depl} \qquad (5.8.8)$$

V_{depl} being defined by the above relationship (5.8.8).

As long as there is a portion of neutral silicon in the film (at a distance y from the source), the depth of the front-gate-related depletion region is given by:

$$X_{d1}(y) = \frac{-\varepsilon_{si}}{C_{ox1}} + \sqrt{\varepsilon_{si}^2/C_{ox1}^2 + 2\,\varepsilon_{si}\,(V_{G1} - V_{fb1} - V(y))/q\,N_a} \qquad (5.8.9)$$

Similarly, under the same conditions, the depth of the depletion zone arising from the back interface is given by:

$$X_{d2}(y) = \frac{-\varepsilon_{si}}{C_{ox2}} + \sqrt{\varepsilon_{si}^2/C_{ox2}^2 + 2\,\varepsilon_{si}\,(V_{G2} - V_{fb2} - V(y))/q\,N_a} \qquad (5.8.10)$$

If $X_{d1}(y) + X_{d2}(y)$ is equal to or larger than t_{si}, the silicon film thickness, the film is locally fully depleted. For the sake of simplicity, it will be assumed that the width of the back depletion region, X_{d2}, is constant and given by Equation (5.8.10) where V(y) is held constant and equal to V(y=0). The resistance of an elementary resistor in the body channel is given by:

$$dR = \frac{dy}{W\,\mu_b\,q\,N_a\,(t_{eff} - X_{d1})} \qquad (5.8.11)$$

where t_{eff} is equal to $t_{si}-X_{d2}$ and μ_b is the bulk hole mobility. Using again the gradual channel approximation, and integrating $dV=IdR$ from source to drain for the various operation modes of the device, and using Equation (5.8.11) to find R and Equation (5.8.9) to calculate X_{d1}, one obtains the following expression:

$$I_{body} \int dy = W\, q\, N_a\, \mu_b \int [t_{eff} + \frac{\varepsilon_{si}}{C_{ox1}} - \sqrt{\frac{\varepsilon_{si}^2}{C_{ox1}^2} + \frac{2\varepsilon_{si}\,(V_{G1}-V_{fb1}-V(y))}{q\, N_a}}]\, dV \quad (5.8.12)$$

The integration of Equation (5.8.12) from source to drain gives an expression in the form:

$$I_{body} = q\, N_a\, \mu_b\, \frac{W}{L}\, \xi$$

where ξ results from the integration of (5.8.12) and varies as a function of applied biases, depending on whether the film is fully depleted or not, whether the neutral zone extends all the way to the drain, etc. We will also define V'_{depl} as the V_{depl} of Equation (5.8.8) where $t_{si}=t_{eff}$, or:

$$V'_{depl} \equiv \frac{q\, N_a\, t_{eff}^2}{2\varepsilon_{si}} + \frac{q\, N_a\, t_{eff}}{C_{ox}}$$

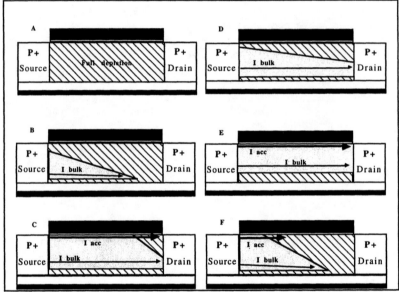

Figure 5.8.2: Cross-section of a p-channel SOI MOSFET illustrating the possible distributions of depletion and neutral zones in the silicon film.

The following cases can be distinguished:

* $\xi = 0$ when $V_{G1} > V_{fb1} + V'_{depl}$, *i.e.* when the film if fully depleted. V'_{depl} is the change in front gate voltage with respect to flatband necessary to obtain $X_{d1} = t_{eff}$ near the source ($V(y=0)$) using Equation (5.8.9). There is no accumulation channel (Figure 5.8.2.A).

* $\xi = t_{eff}\, V_{DS}$ when $V_{G1} - V_{fb1} < 0$ and $V_{G1} - V_{fb1} - V_{DS} < 0$ (Figure 5.8.2.E), *i.e.* when the front interface is in accumulation from source to drain.

*   In the case where $0 < V_{G1}-V_{fb1} < V'_{depl}$ and $0 < V_{G1}-V_{fb1}-V_{DS} < V'_{depl}$, *i.e.* when neither accumulation nor body channel pinchoff occur (Figure 5.8.2.D), Equation (5.8.12) can be written:

$$I_{body} \int_0^L dy = W\, q\, N_a\, \mu_b \int_0^{V_{DS}} [t_{eff} + \frac{\varepsilon_{si}}{C_{ox1}} - \sqrt{\frac{\varepsilon_{si}^2}{C_{ox1}^2} + \frac{2\varepsilon_{si}\,(VG1-Vfb1-V(y))}{q\,N_a}}]\; dV$$

and, therefore:

$$\xi = \quad (t_{eff} + \varepsilon_{si}/C_{ox1})\, V_{DS} + \frac{qN_a}{3\varepsilon_{si}}[(\varepsilon_{si}/C_{ox1})^2 + \frac{2\varepsilon_{si}}{qN_a}(V_{G1}-V_{fb1}-V_{DS})]^{3/2}$$

$$- \frac{qN_a}{3\varepsilon_{si}}[(\varepsilon_{si}/C_{ox1})^2 + \frac{2\varepsilon_{si}}{qN_a}(V_{G1}-V_{fb1})]^{3/2} \tag{5.8.13}$$

*   In the case where $0 < V_{G1}-V_{fb1} < V'_{depl}$ and $V_{G1}-V_{fb1}-V_{DS} > V'_{depl}$, i.e. when body channel is pinched off, but there is no accumulation at the source end nor full depletion of the film (Figure 5.8.2.B), we have:

$$I_{body} \int_0^L dy = W q N_a\, \mu_b \int_0^{V_{G1}-V_{FB1}-V'_{depl}} [t_{eff} + \frac{\varepsilon_{si}}{C_{ox1}} - \sqrt{\frac{\varepsilon_{si}^2}{C_{ox1}^2} + \frac{2\varepsilon_{si}\,(VG1-Vfb1-V(y))}{q\,N_a}}]\; dV$$

which yields:

$$\xi = \quad (t_{eff} + \varepsilon_{si}/C_{ox1})\,(V_{G1}-V_{fb1}-V'_{depl}) + \frac{qN_a}{3\varepsilon_{si}}[(\varepsilon_{si}/C_{ox1})^2 + \frac{2\varepsilon_{si}}{qN_a}V'_{depl}]^{3/2}$$

$$- \frac{qN_a}{3\varepsilon_{si}}[(\varepsilon_{si}/C_{ox1})^2 + \frac{2\varepsilon_{si}}{qN_a}(V_{G1}-V_{fb1})]^{3/2} \tag{5.8.14}$$

*   In the case where $0 > V_{G1}-V_{fb1}$ and $0 < V_{G1}-V_{fb1}-V_{DS} < V'_{depl}$, *i.e.* when body channel is not pinched off and there is accumulation at the source end (Figure 5.8.2.C), we have:

$$I_{body} \int_0^L dy = W q N_a\, \mu_b \int_{V_{G1}-V_{FB1}}^{V_{DS}} [t_{eff} + \frac{\varepsilon_{si}}{C_{ox1}} - \sqrt{\frac{\varepsilon_{si}^2}{C_{ox1}^2} + \frac{2\varepsilon_{si}\,(VG1-Vfb1-V(y))}{q\,Na}}]\; dV + W q N_a\, \mu_b \int_0^{V_{G1}-V_{FB1}} t_{eff}\, dV$$

which yields:

$$\xi = \quad t_{eff}\,(V_{G1}-V_{fb1}) + (t_{eff} + \varepsilon_{si}/C_{ox1})\,(V_{DS}-V_{G1}+V_{fb1})$$

$$+ \frac{qN_a}{3\varepsilon_{si}}[(\varepsilon_{si}/C_{ox1})^2 + \frac{2\varepsilon_{si}}{qN_a}(V_{G1}-V_{fb1}-V_{DS})]^{3/2}$$

$$- \frac{qN_a}{3\varepsilon_{si}}[\varepsilon_{si}/C_{ox1}]^3 \tag{5.8.15}$$

*   In the case where $0 > V_{G1}-V_{fb1}$ and $V_{G1}-V_{fb1}-V_{DS} > V'_{depl}$, *i.e.* when body channel is pinched off and there is accumulation at the source end. (Figure 5.8.2.F), we have:

$$I_{body} \int_0^L dy = WqN_a\mu b \int_{V_{G1}-V_{FB1}}^{V_{G1}-V_{FB1}-V'_{depl}} [t_{eff} + \frac{\varepsilon_{si}}{C_{ox1}} - \sqrt{\frac{\varepsilon_{si}^2}{C_{ox1}^2} + \frac{2\varepsilon_{si}(VG1-Vfb1-V(y))}{q\,Na}}]\ dV\ +\ WqN_a\mu b \int_0^{V_{G1}-V_{FB1}} t_{eff}\,dV$$

which yields:

$$\begin{aligned}
\xi =\ & t_{eff}(V_{G1}-V_{fb1}) - (t_{eff} + \varepsilon_{si}/C_{ox1})(V'_{depl}) \\
& + \frac{qN_a}{3\varepsilon_{si}}[(\varepsilon_{si}/C_{ox1})^2 + \frac{2\varepsilon_{si}}{qN_a}V'_{depl}]^{3/2} \\
& - \frac{qN_a}{3\varepsilon_{si}}[\varepsilon_{si}/C_{ox1}]^3
\end{aligned}$$
(5.8.16)

Finally, the total drain current in the device is given by the sum of the current in the accumulation channel and the current in the body of the device:

$$I_{DS} = I_{acc} + I_{body}$$
(5.8.17)

Figure 5.8.3 presents the linear $I_D(V_{G1})$ characteristics of a thin-film, accumulation-mode, p-channel transistor for different back-gate voltages. The *apparent* front threshold voltage of this device shows a dependence on back-gate bias which is similar to that observed in fully-depleted enhancement-mode devices. Actually, the front accumulation threshold voltage is quite independent of back-gate bias, and the shift of the $I_D(V_{G1})$ characteristics with (more negative) back-gate bias is due to the apparition and increase of the body current. At large negative back biases, even an accumulation channel can be created at the back interface. This case, however, was not dealt with in the above model, since the presence of a back accumulation channel is undesirable for most practical applications.

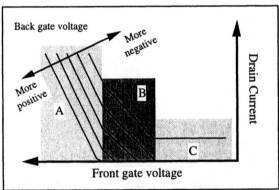

Figure 5.8.3: Linear $I_D(V_{G1})$ characteristics of a thin-film, accumulation-mode, p-channel transistor for different back-gate voltages. The front accumulation current, the body current, and the back channel accumulation current are outlined by the shaded zones A, B, and C, respectively.

Figure 5.8.4 presents the output characteristics of a thin-film, accumulation-mode, p-channel SOI device (t_{si}=100 nm, N_a=4×10^16 cm^-3).

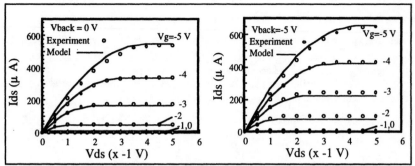

Figure 5.8.4: Output characteristics of a thin-film, accumulation-mode, p-channel SOI device, for two values of back-gate bias.

It is worth noting that in most applications, p-channel SOI MOSFETs are operating with a negative back-gate bias. If one considers, for instance, the case of a CMOS inverter, the source of the p-channel device is connected to $+V_{DD}$ (*e.g.* 5 volts). If the mechanical substrate (the back gate) is grounded, the p-channel transistor is then operating with a back-gate bias equal to $-V_{DD}$ (*i.e.* -5 V).

Good performances are usually obtained from accumulation-mode p-channel MOSFETs. Because of the reduced scattering encountered by the bulk current, a higher mobility is observed in these devices than in enhancement-mode p-channel MOSFETs [137].

Accumulation-mode $(N^+N^-N^+)$ n-channel MOSFETs can also be fabricated, using either p^+ polysilicon or n^+ polysilicon as gate material [138]. In the latter case, however, devices with negative threshold voltage are obtained. Either devices can be simulated using the above model, provided the proper sign changes are made.

5.8.2. Subthreshold slope

In the subthreshold regime (when neither body nor accumulation currents are flowing), the body of the device is fully depleted because of the presence of positive oxide charges in the gate oxide and the buried oxide, and because of the work function difference between the N^+ polysilicon gate and the p-type body. When the body is fully depleted, the classical expressions developed to describe the relationship between the gate voltages and the surface potentials in fully-depleted, enhancement-mode, n-channel MOSFETs can be employed (Equations 5.3.8 and 5.3.9) [139,140]:

$$V_{G1} = V_{FB1} + \Phi_{s1}\left(1 + \frac{C_{it1}}{C_{ox1}} + \frac{C_{si}}{C_{ox1}}\right) - \Phi_{s2}\frac{C_{si}}{C_{ox1}} + \frac{Q_D}{2C_{ox1}}$$

and

$$V_{G2} = V_{FB2} + \Phi_{s2}\left(1 + \frac{C_{it2}}{C_{ox2}} + \frac{C_{si}}{C_{ox2}}\right) - \Phi_{s1}\frac{C_{si}}{C_{ox2}} + \frac{Q_D}{2C_{ox2}}$$

where $V_{FB1} = \Phi_{MS1} - \dfrac{qN_{ox1}}{C_{ox1}}$ is the front flatband voltage, $V_{FB2} = \Phi_{MS2} - \dfrac{qN_{ox2}}{C_{ox2}}$ is the back flatband voltage, and Q_D is the absolute value of the total depletion charge in the silicon film, which is equal to $qN_a t_{si}$. V_{G1}, V_{G2}, Φ_{s1} and Φ_{s2} are the front and back gate voltages (the source being grounded and taken as reference voltage), and the front and back surface potentials in the silicon film, respectively. Using the above equations, the surface potentials can be expressed as a function of the gate voltages:

$$\Phi_{s1} = \frac{V_{G1} - V_{FB1} + \dfrac{C_{si}}{C_{ox1}}\left(\dfrac{V_{G2} - V_{FB2} - \dfrac{Q_D}{2\,C_{ox2}}}{1 + \dfrac{C_{it2}}{C_{ox2}} + \dfrac{C_{si}}{C_{ox2}}}\right) - \dfrac{Q_D}{2C_{ox1}}}{\left(1 + \dfrac{C_{it1}}{C_{ox1}} + \dfrac{C_{si}}{C_{ox1}}\right) - \dfrac{\dfrac{C_{si}}{C_{ox1}}\dfrac{C_{si}}{C_{ox2}}}{1 + \dfrac{C_{it2}}{C_{ox2}} + \dfrac{C_{si}}{C_{ox2}}}} \tag{5.8.18a}$$

and

$$\Phi_{s2} = \frac{V_{G2} - V_{FB2} + \dfrac{C_{si}}{C_{ox2}}\left(\dfrac{V_{G1} - V_{FB1} - \dfrac{Q_D}{2\,C_{ox1}}}{1 + \dfrac{C_{it1}}{C_{ox1}} + \dfrac{C_{si}}{C_{ox1}}}\right) - \dfrac{Q_D}{2C_{ox2}}}{\left(1 + \dfrac{C_{it2}}{C_{ox2}} + \dfrac{C_{si}}{C_{ox2}}\right) - \dfrac{\dfrac{C_{si}}{C_{ox2}}\dfrac{C_{si}}{C_{ox1}}}{1 + \dfrac{C_{it1}}{C_{ox1}} + \dfrac{C_{si}}{C_{ox1}}}} \tag{5.8.18b}$$

Equation (5.3.2) gives the potential distribution through the film:

$$\Phi(x) = \frac{q\,N_a}{2\,\varepsilon_{si}}x^2 + \left(\frac{\Phi_{s2} - \Phi_{s1}}{t_{si}} - \frac{q\,N_a\,t_{si}}{2\,\varepsilon_{si}}\right)x + \Phi_{s1}$$

where x is the depth into the silicon film (x=0 at the silicon-gate oxide interface and x=t_{si} at the silicon-buried oxide interface). Assuming that the drain-to-source voltage drop is small ($V_D \cong V_S$), the hole concentration in the silicon film in any section of the device between the source and the drain is given by:

$$p(y) = N_a \int_0^{t_{si}} \exp\left(\frac{-q\Phi(x)}{kT}\right) dx$$

$$= N_a \exp\left(\frac{-q\Phi_{s1}}{kT}\right) \int_0^{t_{si}} \exp\left[\frac{-q\left(\dfrac{qN_a}{2\varepsilon_{si}}x^2 + \left(\dfrac{\Phi_{s2} - \Phi_{s1}}{t_{si}} - \dfrac{qN_a t_{si}}{2\varepsilon_{si}}\right)x\right)}{kT}\right] dx \tag{5.8.19}$$

The above expression cannot be integrated analytically but it can readily be evaluated using numerical integration. Hence, one can write:

$$p(y) = N_a \, \exp\!\left(\frac{-\Phi_{s1}}{kT}\right) F(t_{si}, \Phi_{s1}, \Phi_{s2}) \tag{5.8.20}$$

where $F(t_{si}, \Phi_{s1}, \Phi_{s2})$ is the result of the numerical integration of Equation (5.8.19).

Measurements [141] and numerical simulations of accumulation-mode devices show that the subthreshold current is independent of the drain voltage, as long as V_{DS} is larger than several kT/q. This suggests that the transport mechanism is due to diffusion rather than drift, as in the case of enhancement-mode MOSFETs.

The diffusion current in the transistor is given by:

$$I_D = - A \, q \, D_p \, \frac{dp(y)}{dy}$$

where D_p is the hole diffusion coefficient, and $A = Wt_{si}$ is the area of the device cross section. If the absolute value of the drain voltage is small enough (with $V_{DS} < 0$), this last expression can be simplified into:

$$
\begin{aligned}
I_D &= - A \, q \, D_p \, \frac{p(L) - p(0)}{L} = \frac{A}{L} \, q \, p(0) \, D_p \left(1 - \exp\!\left(\frac{qV_{DS}}{kT}\right)\right) \\[2mm]
&= \frac{A}{L} q \, N_a \, \exp\!\left(\frac{-q\Phi_{s1}}{kT}\right) F(t_{si}, \Phi_{s1}, \Phi_{s2}) \, D_p \left(1 - \exp\!\left(\frac{qV_{DS}}{kT}\right)\right)
\end{aligned}
\tag{5.8.21}
$$

By definition, the inverse subthreshold slope is given by:

$$S = \frac{dV_{G1}}{d \log(I_D)} = \frac{\ln(10)}{\dfrac{d \ln(I_D)}{dV_{G1}}} \tag{5.8.22}$$

with

$$\frac{d \ln(I_D)}{dV_{G1}} = \frac{1}{I_D} \frac{dI_D}{dV_G} = \frac{1}{I_D} \frac{dI_D}{d\Phi_{S1}} \frac{d\Phi_{S1}}{dV_{G1}}$$

Using Equation (5.8.21), the above expression can be rewritten as:

$$
\frac{d \ln(I_D)}{dV_{G1}} = \frac{d\!\left(\exp\!\left(\dfrac{-q\Phi_{s1}}{kT}\right) F(t_{si}, \Phi_{s1}, \Phi_{s2})\right)}{\exp\!\left(\dfrac{-q\Phi_{s1}}{kT}\right) F(t_{si}, \Phi_{s1}, \Phi_{s2}) \, d\Phi_{S1}} \frac{d\Phi_{S1}}{dV_{G1}}
$$

$$
= \left(\frac{-q}{kT} + \frac{d}{d\Phi_{S1}} \ln(F(t_{si}, \Phi_{s1}, \Phi_{s2}))\right) \frac{d\Phi_{S1}}{dV_{G1}} \tag{5.8.23}
$$

Using expressions (5.8.18a), (5.8.22) and (5.8.23), one finally obtains the inverse subthreshold slope of the device:

177

$$S = \frac{-\ln(10)}{\frac{q}{kT} - \frac{d}{d\Phi_{S1}}\ln(F(t_{si},\Phi_{s1},\Phi_{s2}))} \left[\left(1 + \frac{C_{it1}}{C_{ox1}} + \frac{C_{si}}{C_{ox1}}\right) - \frac{\frac{C_{si}}{C_{ox1}}\frac{C_{si}}{C_{ox2}}}{1 + \frac{C_{it2}}{C_{ox2}} + \frac{C_{si}}{C_{ox2}}} \right] \quad (5.8.24)$$

In most practical cases the difference $\Phi_{s2} - \Phi_{s1}$ is only weakly dependent on Φ_{s1}. In that case, $\Phi_{s2} - \Phi_{s1}$ can be considered as a constant, and the integral term $F(t_{si},\Phi_{s1},\Phi_{s2})$ can be approximated by $F(t_{si})$ which is obtained from Equation (5.8.19), where $\Phi_{s2} - \Phi_{s1}$ is considered to be constant. Using the latter approximation, $F(t_{si})$ being independent of Φ_{s1}, the expression (5.8.24) becomes:

$$S = -\frac{kT}{q}\ln(10) \left[\left(1 + \frac{C_{it1}}{C_{ox1}} + \frac{C_{si}}{C_{ox1}}\right) - \frac{\frac{C_{si}}{C_{ox2}}\frac{C_{si}}{C_{ox1}}}{1 + \frac{C_{it2}}{C_{ox2}} + \frac{C_{si}}{C_{ox2}}} \right] \quad (5.8.25)$$

This last expression is equal, in absolute value, to Equation (5.5.18) and can be evaluated analytically without the need for integrating expression (5.8.19). Numerically, the exact subthreshold slope given by (5.8.24) and the approximate value found using (5.8.25) are equal within a few percent.

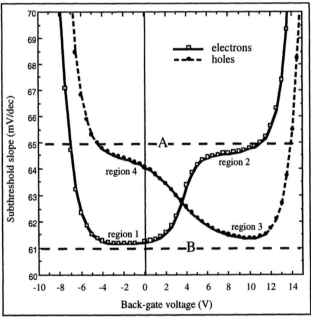

Figure 5.8.5: Subthreshold slope of an enhancement-mode n-channel (electrons) and an accumulation-mode p-channel device (holes). Regions 1, 2, 3 and 4 correspond to a front-surface inversion channel, a back inversion channel, a front body current, and a back body current, respectively.

Figure 5.8.5 presents the subthreshold slope of an enhancement-mode n-channel (electrons) and an accumulation-mode p-channel device (holes), calculated using the FIXFUN program [142]. Regions 1, 2, 3 and 4 correspond to a front-surface inversion channel, a back inversion channel, a front body current, and a back body current, respectively. The lower limit of the subthreshold slope (dotted line B) is given by: $S = \dfrac{kT}{q} \ln (10) (1+ \alpha)$, where $\alpha = \dfrac{C_{si}C_{ox2}}{C_{ox1}(C_{si}+C_{ox2})}$, and its upper limit (dotted line A) is obtained using the same equation with $\alpha = \dfrac{C_{ox2}(C_{ox1}+C_{si})}{C_{ox1}C_{si}}$. For large positive and negative back gate biases accumulation or inversion is reached at the back interface, and high subthreshold swing values are observed.

5.9. Unified body-effect representation

The body-effect coefficient (also called: body factor) of an MOS transistor, which is represented by the letter n, is an image of the efficiency of the coupling between the gate voltage and the channel. It is equal to $1 + \alpha$ and plays a role in each regime of operation of the device. In strong inversion, it shows up in the expression of the drain current (Equation 5.3.31):

$$I_{Dsat} \cong \frac{W \mu_n C_{ox1}}{2 n L} [V_{G1}-V_{th}]^2 \tag{5.9.1}$$

It is also found in the expression of the subthreshold swing (Equation 5.5.19):

$$S = \frac{nkT}{q} \ln (10) \tag{5.9.2}$$

The value of g_m/I_D in weak inversion can be written using Equation 5.4.2:

$$\frac{g_m}{I_D} = \frac{dI_D}{I_D \, dV_G} = \frac{\ln(10)}{S} = \frac{q}{nkT} \tag{5.9.3}$$

In strong inversion, g_m/I_D becomes (Equation 5.4.3):

$$\frac{g_m}{I_D} = \sqrt{\frac{2 \mu C_{ox} W/L}{n I_D}} \tag{5.9.4}$$

In all instances the body factor, n, is given by the following expression:

$$n = 1 + \alpha \quad \text{with} \quad \alpha = \frac{C_{CH-GND}}{C_{G-CH}} \tag{5.9.5}$$

where C_{CH-GND} is the capacitance between the channel (which is a channel "under formation" in the case of the subthreshold regime) and ground (the back gate or the substrate), and C_{G-CH}

is the capacitance between the gate and the channel. The values of C_{G-CH} and C_{CH-GND} are presented in Table 5.9.1.

Equivalent capacitor circuits can be established for each device and are presented in Figure 5.9.1. These equivalent circuits illustrate well the fact that the body factor is an image of the efficiency of the coupling between the gate voltage and the channel. Indeed, C_{G-CH} represents the capacitive coupling between the gate and the channel (a surface inversion channel, except in the case of the AM device in the subthreshold regime, where the subthreshold body current is located at a depth $x(\Phi_{min})$ within the device). The other capacitance, C_{CH-GND}, represents the "force" that tends to prevent the potential in the channel from following gate voltage variations. It is worth noting that the body factors in weak and strong inversion are different in a bulk device, while n remains constant for all regimes of operation in the case of a fully depleted SOI MOSFET.[143] It is worthwhile noting that an additional capacitor can be added below each of these equivalent circuits to represent the influence of a depletion layer in substrate, underneath the buried oxide.[144]

Device	C_{G-CH}	C_{CH-GND}
Bulk MOSFET (strong inversion)	C_{ox}	$\dfrac{\varepsilon_{si}}{x_{dmax}}$
Bulk MOSFET (weak inversion)	C_{ox}	$\dfrac{\varepsilon_{si}}{x_d}$
FD SOI MOSFET	C_{ox1}	$\dfrac{C_{si}\,C_{ox2}}{C_{si}+C_{ox2}}$
FD SOI MOSFET with back accumulation	C_{ox1}	$\dfrac{\varepsilon_{si}}{t_{si}}$
FD SOI MOSFET with back channel	$\dfrac{C_{ox1}C_{si}}{C_{ox1}+C_{si}}$	C_{ox2}
AM SOI MOSFET (subthreshold regime)	$\dfrac{C_{ox1}C_{si1}}{C_{ox1}+C_{si1}}$	$\dfrac{C_{ox2}C_{si2}}{C_{ox2}+C_{si2}}$

Table 5.9.1: Values of C_{G-CH} and C_{CH-GND} for the different types of MOSFETs. In the case of the AM SOI MOSFET $C_{si1}=\varepsilon_{si}/x(\Phi_{min})$ and $C_{si2}=\varepsilon_{si}/(t_{si}-x(\Phi_{min}))$, where $x(\Phi_{min})$ is the depth at which the potential is minimum.

The variation of the value of the subthreshold slope in fully depleted and accumulation-mode SOI MOSFETs (Figure 5.8.5) can be best understood using the equivalent circuits of Figure 5.9.1. One can observe that in an inversion-mode MOSFET there can only be channel at the top or the bottom of the silicon film, such that the subthreshold slope abruptly jumps from a minimum (dotted line B in Figure 5.8.5) to a maximum value (dotted line A) depending whether the channel is at the top or the bottom of the silicon film. In an accumulation-mode device, the body subthreshold current flows within the silicon film, at a depth which varies continuously from 0 to t_{si} as the back-gate voltage is decreased.

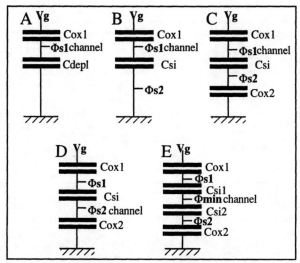

Figure 5.9.1: Capacitance model representing the coupling between the gate voltage and the subthreshold channel: A: bulk device, B: fully depleted SOI MOSFET with back accumulation, C: fully depleted SOI MOSFET, D: fully depleted SOI MOSFET with back inversion channel, E: accumulation-mode SOI MOSFET in the subthreshold regime.

5.10. Microwave MOSFETs

Portable systems are, of course, a market of choice for low-voltage, low-power SOI circuits. Many of these systems will require telecommunication functions, which involves the need for microwave functions. Is it possible to realize these functions in SOI and to integrate them on the same chip as regular CMOS devices?

It has been demonstrated that high-frequency n- and p-channel MOSFETs with transition frequencies above 10 GHz can be fabricated in SOI CMOS. In addition the use of high-resistivity SOI substrates (5,000 Ω.cm or higher) allows for the fabrication of passive elements, such as strip or slot lines with relatively low losses [145,146], and planar inductors with a relatively good quality factor. Rectangular planar inductors with quality factors of 5, 8, and 11 have been obtained when realized on 20, 4,000, and 10,000 Ω.cm substrates, respectively.[147]. The highest-performance SIMOX microwave MOSFETs were fabricated using a dedicated MOS process, called MICROX™, which uses non-standard CMOS features, such as a metal (gold) gate and air-bridge metallization. In other approaches, standard CMOS SOI process was used. The latter devices are, therefore, compatible with lower-frequency (base band) analog and digital circuits fabricated using the same process. For correct microwave operation of a MOSFET the gate sheet resistivity must be low. If this is not the case, the gate behaves like a delay line for the input signal and the part of the transistor which is farthest from the gate contact does not respond to high frequencies.[148] In addition, a low gate resistance is crucial for obtaining a low noise factor. Therefore, silicided gates (or metal-covered gates as in the case of MICROX™) must be used.

SOI material	L (μm)	V_D (V)	I_D (mA)	f_T (GHz)	f_{max} (GHz)	Noise figure / Associated gain (dB) at 2GHz	Ref.
BESOI	1	-	-	-	14	5 / 6.4	149
SIMOX	1	-	-	-	11	5 / 4.4	idem
SOS	0.35	3	10	23	**56**	- / -	150
SIMOX◊	0.32	3	33	14	21	3 / 13.4	151
SIMOX◊	0.25	3	41	23.6	32	1.5 / **17.5**	152
SIMOX	0.75	**0.9**	**3**	10	11	1.5 / 9	153
SIMOX	0.75	**0.9**	10	12.9	**30**	- / 13.9 (10.4¥)	154
SIMOX	0.3	2	-	-	24.3	**0.9** / 14	155
SOS*	0.5	2	**2**	**26**	**60**	1.7 / 16.3	156,157
SIMOX◊	**0.2**	2	125	**28.4**	46	1/15.3†	158

Table 5.10.1: Performances and characteristics of microwave n-channel SOI MOSFETs. (*) T-gate technology is used; (◊) with metal shunt on the gate; (†) at 3 GHz and (¥) at 5 GHz.

Table 5.10.1 presents the high-frequency performances of n-channel, fully-depleted devices. The subthreshold slope, the transconductance at $V_{GS}=V_{DS}=1V$ the OFF current at $V_{GS}=0V$ and $V_{DS}=1V$, and the threshold voltage values are 71 mV/decade, 49 mS/mm, 1.25 nA per micron of channel width and 350 mV, respectively. The silicided gate and silicides source/drain diffusions have a sheet resistance of 6.2 and 6.3 Ω/square, respectively. In order to obtain optimized microwave performances, the 125-μm gate width of the transistors is obtained using a comb-like design of the polysilicon gate, which is composed of 10 fingers having a length of 12.5 μm each. Figure 5.10.1 presents the current gain (H_{21}), the Maximum Available Gain (MAG) and the Unilateral Gain (ULG) as a function of frequency in an n-channel SOI transistor having a length of 0.75 μm ($L_{eff}=0.65$ μm) and a width of 125 μm. These parameters were measured through s-parameter extraction under a supply voltage ($V_{DS} = V_{GS}$) of 0.9 volt. The Maximum Available Gain is 11 dB at 2 GHz. The unit-gain frequency, f_T (found when $H_{21} = 0$ dB) is equal to 10 GHz, and the maximum oscillation frequency, f_{max} (found when ULG = 0 dB), is equal to 11 GHz.

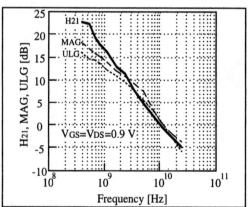

Figure 5.10.1: Current gain (H_{21}), the Maximum Available Gain (MAG) and the Unilateral Gain (ULG) as a function of frequency in an n-channel SOI transistor, having a length of 0.75 μm and a width of 125 μm at $V_{DS} = V_{GS} = 0.9$ V.[159]

The unit-gain frequency, f_T and the maximum oscillation frequency, f_{max}, are presented in Figure 5.10.2 as a function of supply voltage ($V_{DS} = V_{GS}$).

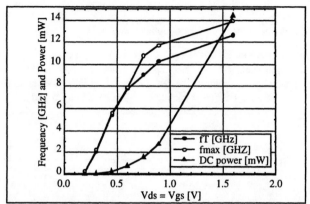

<u>Figure 5.10.2</u>: Unit-gain frequency, f_T , maximum oscillation frequency, f_{max}, and dc power consumption as a function of supply voltage. Same transistor as in Figure 5.10.1.

These frequencies rapidly increase as the supply voltage is increased, up to 1 volt. At higher frequencies, f_T and f_{max} tend to saturate, due to electron velocity saturation in the channel. The values of f_T and f_{max} are 13 and 15.8 GHz, respectively, for a supply voltage of 3 volts. The dc power consumption as a function of supply voltage is also presented. The dc power consumed at Vdd=0.9 V, where f_T = 10 GHz and f_{max} = 11 GHz, is equal to 2.8 mW.

5.11. Quantum effects

Several effects are related to the modification of the band structure in extremely thin SOI films. When the silicon film becomes thin enough the energy minimum of the conduction band (E_C) increases. Its increase is given by relationship (6.5.2): $E_C = E_{C0} + \dfrac{\hbar^2}{2m}\left(\dfrac{\pi}{t_{si}}\right)^2$ where m is the electron effective mass, $\hbar$ is the reduced Planck constant and E_{C0} is the "original" conduction band minimum.[160] The bandgap energy increases as well, and the density of states becomes a staircase function of the energy.[161,162,163,164] As a result, an increase of threshold voltage with reduced film thickness is expected (Figure 5.11.1). It is worthwhile noting that this result cannot be predicted by classical theory, which predicts a decrease of the threshold voltage when the silicon film thickness is decreased (Equation 5.3.12).

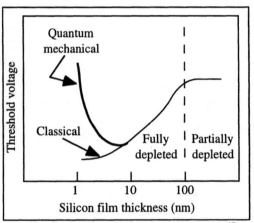

Figure 5.11.1: Variation of threshold voltage with film thickness ($N_a=10^{17}$ cm^{-3}) according to classical and quantum-mechanical models.[165]

An increase of short-channel effect is predicted for very thin silicon films (<10 nm) because of the increase of bandgap energy and the decrease of effective density of states in the conduction band due to 2D quantization.[166,167]

References

1   A.J. Auberton-Hervé, Proceedings of the fourth international Symposium on Silicon-on-Insulator Technology and Devices, ed. by D.N. Schmidt, Vol. 90-6, The Electrochemical Society, p. 455, 1990

2   A.J. Auberton-Hervé, J.M. Lamure, T. Barge, M. Bruel, B. Aspar, and J.L. Pelloie, Semiconductor International, p. 97, October 1995

3   D. Flandre, IEEE Trans. on electron Devices, Vol. 40, No. 10, p. 1789, 1993

4   D. Flandre, F. Van de Wiele, P.G.A. Jespers, and M. Haond, IEEE Electron Device Letters, Vol. 11, No. 7, p. 291, 1990

5   B. Gentinne, D. Flandre, and J.P. Colinge, Solid-State Electronics, Vol. 39, No. 7, p. 1071, 1996

6   D. Flandre and F. Van de Wiele, Proc. of the IEEE SOS/SOI Technology Conference, p. 27, 1989

7   N.J. Thomas and J.R. Davis, Proc. of the IEEE SOS/SOI Technology Conference, p. 130, 1989

8   see, for example, S.M. Sze, Physics of Semiconductor Devices, 2nd Ed., New York: J. Wiley & Sons, 1981

9   H.K. Lim and J.G. Fossum, IEEE Trans. on Electron Devices, Vol. 30, p. 1244, 1983

10   J.H. Choi, Y.J. Park, and H.S. Min, Technical Digest of IEDM, p. 645, 1994

11   J.P. Colinge, IEEE Electron Device Lett., Vol. 6, p. 573, 1985

12   R.S. Muller and T.I. Kamins, Device Electronics for Integrated Circuits, J. Wiley & Sons, p. 436, 1986

13   J. Witfield and S. Thomas, IEEE Electron Dev. Lett., Vol. 7, p. 347, 1986

14   J.P. Colinge, Microelectronic Engineering, Vol. 8, p. 127, 1988

15   D. Hisamoto, T. Kaga, Y. Kawamoto, and E. Takeda, Techn. Digest of International Electron Devices Meeting, p. 833, 1989

16   R.S. Muller and T.I. Kamins, Device Electronics for Integrated Circuits, J. Wiley & Sons, p. 487, 1986

17   S.Veeraraghavan and J.G. Fossum, IEEE Trans. on Electron Devices, Vol. 35, p. 1866, 1988

18   T.W. MacElwee and D.I. Calder, Proceedings of the second international Symposium on Ultra Large Scale Integration Science and Technology, Ed. by C.M. Osburn and J.M. Andrews, Vol. 89-9, The Electrochemical Society, p. 693, 1989

19   T.W. MacElwee and I.D. Calder, Proceedings IEEE SOS/SOI Technology Conference, p. 171, 1989

20   J.G. Fossum, Proceedings of the fourth international Symposium on Silicon-on-Insulator Technology and Devices, ed. by D.N. Schmidt, Vol. 90-6, The Electrochemical Society, p. 491, 1990

21   S.Veeraraghavan and J.G. Fossum, IEEE Trans. on Electron Devices, Vol. 36, p. 522, 1989

22   T. Sekigawa and Y. Hayashi, Solid-State Electronics, Vol. 27, pp. 827, 1984

23   D. Hisamoto, T. Kaga, Y. Kawamoto, and E. Takeda, Techn. Digest of International Electron Devices Meeting (IEDM), p. 833, 1989

24   J.P. Colinge, Techn. Digest of International Electron Devices Meeting (IEDM), p. 817, 1989

25   S.Veeraraghavan and J.G. Fossum, IEEE Trans. on Electron Devices, Vol. 35, p. 1866, 1988

26   S.Veeraraghavan and J.G. Fossum, IEEE Trans. on Electron Devices, Vol. 36, p. 522, 1989

27   H.K. Lim and J.G. Fossum, IEEE Trans. on Electron Devices, Vol. 31, p. 401, 1984

28   S.Veeraraghavan and J.G. Fossum, IEEE Trans. on Electron Devices, Vol. 35, p. 1866, 1988

29   J.G. Fossum, J.Y. Choi, and R. Sundaresan, IEEE Trans. on Electron Devices, Vol. 37, p. 724, 1990

30   P. Antognetti and G. Massobrio, Semiconductor Device Modeling with Spice, McGraw-Hill, p. 185, 1988

31   J.C. Sturm and K. Tokunaga, Electronics Letters, Vol. 25, p. 1233, 1989

32   J.G. Fossum and S. Krishnan, IEEE Transactions on Electron Devices, Vol. 39, p. 457, 1993

33   H.K. Lim and J.G. Fossum, IEEE Trans. on Electron Devices, Vol. 32, p. 446, 1985

34   S.Veeraraghavan and J.G. Fossum, IEEE Trans. on Electron Devices, Vol. 35, p. 1866, 1988

35   C.C. Enz, in Low-power HF microelectronics: a unified approach, edited by G.A.S. Machado, IEE circuits and systems series 8, the Institution of Electrical Engineers, p. 247, 1996

36   C. Enz, F. Krummenacher, and E.A. Vittoz, Analog Integrated Circuit and Signal Processing, Vol. 8, No1, p. 83, 1995

37   A.M. Ionescu, S. Cristoloveanu, A. Rusu, A. Chovet, and A. Hassein-Bey, Proceedings of the International SOI Conference, p. 144, 1993

38   B. Iñiguez, L.F. Ferreira, B. Gentinne, and D. Flandre, IEEE Transactions on Electron Devices, Vol. 43, No. 4, p. 568, 1996

39   C. Mallikarjun and K.N. Bhat, IEEE Transactions on Electron Devices, Vol. 37, No. 9, p. 2039, 1990

40   E.A. Vittoz, Proceedings of the 23rd European Solid-State Device Research Conference (ESSDERC), Ed. by. J. Borel, P. Gentil, J.P. Noblanc, A. Nouailhat and M. Verdone, Editions Frontières, p. 927, 1993

41   K. Lee, M. Shur, T.A. Fjeldly, and T. Ytterdal, Semiconductor Device Modelling for VLSI, Englewood Cliffs, NJ, Prentice-Hall, 1993

42   B. Iñiguez, L.F. Ferreira, B. Gentinne, and D. Flandre, IEEE Transactions on Electron Devices, Vol. 43, No. 4, p. 568, 1996

43   Grove, A.S., Physics and Technology of Semiconductor Devices, J. Wiley & Sons, pp. 326, 1967

44   J.C. Sturm and K. Tokunaga, Electronics Letters, Vol. 25, p. 1233, 1989

45   F. Silveira, D. Flandre, and P.G.A. Jespers, IEEE Journal of Solid-State Circuits, Vol. 31, No. 9, p. 1314, 1996

46   E.A. Vittoz, Tech. Digest of Papers, ISSCC, p. 14, 1994

47   D. Flandre, L.F. Ferreira, P.G.A. Jespers, and J.P. Colinge, Solid-State Electronics, Vol. 39, No. 4, p. 455, 1996

48   H.K. Lim and J.G. Fossum, IEEE Trans. on Electron Devices, Vol. 31, p. 401, 1984

49   S.C. Sun and J.D. Plummer, IEEE Trans. on Electron Devices, Vol. 27, pp. 1497, 1980

50   M. Yoshimi, H. Hazama, M. Takahashi, S. Kambayashi, T. Wada, K. Kato, and H. Tango, IEEE Trans. on Electron Devices, Vol. 36, pp. 493, 1989

51   A. Yoshino, Proceedings of the fourth international Symposium on Silicon-on-Insulator Technology and Devices, ed. by D.N. Schmidt, Vol. 90-6, The Electrochemical Society, p. 544, 1990

52   F. Balestra, S. Cristoloveanu, M. Benachir, J. Brini, and T. Elewa, IEEE Electron Device Lett., Vol. 8, pp. 410, 1987

53   R.J. Van Overstraeten, G.J. Declerck, and P.A. Muls, IEEE Trans. Electron Devices, Vol. ED-20, p. 1150, 1973

54   S.M. Sze, Physics of Semiconductor Devices, Wiley & Sons, p. 446, 1981

55   D.J. Wouters, J.P. Colinge, and H.E. Maes, IEEE Trans. Electron Devices, Vol. ED-37, p. 2022, 1990

56   J.P. Colinge, Techn. Digest of International Electron Devices Meeting (IEDM), p. 817, 1989

57   J.P. Colinge, IEEE Electron Device Lett., Vol. EDL-7, p. 244, 1986

58   J.P. Colinge, Ext. Abstracts of 5th Internat. Workshop on Future Electron Devices, Miyagi-Zao, Japan, p. 105, 1988

59   K.Asada, H. Miki, M. Kumon, and T. Sugano, Ext. Abstracts of 8th Internat. Workshop on Future Electron Devices, Kochi, Japan, p. 165, 1990

60   H.O. Joachim, Y. Yamaguchi, K. Ishikawa, Y. Inoue, and T. Nishimura, IEEE Trans. on electron Devices, Vol. 40, No. 10, p. 1812, 1993

61   B. Dierickx, L. Warmerdam, E. Simoen, J. Vermeiren, and C. Claeys, IEEE Trans. on Electron Devices, Vol. 35, p. 1120, 1988

62   J. Tihanyi and H Schlötterer, IEEE Trans. on. Electron Devices, Vol. 22, p. 1017, 1975

63   G. Merckel, NATO Course on Process and Device Modeling for Integrated Circuit Design, Ed. by F. Van de Wiele, W. Engl and P. Jespers, Groningen, The Netherlands, Noordhoff, p. 725, 1977

64   J.P. Colinge, IEEE Electron Device Lett., Vol. 9, p. 97, 1988

65   J.G. Fossum, R. Sundaresan, and M. Matloubian, IEEE Transactions on Electron Devices, Vol. 34, p. 544, 1987

66   K.M. Cham, S.Y. Oh, D. Chin, and J.L. Moll, Computer Aided Design and VLSI Device Development, Hingham, MA, Kluwer Academic Publishers, p. 240, 1986

67   P.K. Ko, S. Tam, C. Hu, S.S. Wong, and C.G. Sodini, Techn. Digest of International Electron Devices Meeting (IEDM), p. 88, 1984

68   C. Hu, Techn. Digest of International Electron Devices Meeting (IEDM), p. 176, 1983

69   J.G. Fossum, J.Y. Choi, and R. Sundaresan, IEEE Trans. on Electron Devices, Vol. 37, p. 724, 1990

70   C.Hu, S.C. Tam, F.C. Hsu, P.K. Ko, T.Y. Chan, and T.W. Terrill, IEEE Trans. on. Electron Devices, Vol. 32, p. 375, 1985

71   J.G. Fossum, J.Y. Choi, and R. Sundaresan, IEEE Trans. on Electron Devices, Vol. 37, p. 724, 1990

72   J.P. Colinge, IEEE Trans. on Electron Devices, Vol. 34, p. 2173, 1987

73   J.G. Fossum, J.Y. Choi, and R. Sundaresan, IEEE Trans. on Electron Devices, Vol 37, p. 724, 1990

74   J.P. Colinge, IEEE Trans. on Electron Devices, Vol. 34, p. 2173, 1987

75   L.T. Su, H. Fang, J.E. Chung, and D.A. Antoniadis, Technical Digest of IEDM, p. 349, 1992

76   G. Reimbolt and A.J. Auberton-Hervé, IEEE Trans. on Electron Devices, Vol. 14, p. 364, 1993

77   S.M. Guwaldi, A. Zaleski, D.E Ioannou, S. Cristoloveanu, G.J. Campisi, and H.L. Hughes, Proceedings of the Fifth International Symposium on Silicon-on-Insulator Technology and Devices, Ed. by. W.E. Bayley, Proc. vol. 92-13, The Electrochemical Society, p. 157, 1992

78   B. Zhang, A. Yoshino, and T.P. Ma, Proceedings of the Fifth International Symposium on Silicon-on-Insulator Technology and Devices, Ed. by. W.E. Bayley, Proc. vol. 92-13, The Electrochemical Society, p. 163, 1992

79   B. Yu, Z.J. Ma, G. Zhang, and C. Hu, Solid-State Electronics, Vol. 39, No. 12, p. 1791, 1996

80   P.H. Woerlee, C. Juffermans, H. Lifka, W. Manders, F. M. Oude Lansink, G.M. Paulzen, P. Sheridan, and A. Walker, Technical Digest of IEDM, p. 583, 1990

81   T. Ouisse, S. Cristoloveanu, and G. Borel, Proceedings of IEEE SOS/SOI Technology Conference, p. 38, 1990

82   D.E. Ioannou, in "Physical and Technical Problems of SOI Structures and Devices", Kluwer Academc Publishers, NATO ASI Series - High Technology, Vol. 4, Ed. by J.P. Colinge, V.S. Lysenko and A.N. Nazarov, p. 199, 1995

83   S.P. Sinha, F.L. Duan, and D.E. Ioannou, Proceedings of the IEEE International SOI Conference, p. 18, 1996

84   Y. Omura and K. Izumi, Extended Abstracts of the Solid-State Devices and Materials Conference (SSDM), p. 496, 1992

85   L.T. Su, H. Fang, J.E. Chung and D.A. Antoniadis, Technical Digest of IEDM, p. 349, 1992

86   J.R. Davis, A.E. Glaccum, K. Reeson, and P.L.F. Hemment, IEEE Electron Device Letters, Vol. 7, p. 570, 1986

87   J.G. Fossum, R. Sundaresan, and M. Matloubian, IEEE Transactions on Electron Devices, Vol. 34, p. 544, 1987

88   J.Y. Choi and J.G. Fossum, Proceedings IEEE SOS/SOI Technology Conference, p. 21, 1990

89   C.E.D. Chen, M. Matloubian, R. Sundaresan, B.Y. Mao, C.C. Wei, and G.P. Pollack, IEEE Electron Device Letters, Vol. 9, p. 636, 1988

90   R. Sundaresan and C.E.D. Chen, Proceedings of the fourth international Symposium on Silicon-on-Insulator Technology and Devices, Ed. by D.N. Schmidt, Vol. 90-6, The Electrochemical Society, p. 437, 1990

91   A.J. Auberton-Hervé, Proceedings of the fourth international Symposium on Silicon-on-Insulator Technology and Devices, ed. by D.N. Schmidt, Vol. 90-6, The Electrochemical Society, p. 455, 1990

92   H.S. Sheng, S.S. Li, R.M. Fox, and W.S. Krull, IEEE Trans. on Electron Devices, Vol. 36, no. 3, p. 488, 1989

93   A.S. Grove, <u>Physics and Technology of Semiconductor Devices</u>, J. Wiley & Sons, p. 230, 1967

94   J.Y. Choi and J.G. Fossum, Proceedings IEEE SOS/SOI Technology Conference, p. 21, 1990

95   M. Haond and J.P. Colinge, Electronics Letters, Vol. 25, p.1640 , 1989

96   K.K. Young and J.A. Burns, IEEE Transactions on Electron Devices, Vol. 35, p. 426, 1988

97   F.L. Duan, S.P. Sinha, D.E. Ioannou, and F.T. Brady, IEEE Transactions on Electron Devices, Vol. 44, No. 6, p. 972, 1997

98   P.H. Woerlee, C. Juffermans, H. Lifka, W. Manders, F.M. Oude Lansink, G.M. Paulzen, P. Sheridan, and A. Walker, Technical Digest of IEDM, p. 583, 1990

99   F. Deng, R.A. Johnson, W.B. Dubbelday, G.A. Garcia, P.M. Asbeck, and S.S. Lau, Proceedings of the IEEE International SOI Conference, p. 78, 1996

100   M. Yoshimi, M. Terauchi, M. Takahashi, K. Matsuzawa, N. Shigyo,and Y. Ushiku, Extended Abstracts of the International Electron Device Meeting, p. 429, 1994

101   A. Nishiyama, O. Arisumi, and M. Yoshimi, Proceedings of the IEEE International SOI Conference, p. 68, 1996

102   S. Krishnan, J.G. Fossum, and M.M. Pelella, Proceedings of the IEEE International SOI Conference, p. 140, 1996

103   J.G. Fossum, Proceedings of the fourth international Symposium on Silicon-on-Insulator Technology and Devices, ed. by D.N. Schmidt, Vol. 90-6, The Electrochemical Society, p. 491, 1990

104   J.Y. Choi and J.G. Fossum, Proceedings IEEE SOS/SOI Technology Conference, p. 21, 1990

105   M. Pelella, J.G. Fossum, D. Suh, S. Krishnan, and K.A. Jenkins, Proceedings of the IEEE International SOI Conference, p. 8, 1995

106   A. Wei and D.A. Antoniadis, IEEE Electron Device letters, Vol. 15, No. 5, p. 193, 1996

107   F. Assaderaghi, G. Shahidi, L. Wagner, M. Hsieh, M. Pelella, S. Chu, R.H. Dennard, and B. Davari, IEEE Electron Device Letters, Vol. 18, No. 6, p. 241, 1997

108   J.A. Mandelman, J.E. Barth, J.K. DeBrosse, R.H. Dennard, H.L. Kalter, J. Gautier, and H.I. Hanafi, Proceedings of the IEEE International SOI Conference, p. 136, 1996

109   M. Terauchi and M. Yoshimi, Proceedings of the IEEE International SOI Conference, p. 138, 1996

110   J. Gautier, K.A. Jenkins, and Y.C. Sun, Technical Digest of the IEDM, p. 623, 1995

111   T. Saraya, M. Takamiya, T.N. Duyet, T. Tanaka, H. Ishikuro, T. Hiramoto, and T. Ikoma, Proceedings of the IEEE International SOI Conference, p. 70, 1996

112   K.A. Jenkins, Y. Taur, and J.Y.C. Sun, Proceedings of the IEEE International SOI Conference, p. 72, 1996

113   R.A. Schiebel, T.W. Houston, R. Rajgopal, K. Joyner, J.G. Fossum, D. Suh, and S. Krishnan, Proceedings of the IEEE International SOI Conference, p. 125, 1995

114   K.A. Jenkins, J.Y.C. Sun, and J. Gautier, IEEE Electron Device Letters, Vol. 17, No. 11, p. 7, 1996

115   F. Assaderaghi, G.G. Shahidi, M. Hargrove, K. Hathorn, H. Hovel, S. Kulkarni, W. Rausch, D. Sadana, D. Schepis, R. Schulz, D. Yee, J. Sun, R. Dennard, and B. Davari, Digest of Technical Papers, Symposium on VLSI Technology, p. 122, 1996

116   A. Wei and D.A. Antoniadis, Proceedings of the IEEE International SOI Conference, p. 74, 1996

117   H.K. Lim and J.G. Fossum, IEEE Transactions on Electron Devices, Vol. 31, No. 9, p. 1251, 1985

118   K. Ueda, H. Morinaka, Y. Yamaguchi, T. Iwamatsu, I.J. Lim, Y. Inoue, K. Mashiko, and T. Sumi, Proceedings of the IEEE International SOI Conference, p. 142, 1996

119   S.C. Chin, Y.C. Tseng, and J.C.S. Woo, Proceedings of the IEEE International SOI Conference, p. 144, 1996

120   K. Verhaege, G. Groeseneken, J.P. Colinge, and H.E. Maes, IEEE Electron Device Letters, Vol. 14, No. 7, p. 326, 1993

121   K. Verhaege, G. Groeseneken, J.P. Colinge, and H.E. Maes, Microelectronics and Reliability, Special Issue 'Reliability Physics of Advanced Electron Devices', Vol. 35, No. 256, Issue 3, p. 555, 1994

122   J.S.T. Huang, Proceedings of the IEEE International SOI Conference, p. 122, 1993

123   J. Jomaah, G. Ghibaudo, F. Balestra, and J.L. Pelloie, Proceedings of the IEEE International SOI Conference, p. 114, 1995

124   J.S. Brodsky, R.M. Fox, D.T. Zweidinger, and S. Veeraraghavan, IEEE Transactions on Electron Devices, Vol. 44, No. 6, p. 957, 1997

125   D. Yachou and J. Gautier, Proceedings of the 24th European Solid State Device Research Conference (ESSDERC), Ed. by. C. Hill and P. Ashburn, Editions Frontières, p. 787, 1994

126   L.J. Mc Daid, S. Hall, P.H. Mellor, W. Eccleston, and J.C. Alderman, Electronics Letters, Vol. 25, p. 827, 1989

127   L.T. Su, K.E. Goodson, D.A. Antoniadis, M.I. Flik, and J.E. Chung, Technical Digest of IEDM, p. 357, 1992

128   M. Berger and Z. Chai, IEEE Transactions on Electron Devices, Vol. 38, p. 871, 1991

129   Y. Cheng and T.A. Fjeldly, IEEE Transactions on Electron Devices, Vol. 43, No. 8, p. 1291, 1996

130   O. Le Neel and M. Haond, Electronics Letters, Vol. 26, p. 74, 1991

131   D. Yachou, J. Gautier and C. Raynaud, Proceedings of the IEEE International SOI Conference, p. 148, 1993

132   B.M. Tenbroek, M.S.L. Lee, W. Redman-White, R.J.T. Bunyan, and M.J. Uren, IEEE Transactions on Electron Devices, Vol. 43, No. 12, p. 2240, 1996

133   B.M. Tenbroek, W. Redman-White, M.S.L. Lee, R.J.T. Bunyan, M.J. Uren, and K.M. Brunso IEEE Transactions on Electron Devices, Vol. 43, No. 12, p. 2227, 1996

134   S.Veeraraghavan and J.G. Fossum, IEEE Trans. on Electron Devices, Vol. 36, p. 522, 1989

135   J.P. Colinge, IEEE Trans. on Electron Devices, Vol. 37, p. 718, 1990

136 K.W. Su and J.B. Kuo, IEEE Transactions on Electron Devices, Vol. 44, No. 5, p. 832, 1997

137 T.W. MacElwee and D.I. Calder, Proceedings of the second international Symposium on Ultra Large Scale Integration Science and Technology, Ed. by C.M. Osburn and J.M. Andrews, Vol. 89-9, The Electrochemical Society, p. 693, 1989

138 N.J. Thomas and J.R. Davis, Proc. of the IEEE SOS/SOI Technology Conference, p. 130, 1989

139 H.K. Lim and J.G. Fossum, IEEE Trans. on Electron Devices, Vol. 30, p. 1244, 1983

140 D.J. Wouters, J.P. Colinge, and H.E. Maes, IEEE Trans. on Electron Devices,Vol. 37, p. 2022, 1990

141 J.P. Colinge, D. Flandre and F. Van de Wiele, Solid-State Electronics, Vol. 37, No 2, p. 289, 1994

142 F. Van de Wiele and P. Paelinck, Solid-State Electronics, Vol. 32, p. 567, 1989

143 M. Bucher, C. Lallement, and C.C. Enz, Proceedings of the 1996 IEEE International Conference on Microelectronic Test Structures, Vol. 9, p. 145, 1996

144 M.A. Pavanello, J.A. Martino, and J.P Colinge, Solid-State Electronics, Vol. 41, No. 1, p. 111, 1997

145 R.N. Simons, Electronics Letters, Vol. 30, p. 654, 1994

146 J.P. Raskin, I. Huynen, R. Gillon, D. Vanhoenacker, and J.P. Colinge, Proceedings of the IEEE International SOI Conference, p. 28, 1996

147 A. Hürrich, P. Hübler, D. Eggert, H. Kück, W. Barthel, W. Budde, and M. Raab, Proceedings IEEE International SOI Conference, p. 130, 1996

148 R. Gillon, J.P. Raskin, D. Vanhoenacker, and J.P. Colinge, Proceedings of the European Microwave conference, Bologna, Italy, p. 543, 1995

149 A.L. Caviglia, R.C. Potter, and L.J. West, IEEE Electron Device Letters, Vol. 12, No. 1, p. 26, 1991

150 A.E. Schmitz, R.H. Walden, L.E. Larson, S.E. Rosenbaum, R.A. Metzger, J.R. Behnke, and P.A. Macdonald, IEEE Electron Device Letters, Vol. 12, No. 1, p. 16, 1991

151 A.K. Agarwal, M.C. Driver, M.H. Hanes, H.M. Hobgood, P.G. McMullin, H.C. Nathanson, T.W. O'Keefe, T.J. Smith, J.R. Szedon and R.N. Thomas, Technical Digest of IEDM, p. 687, 1991

152 M.H. Hanes, A.K.Agarwal, T.W. O'Keefe, H.M. Hobgood, J.R. Szedon, T.J. Smith, R.R. Siergiej, P.G. McMullin, H.C. Nathanson, M.C. Driver, and N.R. Thomas, IEEE Electron Device Letters, Vol. 7, p. 219, 1993

153 J.P. Colinge, J. Chen, D. Flandre, J.P. Raskin, R. Gillon, and D. Vanhoenaecker, Proceedings IEEE International SOI Conference, p. 28, 1996

154 J. Chen, J.P Colinge, D. Flandre, R. Gillon, J.P. Raskin, and D. Vanhoenacker, to be published in the Journal of the Electrochemical Society, 1997

155 A. Hürrich, P. Hübler, D. Eggert, H. Kück, W. Barthel, W. Budde, and M. Raab, Proceedings IEEE International SOI Conference, p. 130, 1996

156 R.A. Johnson, C.E. Chang, P.R. de la Houssaye, G.A. Garcia, I. Lagnado, and P.M. Asbeck, Proceedings of the IEEE International SOI Conference, p. 18, 1995

157 P.R. de la Houssaye, C.E. Chang, B. Offord, G. Imthurn, R. Johnson, P.M. Asbeck, G.A. Garcia, and I. Lagnado, IEEE Elctron Device Letters, Vol. 16, No. 6, 289, 1995

158 D. Eggert, P. Huebler, A. Huerrich, H. Kueck, W. Budde, and M. Vorwerk, to be published in IEEE Transactions on Electron Devices, 1997

159 J.P. Colinge, J. Chen, D. Flandre, J.P. Raskin, R. Gillon, and D. Vanhoenaecker, Proceedings IEEE International SOI Conference, p. 28, 1996

160 J.P. Colinge and F. Van de Wiele, <u>Physique des dispositifs semi-conducteurs</u>, De Boeck Université, p. 285, 1996 (in French)

161 L. Esaki, <u>Electronic properties of multilayers and low-dimensional semiconductor structures</u>, NATO ASI Series, Plenum Press, Series B: Physics Vol. 321, pp. 1-24, 1990

162 T. Ando, A.B. Fowler, and F. Stern, Review of Modern Physics, Vol. 54, p. 437, 1982

163 P.N. Butcher, <u>Physics of low-dimensional semiconductor structures</u>, Ed. by P.N. Butcher, N.H. March and M.P. Tosi, Plenum Press, p. 95, 1993

164 F. Stern, <u>Physics of low-dimensional semiconductor structures</u>, Ed. by P.N. Butcher, N.H. March and M.P. Tosi, Plenum Press, p. 177, 1993

165 Y. Omura, S. Horiguchi, M. Tabe and K. Kishi, IEEE Electron Device Letters, Vol. 14, No. 12, p. 569, 1993

166 Y. Omura and K. Izumi, IEEE Electron Device Letters, Vol. 17, no. 6, p. 300, 1996

167 Y. Omura, T. Ishiyama, M. Shoji, and K. Izumi, in "Silicon-on-Insulator Technology and Devices VII", Ed. by. P.L.F. Hemment, S. Cristoloveanu, K. Izumi, T. Houston, and S. Wilson, Proceedings of the Electrochemical Society, Vol. 96-3, p. 199, 1996

CHAPTER 6 - Other SOI Devices

Although CMOS remains the most obvious field of application for SOI, the ease of processing SOI substrates, the full dielectric isolation of the devices and the possibility of using a back gate have sparked a large research activity in the field of novel SOI devices. Indeed, different novel bipolar and MOS structures have been proposed such as lateral bipolar and bipolar-MOS devices, vertical bipolar transistors with back gate-induced collector, high-voltage lateral devices of various kinds and double-gate MOS devices. This Chapter will review these devices, qualitatively explain their physics and explore their possible fields of application.

6.1. COMFET

SOI lateral COMFETs (COnductivity Modulation FET) can be realized in relatively thin SOI material using a standard SOI CMOS process (Figure 6.1.1)[1]. In spite of the use of such a thin gate oxide as 23 nm, forward and reverse breakdown voltages of 80 volts have been demonstrated. When the device is turned off (low gate voltage), the reverse-biased diode between the P-type region under the gate and the N⁻ drift region prevent current from flowing through the structure.

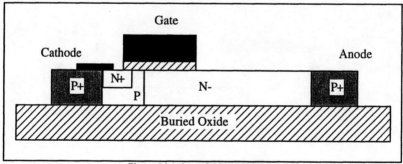

Figure 6.1.1: Lateral SOI COMFET.

The low doping concentration in the N⁻ region ensures a relatively high breakdown voltage. When the gate is positively biased, electrons are injected from the N⁺ part of the cathode into the N⁻ region through a channel created underneath the gate oxide in the p-type region. These electrons can recombine with holes injected in the N⁻ region from the positively biased anode.

The devices then basically operates as a forward-biased PIN diode, and current flows through the structure. The limitation of these thin-film devices (a silicon film thickness of 200 nm was used in the reported case) lies in the fact that rather high current densities are readily reached in the silicon film, limiting the usefulness of such devices for power applications. Their ability to control high voltages and their compatibility with CMOS processing, on the other hand, may render them useful for applications where high-speed SOI CMOS logic and relatively high-voltage interface devices are required.

6.2. High-voltage and power devices

Double implantation of oxygen ions has been used to produce CMOS and high-voltage, offset-gate MOS devices (Figure 6.2.1). The double implantation is performed as follows: after the formation of classical SIMOX material, epitaxy is used to increase the thickness of the silicon overlayer. The epitaxial layer then serves as substrate for the formation of a second SIMOX structure, in such a way that a "silicon-on-oxide-on-silicon-on-oxide" structure is obtained. The buried silicon layer can be connected to ground and serves as a buried back gate which acts as a shield from electrical interference from high-voltage devices realized in the substrate.

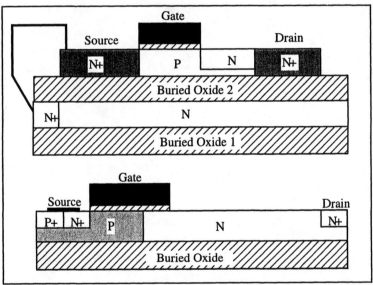

Figure 6.2.1: High-voltage SOI MOSFET (top) [2] and SOI LDMOS (bottom).[3,4,5,6]

A second role of the buried gate is to reduce the maximum electric field in the device (RESURF device) by providing a grounded field plate under the transistor.[7] As in most high-voltage lateral MOSFETs, the function of the N-type drift region near the drain is to reduce the drain electric field between and, hence, to improve the breakdown voltage. Breakdown voltages of 90 volts are indeed obtained for an active silicon thickness of 400 nm and a gate oxide thickness of 100 nm. Higher breakdown voltages can be reached when thicker silicon films are used. One can estimate that for lateral double-diffused (LDMOS) devices the breakdown voltage increases by approximately 20V for every additional micrometer of silicon. LDMOS devices made in 20 µm-thick silicon films and having a breakdown voltage of 560 V have been reported.[8] Masked implantation can also be used to create devices in both the bulk silicon (which is locally protected from oxygen implantation by a mask) and in the SOI overlayer.[9]

Other lateral high-voltage devices have been fabricated in SOI as well. These are: the lateral insulated gate bipolar transistor (LIGBT) [10,11,12], the IGBT mode turn-off thyristor (IGTT) [13], the lateral insulated-gate p-i-n transistor (LIGPT) [14], the lateral high-voltage P-i-N rectifier [15], the base resistance controlled thyristor (BRT) [16], and the VDMOS.[17]

<u>6.3. JFET</u>

Junction field-effect transistors (JFETs) can also be fabricated on SOI substrates. JFETs are particularly interesting for applications where low noise, high input impedance and good radiation hardness are required. Figure 6.3.1 shows the cross-section of an SOI p-channel JFET [18]. The structure is fabricated as follows. Buried N+ diffusions are created in the silicon overlayer of a SIMOX wafer, and p-type silicon epitaxy is carried out to increase the silicon film thickness from 0.2 to 1 µm. These buried N+ diffusions will be utilized as back gate. The silicon islands are defined by means of a mesa isolation process, and ion implantation is used to form the top front-gate (N+) and source and drain (P+) electrodes (Figure 6.3.1). Modulation of the source-drain current can be achieved by applying a voltage to either the front gate or the back gate while keeping the other gate grounded, but the highest transconductance is obtained by connecting G1 and G2 together and using them as a single input terminal.

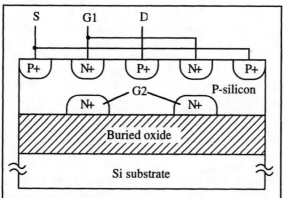

<u>Figure 6.3.1</u>: P-channel JFET. G1 is the front gate, G2 is the back gate.

SOI JFETs show good radiation hardness characteristics. Indeed, upon total dose irradiation (X-rays), a saturation current degradation of 20% is observed for a 30 krad(Si) irradiation. Above 30 krad(Si) and up to 1 Mrad(Si), no further shift is observed. The current saturation degradation caused by a 1 MeV neutron fluence of 10^{14} cm^{-2} is below 10%. Such JFETs can be integrated with complementary bipolar transistors and CMOS to provide a rad-hard analog/digital SOI technology (*Durcie Mixte sur Isolant Logico-Linéaire* or DMILL, in French).[19,20]

<u>6.4. Bipolar-MOS "hybrid" device</u>

As seen in Section 5.7, every enhancement-mode SOI MOSFET contains a parasitic bipolar transistor. The hybrid bipolar-MOS device controls the bipolar effect and makes use of the combined current drive capabilities of both the bipolar and the MOS parts of a "normally MOS" device [21]. This is achieved by connecting the gate, which controls the current flow in the MOS part of the device, to the floating substrate, which acts as the base of the lateral bipolar transistor (Figure 6.4.1). The source and the drain of the MOS transistor are also the emitter and the collector of the bipolar device, respectively.

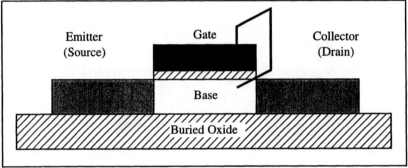

Figure 6.4.1: Bipolar-MOS hybrid device

We will now consider the case of an n-channel (NPN) device, although p-channel (PNP) devices can be realized as well. When the device is OFF ($V_G=V_B=0$), the potential of the MOS substrate (= the base) is low, which maximizes the value of the threshold voltage (and thereby minimizes the OFF current). When a gate bias is applied, on the other hand, the potential of the MOS substrate is increased, which decreases the threshold voltage according to the expressions developed in section 5.3.2 for the body effect (bulk case). This lowering of the threshold voltage increases the current drive for a given gate voltage, compared to an MOS transistor without gate-to-body connection. Similarly, the application of a gate voltage when the device is "ON" increases the collection efficiency of the bipolar device, and reduces the effective neutral base width of the device, thereby improving the bipolar gain. This effect is illustrated in Figure 6.4.2. In absence of any gate bias (we assume a flat-band condition at the Si-SiO2 interfaces), the minority carrier (electrons) current injected from the emitter has to diffuse across the width of the neutral base before being collected by the depletion zone near the collector (Figure 6.4.2.A). According to equation 5.7.2, the gain of the bipolar transistor, β, can be approximated by: $\beta \cong 2(L_n/L_B)^2 -1$, where L_n is the minority carrier (electrons in this case) diffusion length in the base, and L_B is the base width. When a positive gate voltage is applied, an inversion channel is created. This channel allows for the MOS current component , I_{ch}, to flow from source to drain. Two components of bipolar current can be distinguished: the electrons injected by the emitter can either be attracted by the surface potential created by the gate bias and join the channel electrons (I_{C1} component) or diffuse directly from emitter to collector through the base (I_{C2} component). In both cases, the path traveled by the electrons is shorter than if no gate bias were applied (Figure 6.4.2.B). Indeed, the widening of the depletion zone caused by the gate bias reduces the effective neutral base length. The even shorter path traveled by the electrons injected from the emitter and joining the channel current is also obvious (Figure 6.4.2.B).

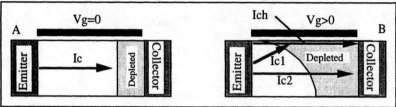

Figure 6.4.2: Emitter-to-collector current (I_c) flowing from emitter to collector in the lateral bipolar transistor with no applied gate bias (A), and the different current components flowing from source (emitter) to drain (collector) when a gate bias is applied to both gate and base: the MOS channel current (I_{ch}), a bipolar current component collected by the channel, (I_{c1}), and a bipolar current component flowing from emitter to collector (I_{c2}) (B).

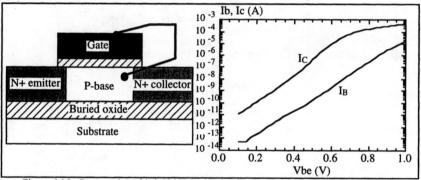

Figure 6.4.3: Cross section of hybrid bipolar-MOS fully depleted SOI NPN device (left) and measured Gummel plot (right). W/Leff = 20μm/0.6μm, t_{si} = 80 nm, t_{ox} = 30 nm. β_F = 5,000 @ I_C = 10 μA.[22,23]

As a result from these considerations and of equation 5.7.2, the gain of the bipolar device is increased when a positive gate voltage is applied, which is always the case in the bipolar-MOS device, since the gate is connected to the base. Recently, this principle of gain enhancement has been applied to bulk lateral bipolar transistors.[24] In summary, the presence of a gate improves the gain of the bipolar transistor, and the presence of a base contact improves both the ON and OFF characteristics of the device. Such a mutually beneficial phenomenon could almost be called "symbiosis". An analytical model of the device can be found in the literature.[25]

The electrical characteristics of the bipolar-MOS device are presented in Figures 6.4.3 (NPN device) and 6.4.4 (quasi-PNP device). Such devices can provide common-emitter current gains of 10,000 for L= 0.3 μm.[26]

197

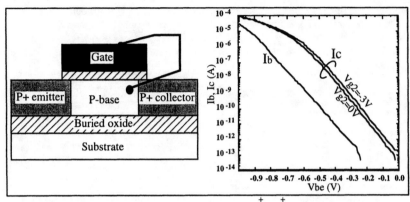

Figure 6.4.4: Cross section of hybrid bipolar-MOS lateral P+-P-P+ device (left) and measured Gummel plot (right) for back-gate voltage values of 0 and -3V. Gate is tied to base. V_{CE}=-1V and W/L_{eff} = 20μm/1.1μm.[27]

Low-Power, Low-Voltage CMOS ring oscillators have been fabricated using complementary hybrid MOS-bipolar devices. These were shown to operate with supply voltages ranging from 0.5 to 1 V. Very low power dissipation is observed.

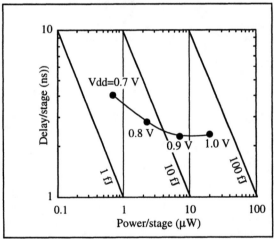

Figure 6.4.5: Delay per stage vs. power per stage in a hybrid complementary MOS-bipolar complementary ring oscillator. $L_{eff,N}$ = 0.45 μm, $L_{eff,P}$ = 0.58 μm.[28]

More recently, a 0.5 V SIMOX-CMOS circuit (8k-gate ALU) has been realized using hybrid MOS-bipolar transistors (called multi-threshold CMOS (MTCMOS in Refs [29,30]). The gate delay and the clock frequency are 200 ps and 40 MHz, respectively, for a supply voltage of 0.5 V. The standby (sleep mode) power dissipation of the ALU is 5 nW and the power operation during 40-MHz operation is 350 nW.

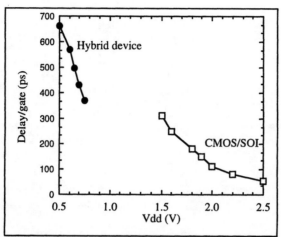

Figure 6.4.6: Delay per stage *vs.* V_{dd} in a hybrid complementary MOS-bipolar complementary ring oscillator. Leff,N = Leff,P = 0.3 µm. The data for an equivalent SOI CMOS ring oscillator are shown for comparison.[31]

6.5. Dual-gate MOSFET

SOI MOSFETs have always two gates. The top gate is used to control the carrier concentration in the top channel, while the back gate is usually grounded. Usually, back-channel conduction is avoided, and the back interface is kept in depletion or accumulation. The advantages of fully-depleted SOI devices are well known. Most of them derive from the excellent coupling between the front gate and the front surface potential. This excellent coupling allows for a reduced slope factor (also called "body-effect factor"). The slope factor, noted *n*, is equal to ...1.5... in a bulk device, and to 1.05...1.1 in a fully depleted SOI MOSFET. The lower the slope factor, the more ideal the characteristics of the device, the lowest possible value being unity. This value is not completely reached in fully depleted SOI transistors because of the capacitor divider formed by the gate oxide, the silicon film, and the buried oxide capacitances. The only ways to obtain a perfect coupling between the surface potential in the channel region and the gate is either to use an infinitely thick buried dielectric or to have the back gate connected to the front gate, with a back-gate oxide thickness equal to the front one. The latter solution defines the concept of the dual-gate SOI MOSFET. Figure 6.5.1 describes this device: the front and back gates are symmetrical (the same gate oxide thickness is used), and tied together electrically. One should not be mistaken by other devices which bear names similar the dual-gate MOSFET, such as the twin-gate MOSFET and dual-poly gates MOSFETs. The pillar-shaped vertical transistor with a surrounding gate has the same physics as the dual-gate SOI transistor, but it is not an SOI device.

One of the first publication on the dual-gate transistor concept dates back to 1984.[32] It shows that one can obtain significant reduction of short-channel effects in a device, called

199

XMOS, where an excellent control of the potential in the silicon film is achieved by using a top-and-bottom gate. The name of the device comes from its resemblance with the Greek letter Ξ. Using this configuration, a better control of the channel depletion region is obtained than in a "regular" SOI MOSFET, and, in particular, the influence of the source and drain depletion regions are kept minimal, which reduces the short-channel effects through screening the source and drain field lines away from the channel.[33] More complete modeling, including Monte-Carlo simulations, is presented in [34] in which the ultimate scaling of silicon MOSFETs is explored. According to that paper, the ultimate silicon device is a dual-gate SOI MOSFET with a gate length of 30 nm, an oxide thickness of 3 nm, and a silicon film thickness of 5 to 20 nm. Such a (simulated) device shows no short-channel effects for gate lengths larger than 70 nm, and provides transconductance values up to 2300 mS/mm.

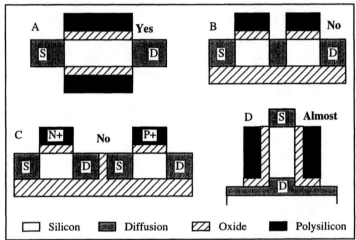

Figure 6.5.1: Who's a dual-gate SOI MOSFET? A: a genuine SOI dual-gate MOSFET; B: the twin-gate MOSFET [35] is not a dual-gate MOSFET; C: the SOI CMOS inverter with dual polysilicon gate (N^+ and P^+ doping) is not a dual-gate MOSFET; D: the pillar-shaped vertical transistor with surrounding gate has the same physics as the dual-gate transistor, but it is not an SOI device.[36, 37, 38,39,40]

One problem with such thin devices is the splitting of the conduction band into subbands. The energy minimum of the first subband controls the threshold voltage and is dependent on the silicon film thickness, according to the equation: $\sigma_{VT} = \dfrac{\hbar^2 \pi^2}{q\, m^*\, t_{si}^3}\, \sigma_{tsi}$, where σ_{VT} is the threshold uncertainty, and σ_{tsi}, is the silicon film thickness uncertainty. The latter result is similar to the quantum-mechanical increase of threshold voltage in ultra-thin SOI devices.[41].

Fully-depleted SOI MOSFETs are known to have a near-ideal subthreshold slope. For a slope factor of 1.05, a subthreshold slope of 63 mV/decade is indeed expected, but the presence of interface states brings it up to values around 66-68 mV/decade. In dual-gate SOI MOSFETs, values very close to the theoretical limit of 60 mV/ decade are expected.

Furthermore, this low value can be obtained for very short channel lengths, provided that the gate oxide thickness and the silicon film thickness are scaled accordingly.

Theoretical investigation shows a subthreshold slope with values lower than 63 mV/ decade can be obtained for a gate length down to 0.1 µm if the device is designed using a scaling parameter $\upsilon = L/2\lambda$, where L is the gate length, and λ, called the "natural length", is equal to:

$$\lambda = \sqrt{\frac{\varepsilon_{si}t_{si}t_{ox}}{2\varepsilon_{ox}}\left(1 + \frac{\varepsilon_{ox}t_{si}}{4\varepsilon_{si}t_{ox}}\right)}.$$

In that case, the subthreshold swing is given by $S = \frac{kT}{q}\ln(10)\frac{1}{1-2\exp(-\upsilon)}$. As long as the value of υ is larger than 3, the device is free from punchthrough, threshold voltage roll-off, and subthreshold slope degradation.[42,43,44] It is worthwhile noting that similar improvements are obtained in polysilicon TFTs when a dual gate is used.[45]

The most innovative device property of the dual-gate SOI MOSFET is the possibility of forming not only inversion layers at the top and the bottom of the channel region, but inverting the entire film thickness. This effect, which appears if the silicon film is thin enough, is called "volume inversion".[46,47]

6.5.1. Threshold voltage

One of the most common criteria for defining the threshold voltage of a MOSFET uses the relationship $\Phi_S = 2\Phi_F$, *i.e.* strong inversion and, therefore, threshold are reached when the surface potential reaches twice the Fermi potential, which depends on the doping level. This definition is inadequate for thin-film, dual-gate devices, where current appears following a weak inversion mechanism.

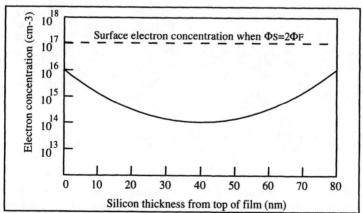

Figure 6.5.2: Minority carrier (electron) concentration across a dual-gate, inversion-mode, n-channel transistor, at threshold (t_{si}=75 nm, t_{ox}=55 nm, N_a=10^{17} cm^{-3}, V_{th}=0.71 V).

Indeed, Figure 6.5.2 presents the electron concentration in the silicon film as current starts to flow in the device, which corresponds to the practical definition of strong inversion threshold. Therefore, the necessity arises to adopt a mathematical definition of the threshold voltage which differs from the classical $\Phi_S = 2\Phi_F$ relationship.

Francis *et al.* have developed an extensive model of the dual-gate, inversion-mode SOI transistor [48,49,50,51] where the threshold voltage is defined by the transconductance change (TC) method. According to this method, the threshold voltage can be defined as the gate voltage where the derivative of the transconductance reaches a maximum, or, in mathematical terms, when $d^3I_D/dV_G^3 = 0$.[52]

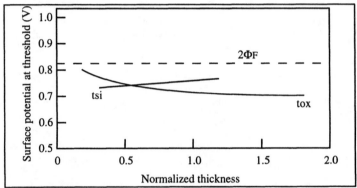

Figure 6.5.3: Surface potential evolution as a function of normalized silicon film and gate oxide thickness ($t_{si}/75$ nm and $t_{ox}/55$ nm). $N_a = 10^{17}$ cm^{-3}.

Using this condition, the surface potential at threshold can be obtained:

$\Phi_S^* = 2\Phi_F + \dfrac{kT}{q} \ln\left[\delta\, \dfrac{1}{1 - \exp(-\alpha)}\right]$ where $\alpha = \dfrac{q}{kT}\dfrac{Q_D}{8C_{si}}$ and $\delta = \dfrac{C_{ox}}{4C_{si}}$, all other symbols having

their usual meaning. The last term of the surface potential at threshold is negative, such that Φ_S^* is smaller than $2\Phi_F$ by a value of 10 to 90 mV, which justifies the previous assumption of having a weak inversion current at threshold. The threshold voltage can be obtained analytically:

$$V_{th} = \Phi_S^* + V_{FB} + \dfrac{kT}{q}\dfrac{\alpha}{\delta}\sqrt{1 + \dfrac{\delta}{\alpha}}$$

The difference between the surface potential at inversion and $2\Phi_F$ depends on the silicon film thickness, the gate oxide thickness, and the doping concentration. The silicon film and gate oxide film thickness dependence is illustrated in Figure 6.5.3.

Thin-film, dual-gate transistors have a low threshold voltage (around 0 volt) when the p-type channel doping is low and the top and bottom gates are made of N$^+$ polysilicon. To obtain a higher threshold voltage is possible to use P$^+$ polysilicon for both gates. In that case a threshold voltage around 1 volt is obtained, which is too high for most applications. An intermediate solution consists into using either midgap or dual-type gate material (one gate is N^+-doped and the other one is P^+-doped). Figure 6.5.4 presents the threshold voltage in an n-channel dual-gate MOSFET with low channel doping concentration ($N_a=10^{15}$ cm^{-3}), as a function of the silicon film thickness. Another solution is, of course, to use two N^+-poly gates and to increase the threshold voltage by increasing the channel doping level. This solution, however, has the disadvantage of decreasing the mobility through increasing impurity scattering.

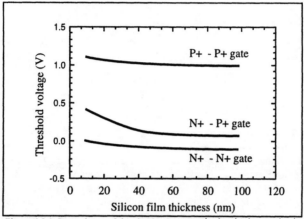

Figure 6.5.4: Dependence of threshold voltage in P^+-P^+, N^+-P^+ and N^+-N^+ dual-gate SOI MOSFETs on silicon film thickness ($N_a=10^{15}$ cm^{-3}). [53,54]

6.5.2. Fabrication process

The fabrication steps necessary to produce a dual-gate MOSFET are not easily derived from a classical SOI process. The first technique that was reported makes use of thin, vertical silicon walls etched in bulk silicon. A local oxidation step is then used to oxidize the silicon between these walls. A bird's beak-like effect allows for some oxide to grow underneath the silicon walls and to produce vertical silicon islands which are dielectrically insulated from the substrate. Gate oxide is then growth, polysilicon is deposited and etched to form the gate, and implantation is used to form the sources and drains. This device is called "fully DEpleted Lean-channel TrAnsistor (DELTA)" and it is presented in Figure 6.5.5.[55] In accordance with the device physics issues discussed above, dual-gate device effects such as volume inversion and increased transconductance are not observed when thick (500 nm) silicon walls are produced, but these effects appear clearly for thin (150 nm) silicon walls.

Silicon wafer bonding has also been used to fabricate dual-gate transistors. In this case, the bottom gates are realized on the handle wafer. After oxidation and planarization of the oxide, the SOI wafer is bonded to the handle wafer, thinned down, and SOI MOSFETs are fabricated on top of the bottom gates. This technique has been used to realize ultra-fast devices with P+ and N+ dual gates.[56] Dual-gate SOI transistors can also be fabricated using epitaxial lateral overgrowth (ELO) of silicon over an oxidized polysilicon gate. A second gate is then fabricated on top of the device.[57,58,59,60,61]

203

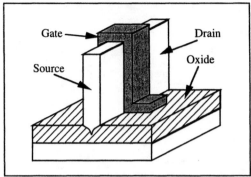

Figure 6.5.5: Fully depleted lean-channel transistor (DELTA)

Another embodiment of the dual-gate SOI transistor makes use of regular SIMOX wafers and adopts a process sequence which is similar to that used for regular SOI MOSFET fabrication, with only one additional mask step and a wet etch step in buffered HF. The device is called the Gate-All-Around (GAA) MOSFET.[62] Device fabrication makes use of SIMOX substrates where the original thickness of the silicon film is 120 nm. A thin pad oxide is grown, and silicon nitride is deposited. Using a mask step, the nitride and the silicon are etched to define the active areas. A local oxidation step is used to round off the edges of the silicon islands, after which the nitride and the pad oxide are stripped. A mask step is then used to cover the entire wafer with resist except areas which correspond to an oversize of the intersection between the active area and the poly gate layers. The wafers are then immersed in buffered HF (BHF). At this step the oxide on the sidewalls of the silicon islands as well as the buried oxide are etched, and a cavity is created underneath the center part of the silicon islands, as shown in Figure 6.5.6.

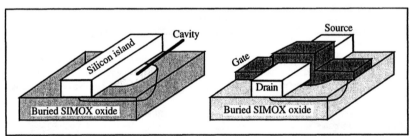

Figure 6.5.6: Cavity etch underneath the silicon island (left) and completed GAA device (right)

The wafers are removed from BHF once the cavity etch is completed. At this point, the device looks like a silicon bridge supported by its extremities (which will later on become source and drain), which is hanging over an empty cavity. Gate oxidation is then carried out. In this step, a 30 nm-thick gate oxide is grown over all the exposed silicon (top, bottom and edges of the active silicon, as well as on the silicon substrate in the bottom of the cavity). Boron is implanted to adjust the threshold voltage, and polysilicon gate material is then deposited and doped n-type. Because of the extremely good step coverage of LPCVD polysilicon, the gate oxide over the cavity is completely coated with polysilicon, and a gate is formed on the top, the sides and the bottom of the channel area (hence the name of Gate-All-Around (GAA) device). The polysilicon gate is then patterned using conventional lithography and anisotropic plasma etch. Source and drain are formed using phosphorous implantation followed by an annealing step. CVD oxide is deposited, and contact holes are opened. An aluminum metallization step completes the process. The final silicon thickness is 80 nm. A schematic cross section of the device is presented in Figure 6.5.7. Accumulation-mode p-channel GAA devices can be obtained just as easily by using a lower-dose boron channel implant and P^+ doping of the sources and drain.

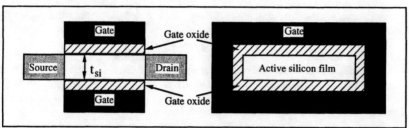

Figure 6.5.7: GAA device cross-section: parallel (left) and perpendicular (right) to the current flow direction.

6.5.3. Device properties

The GAA transistor is in volume inversion when the gate voltage is close the threshold voltage. This can clearly be observed in Figure 6.5.8 where the transconductance of a conventional device and that of a GAA device are compared. An additional curve labeled "SOI x 2" presents the transconductance of the SOI MOSFET multiplied by two to account for both the presence of two channels in the GAA device. The gray area represents the extra drive of the GAA device, which is attributed to volume inversion.

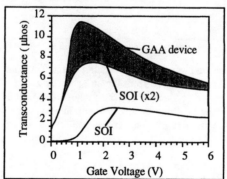

Figure 6.5.8: Transconductance (dI_d/dV_G) at Vds=100 mV in a conventional SOI MOSFET and a GAA device. $(W/L)_{mask}$ = 3μm / 3μm.

The contribution of volume inversion to drain current is more pronounced right above threshold, where the inversion layer is distributed across the entire silicon film and where the effects of bulk mobility (in contrast to surface mobility) can be felt. At higher gate voltages, there is still inversion in the center of the silicon film, but the carriers are now mostly

205

localized in inversion layers near the interfaces. As a result, more scattering occurs, and the transconductance tends to be equal to twice that of that a conventional device. Because of the excellent coupling between the surface potentials and the gate voltage, a subthreshold slope of 60 mV/decade is obtained at room temperature.

6.5.4. Temperature and radiation hardness

Beside the beneficial effects related to volume inversion, GAA devices present some interesting features in the fields of high-temperature applications. Being thin-film SOI devices with minimal junction area, they possess the classical SOI low leakage current features of SOI. Being fully depleted, the variation of their threshold voltage with temperature is also minimal. But, because of the dual-gate control of the channel region, the GAA MOSFET remains fully depleted at higher temperatures than a fully depleted SOI MOSFET having the same silicon film thickness. Indeed, reduced dV_{th}/dT is observed up to only 200°C in 80 nm-thick FDSOI devices, while it is observed up to 300°C in GAA devices having the same silicon film and gate oxide thickness. Figure 6.5.9 presents the transfer characteristics of a GAA inverter at 25, 100, 200 and 300°C. Virtually no difference is observed between the different curves.

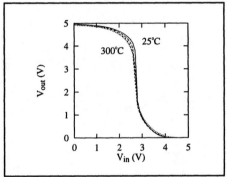

Figure 6.5.9: Transfer characteristics of a GAA inverter at 25, 100, 200 and 300°C.[63]

GAA devices are less prone to the self-heating effect than regular SOI devices. Indeed, SOI MOSFETs are thermally isolated from the substrate by a relatively thick buried oxide which presents a low thermal conductivity. In the case of a GAA device, the channel region is thermally isolated from the substrate by a polysilicon layer and two thin gate oxide layers, which presents a much better thermal conductivity path to the substrate than a buried oxide. Thus significantly less self-heating is observed in GAA than in regular SOI MOSFETs.[64]

GAA devices also offer excellent total-dose and SEU (Single-event Upset, see Section 7.1.1) radiation hardness. Indeed, the active area of the device is completely surrounded by thin thermal gate oxide and by the polysilicon gate. Therefore, the device has no edges, in that sense that there is no field nor buried oxide in contact with the channel region. As a result, no dose-induced edge leakage is observed.[65] Figure 6.5.10 presents the $I_D(V_G)$ characteristics of an n-channel GAA transistor irradiated with different doses of ^{60}Co gamma rays. No edge leakage is observed. The only observed degradation mechanism is due to the inevitable creation of interface states, which degrade both the subthreshold swing and the surface mobility. Figure 6.5.11 presents the threshold voltage variation observed in n- and p-channel GAA transistors as a function of the irradiation dose.

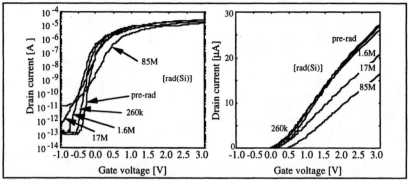

Figure 6.5.10: Drain current *vs.* gate voltage curves of a 3µm × 3µm nMOS/GAA device exposed to different irradiation doses. V_G = 3 V during irradiation, and V_{DS} = 0.1 V during parameter extraction.

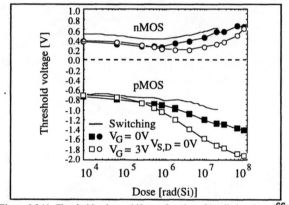

Figure 6.5.11: Threshold voltage shift as a function of irradiation dose. [66]

Classical SEU hardening techniques involve either adding capacitors to the SRAM cells (to absorb the current pulses) or using Silicon-on-Insulator (SOI) technology. Indeed, SOI transistors are made in a thin silicon film, and heavy ions generate less electrical charges in a thin-film device than in a bulk transistor. Figures 6.5.12 and 6.5.13 present the SEU hardness of a 1-kbit GAA (Gate-all-Around) SRAM.[67]

The devices were irradiated with 459 MeV-Xenon ions using a cyclotron. The values of the LET were varied by tilting the samples in order to increase the length of the ion track into the devices. Not a single bit flip were observed when the devices were operated with a standard supply voltage of 3 V. The supply voltage had to be reduced below 2.5 V to reduce the noise margin of the transistors in order to observe some upsets (Figure 6.5.12). It was necessary to further reduce the supply voltage to 1.9 V to observe cross sections comparable to data found in the literature (Figure 6.5.13).

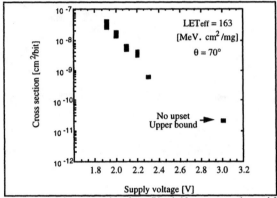

<u>Figure 6.5.12</u>: Upset cross section vs. supply voltage (V_{DD}). No upset was observed for V_{DD}>2.5 V.

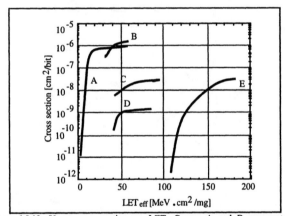

<u>Figure 6.5.13</u>: Upset cross section vs. LET. Curves A and B represent bulk SRAMs with V_{DD}=5V, curves C and D represent SOI SRAMs for V_{DD}=5V, curve E represents a GAA SRAM for V_{DD}=1.9V.

6.5.5. 2DEG effects

When a dual-gate MOSFET operates in the volume inversion regime, the electrons form a two-dimensional electron gas (2DEG), the thickness of which is equal to the silicon film thickness. One can, therefore expect to observe splitting of the conduction band into subbands, since the electrons are confined in one spatial direction, along which the electron wavefunctions form standing waves. The wavefunctions and the corresponding electron energy can be found solving the Schrödinger equation and the Poisson equation in a self-consistent manner:

$$-\frac{\hbar^2}{2m}\frac{d^2\Psi(x)}{dx^2} - q\Phi(x)\,\Psi(x) = E\,\Psi(x) \quad \text{and} \quad \frac{d^2\Phi(x)}{dx^2} = \frac{q(N_a+n(x))}{\varepsilon_{si}} \tag{6.5.1}$$

The general evolution of the energy levels and wave functions is presented in Figure 6.5.14, for a single effective mass value. Threshold is reached when the lowest subband becomes populated. If the depletion and inversion charges are low enough (undoped device right above threshold), the potential well into which the electrons are confined can be considered as square. The position of the energy levels are then given by the equation

$$E_{n-1} = \frac{\hbar^2}{2m^*}\left(\frac{\pi\,n}{t_{si}}\right)^2 \tag{6.5.2}$$

where $n = 1,2,3,\ldots$ and the normalized wave functions are given by $\Psi(x)=\sin(n\pi x/t_{si})$, where t_{si} is the silicon film thickness. When the inversion charge in the film increases (right above threshold), the potential well becomes parabolic and the electrons become more attracted by the gates. The energy levels shift upwards and tend to pair up (E_0 pairs with E_1, E_2 with E_3, etc...). Finally, for higher gate voltages the energy levels become degenerate through pairing, and the wave functions are located mostly at the interfaces, close to the gate oxides (formation of two inversion channels). The observed transconductance peaks are observed for the "square well" and "weak parabolic well" situations, where true volume inversion is present.

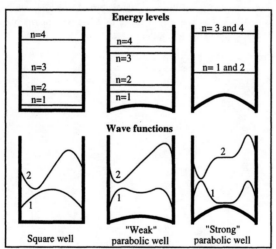

Figure 6.5.14: Four first energy levels and two first wave functions in square, weak parabolic and strong parabolic potential wells.

209

Figure 6.5.15 presents the evolution of the electron concentration in the 40 nm-thick device as a function of gate voltage. One can clearly see the volume inversion right above threshold and the formation of two inversion channels at higher gate voltages.

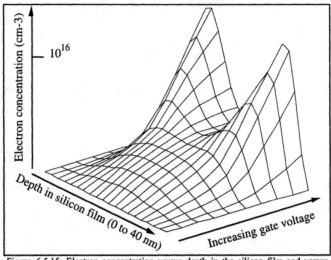

Figure 6.5.15: Electron concentration versus depth in the silicon film and versus $V_G - V_{TH}$ (40 nm-thick device).

Figure 6.5.16 presents the evolution of the subband energy levels as a function gate voltage. The plot is realized in such a way that the Fermi level remains constant. Every time an energy level falls below the Fermi level, the corresponding subband becomes populated.

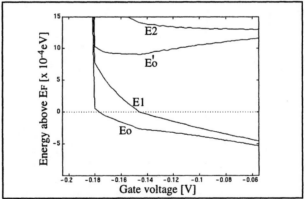

Figure 6.5.16: Evolution of energy levels (relative to E_F) as a function of gate voltage (40 nm device).

Due to intersubband scattering, mobility decreases every time a new subband becomes populated, which generates the humps in the transconductance curve of the device, as presented in the experimental curves of Figure 6.5.17.

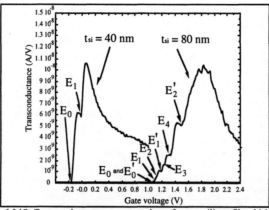

<u>Figure 6.5.17</u>: Transconductance vs. gate voltage for two silicon film thicknesses. The energy levels corresponding to the transconductance humps are indicated (E_0 corresponds to the first subband ($m^*=m_l$), E_1 corresponds to the second subband ($m^*=m_l$), E'_0 corresponds to the first subband with $m^*=m_t$, etc...). Classical effective masses in silicon are used ($m_l = 0.98\ m_o$ and $m_t = 0.19\ m_o$).[68,69]

6.6. Bipolar transistors

Different types of bipolar transistors can be realized in SOI, depending on the application and the silicon film thickness under consideration. If a pure bipolar circuit is contemplated (e.g. ECL), a "thick" silicon layer can be employed and classical vertical transistors can be fabricated.[70,71,72,73,74] The advantages of using an SOI substrate resides in the reduction of collector-substrate capacitances, the full dielectric isolation and the reduction of sensitivity to alpha particles (reduction of the soft-error rate).[75]

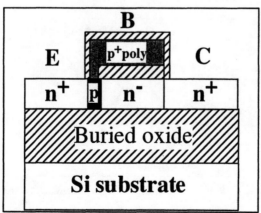

<u>Figure 6.6.1</u>: Lateral SOI bipolar transistor with top base contact.

If the bipolar transistors have to be integrated with thin-film CMOS devices (BiCMOS and CBiCMOS), pretty efficient lateral bipolar transistors can be realized as well. Early SOI lateral bipolar transistors suffered from high base resistance problems due to the fact that the base contact was taken laterally (MOSFET body contact).[76] More recent devices use base contacts made at the top of the device, which improves the base resistance but usually complicates the fabrication process (Figure 6.6.1).[77,78,79,80,81,82]

Lateral polysilicon emitters can also be realized, also at the expense of process complexity.[83] It is, however, worthwhile to mention that the increase of process complexity necessary to upgrade a CMOS SOI process to (C)BiCMOS is much more modest than for a bulk process, simply because of the inherent dielectric isolation of SOI devices. Table 6.6.1 presents the performances of some recent SOI bipolar transistors. It is worthwhile noting that the maximum gain of lateral devices is usually obtained for relatively low collector current values. Indeed, high current densities and high-injection phenomena are reached much more quickly in thin-film lateral devices than in vertical devices because of the small silicon thickness.

Type	L(μm)	β	BV_{CE0}	f_T(GHz)	Company	Ref.
Lateral	0.2	120	2.5 V	4.5 GHz	UCB	[84]
Hybrid	0.3	10000	2.5 V	-	UCB	[85]
Lateral	-	90	3 V	15.4 GHz	Philips	[86]
Lateral	-	80	>3 V	10 GHz	Motorola	[87]
Lateral	-	30	2.8 V	20 GHz	IBM	[88]
Vertical	-	50	8 V	12.4 GHz	SIM	[89]
Vertical	-	110	12 V	4.5 GHz	Analog Devices	[90]

Table 6.6.1: Performances of SOI bipolar transistors

Because of the need for making a buried collector layer, it may seem impossible to fabricate vertical bipolar transistors in thin SOI films. An original solution to this problem has, however, been proposed, and vertical NPN transistors have been realized in 400 nm-thick SOI films (and the use of even thinner films is quite possible) [91].

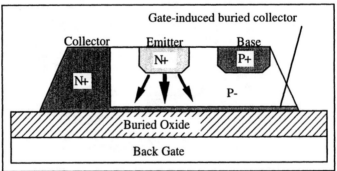

Figure 6.6.2: Vertical bipolar transistor with gate-induced buried collector.

An ingenious solution consists in fabricating a vertical bipolar device with no buried collector diffusion. The bottom of the P- intrinsic base is directly in contact with the buried

212

oxide. If no back gate bias is applied, very few of the electrons injected by the emitter into the base can reach the lateral collector, and the current gain of the device is extremely small. When a positive bias is applied to the back gate, however, an inversion layer is induced at the silicon-buried oxide interface, at the bottom of the base, and acts as a buried collector (Figure 6.6.2). Furthermore, the band bending in the vicinity of the bottom inversion layer attracts the electrons injected by the emitter into the base. As a result, collection efficiency is drastically increased, and the current gain can reach useful values. The use of a field-effect-induced buried collector makes it thus possible to obtain significant collection efficiency without the need for the formation of a diffused buried layer and an epitaxy step. It also solves the high-injection (high current densities) problems inherent to thin-film lateral devices. Furthermore, because the neutral base width does not depend on the collector voltage, but on the back-gate voltage, the Early effect is almost totally suppressed.[92] Since the amplification factor of a bipolar transistor is given by the Early voltage × current gain product, this feature can be extremely attractive for high-performance analog applications. Vertical SOI bipolar transistor with field-induced collector with an Early voltage over 50,000 volts and a current gain of 200 have been demonstrated. Such devices offer thus an Early voltage × current gain product larger than 1,000,000.[93]

6.7. Optical modulator

The presence of a buried layer underneath an active silicon layer makes it possible to use the SOI structure to realize optical modulators. Indeed, an SOI structure capped by an oxide and polysilicon can form a Fabry-Perot cavity [94]. Infrared laser light (λ=1.3 μm) can be introduced into the cavity by means of a single-mode fiber optics. The light bounces forth and back inside the cavity, partially reflected through, and partially reflected back. Resonance can be obtained by optimizing the thickness of the different film thicknesses of the multilayer structure. At resonance, the reflectance is minimized (Figure 6.7.1).

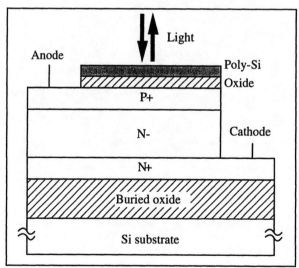

Figure 6.7.1: SOI infrared optical modulator.

Forward biasing the PIN diode of the structure generates carriers inside the intrinsic (actually, N⁻) region which modulates the phase of the laser light in the cavity and shifts the resonance of the Fabry-Perot cavity to convert the phase modulation into intensity modulation. At resonance, a 0.6% change of the refractive index in the "intrinsic" portion of the PIN diode can theoretically induce a 65% reduction of the reflectance.

Using such a structure, a 40% modulation depth of 1.3 µm infrared laser light has been demonstrated, the 40% optical output reduction being obtained for a 2.25 A/cm² current density in the PIN diode.

6.8. Quantum-effect devices

Two-dimensionally confined injection mechanisms have been observed in insulated-gate PN junctions made in sub-10-nm-thick SOI films. Transconductance oscillations are observed. The origin of these fluctuations is found in the splitting of the conduction and valence bands into subbands.[95]

Short quantum wire transistors have been fabricated in SIMOX.[96,97,98] The conductance of these devices increases in a staircase manner as a function of gate voltage. After correcting for the source and drain resistance the transconductance was found to increase in multiples of $4q^2/h$, which agrees with the Landauer formula for universal transconductance fluctuations.[99] Longer quantum wire transistors have been realized as well, and transconductance fluctuations due to subband splitting have been observed as well.[100]

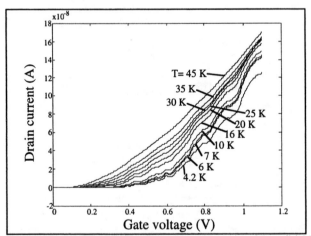

Figure 8.8.1: Current as a function of gate voltage in a series of 7 parallel SOI quantum wire MOSFETs at different temperatures. V_{DS}=10 mV. The device thickness and width are 86 and 100 nm, respectively.[101,102]

The fabrication of a single-electron transistor (SET) on SIMOX has been reported as well. This device is basically a short quantum wire connected to source and drain through constrictions. The higher energy levels in the constrictions isolate the wire from the outside world. When a gate voltage is applied current can flow through the quantum wire by a Coulomb blockade mechanism which is made possible by the very small capacitances (2 aF) involved in the structure. Conductance oscillations as a function of gate voltage have been observed at temperatures as high as 300 K.[103,104] Coulomb blockade oscillations have been observed at room temperature in SOI quantum-wire MOSFETs without constrictions as well. In these devices small interconnected quantum dot domains are formed, probably due to the fluctuations of oxide charge density, interface states density or gate oxide thickness along the wire's length.[105,106]

The fabrication and successful operation of SOI single-electron memory cells with floating dot gate has been reported as well. In such a device, which basically behaves as a miniature EEPROM cell, a single electron can be injected into an electrically floating dot. Because the dot has a very small capacitance the injection of such a minute charge as that of a single electron gives the dot to a potential which is sufficiently high (several volts) to modify the threshold voltage of a quantum-wire MOSFET. This MOSFET can, in turn, be used as a memory element, since it will present different threshold voltage values depending whether or not there is an electron stored in the floating dot.

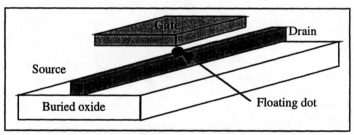

Figure 8.8.2: SOI single-electron memory cell MOSFET. [107,108,109]

215

References

1   J.P. Colinge and S.Y. Chiang, IEEE Electron Device Letters, Vol. 7, p. 697, 1986

2   T. Ohno, S. Matsumoto, and K. Izumi, Electronics Letters, Vol. 25, p. 1071, 1989

3   E. Arnold, Proceedings of the Second International Symposium on "Semiconductor Wafer Bonding: Science, Technology, and Applications", Ed. by M.A. Schmidt, C.E. Hunt, T. Abe, and H. Baumgart, The Electrochemical Society, Proceedings Vol. 93-29, p. 161, 1993

4   E. Arnold, H. Pein, and S.P. Herko, Technical Digest of IEDM, p. 813, 1994

5   Y. Suzuki, T. Kishida, H. Takano, Y. Shirai, and M. Suzumura, Proceedings of the IEEE International SOI Conference, p 134, 1996

6   Y.K. Leung, S.C. Kuehne, V.S.K. Huang, C.T. Nguyen, A.K. Paul, J.D. Plummer, and S.S. Wong, IEEE Electron Device Letters, Vol. 18, No. 1, p. 13, 1997

7   S.K. Chung, S.Y. Han, J.C. Shin, Y.I. Choi, and S.B. Kim, IEEE Electron Device Letters, Vol. 17, No. 1, p. 22, 1996

8   W. Wondrak, E. Stein, and R. Held, in "Semiconductor Wafer Bbonding: Science, Technology, and Applications, Ed. by U. Gösele, T. Abe, J. Haisma, and M.A.A. Schmidt, Proceedings of the Electrochemical Society, Vol. 92-7, p. 427, 1992

9   J. Weyers, H. Vogt, M. Berger, W. Mach, B. Mütterlein, M. Raab, F. Richter, and F. Vogt, Proceedings of the 22nd ESSDERC, Microelectronic Engineering, Vol. 19, p 733, 1992

10   Y.K. Leung, S.C. Kuehne, V.S.K. Huang, C.T. Nguyen, A.K. Paul, J.D. Plummer, and S.S. Wong, Proceedings of the IEEE International SOI Conference, p 132, 1996

11   A. Nakagawa, N. Yasuhara, I. Omura, Y. Yamaguchi, T. Ogura, and T. Matsudai, Extended Abstracts of the IEDM, p. 229, 1992

12   A. Nakagawa, Y. Yamaguchi, N. Yasuhara, K. Hirayama, and H. Funaki, technical Digest of IEDM, p. 477, 1996

13   T. Ogura and A. Nakagawa, Technical Digest of IEDM, p. 241, 1992

14   A.Q. Huang, IEEE Electron Device Letters, Vol. 17, No. 6, p. 297, 1996

15   S. Sridhar, Y.S. Huang, and B.J. Baliga, Extended Abstracts of the IEDM, p. 245, 1992

16   S. Sridhar and B.J. Baliga, IEEE Electron Device Letters, Vol. 17, No. 11, p. 512, 1966

17   C. Harendt, U. Apel, T. Ifström, H.G. Graf, and B. Höfflinger, Proceedings of the Second International Symposium on "Semiconductor Wafer Bonding: Science, Technology, and Applications", Ed. by M.A. Schmidt, C.E. Hunt, T. Abe, and H. Baumgart, The Electrochemical Society, Proceedings Vol. 93-29, p. 129, 1993

18   J.P. Blanc, J. Bonaime, E. Delevoye, J. Gauthier, J. de Pontcharra, R. Truche, E. Dupont-Nivet, J.L. Martin, and J. Montaron, Proceedings IEEE SOS/SOI Technology Conference, p. 85, 1990

19   M. Dentan, E. Delagnes, N. Fourches, M. Rouger, M.C. Habrard, L. Blanquart, P. Delpierre, R. Potheau, R. Truche, J.P. Blanc, E. Delevoye, J. Gautier, J.L. Pelloie, J. de Pontcharra, O. Flament, J.L. Leray, J.L. Martin, J. Montaron, and O. Musseau, IEEE Transactions on Nuclear Science, Vol. 40, No. 6, p. 1555, 1993

20   O. Flament, J.L. Leray, and O. Musseau, in <u>Low-power HF microelectronics: a unified approach</u>, edited by G.A.S. Machado, IEE circuits and systems series 8, the Institution of Electrical Engineers, p. 185, 1996

21   J.P. Colinge, IEEE Transactions on Electron Devices, Vol. 34, No. 4, p. 845, 1987

22   J.P. Colinge, in <u>Low-power HF microelectronics: a unified approach</u>, edited by G.A.S. Machado, IEE circuits and systems series 8, the Institution of Electrical Engineers, p. 139, 1996

23   J.P. Colinge, Technical Digest of IEDM, p. 817, 1994

24   S. Verdonck-Vandenbroek, S.S. Wong, and P.K. Ko, Technical Digest of IEDM, p 406, 1988

25   R. Huang, Y. Y. Wang, and R. Han, Solid-State Electronics, Vol. 39, No. 12, p. 1816, 1996

26   S.A. Parke, C. Hu, and P.K. Ko, IEEE Electron Device Letters14, p. 234, 1993

27   J.P. Colinge, D. Flandre, D. De Ceuster, Electronics Letters, Vol. 30, p. 1543, 1994

28   J.P. Colinge, Electronics Letters, Vol. 23, No. 19, p. 1023-1025, 1987

29   T. Douseki, S. Shigematsu, Y. Tanabe, M. Harada, H. Inokawa, and T. Tsuchiya, Digest of Technical Papers, IEEE International Solid-State Circuits Conference, p. 84, 1996

30   M. Harada, T. Douseki, and T. Tsuchiya, Digest of Technical Papers, Symposium on VLSI Technology, p. 96, 1996

31   F. Assaderaghi, D. Sinitsky, S. Parke, J. Bokor, P.K. Ko, and C. Hu, Technical Digest of IEDM, p. 809, 1994

32   T. Sekigawa and Y. Hayashi, Solid-State Electronics, Vol. 27, p. 827, (1984)

33   B. Agrawal, V.K. De, and J.D. Meindl, Proceedings of 23rd ESSDERC, Ed. by J. Borel, P. Gentil, J.P. Noblanc, A. Nouhaillat, and M. Verdone, Editions Frontières, p. 919, 1993

34   D.J. Frank, S.E. Laux and M.V. Fischetti, Technical Digest of IEDM, p. 553, (1992)

35   M.H. Gao, J.P. Colinge, L. Lauwers, S. Wu, and C. Claeys, Solid-State Electronics, Vol. 35, No 4, p. 505, (1992)

36   H. Takato, K. Sunouchi, N. Okabe, A. Nitayama, K. Hieda, F. Horiguchi and F. Masuoka, Technical Digest of IEDM, p. 222, (1988)

37   A. Nitayama, H. Takato, N. Okabe, K. Sunouchi, K. Hieda, F. Horiguchi and F. Masuoka, IEEE Trans. on Electron Devices, Vol. 38-3, p. 579, 1991

38   S. Miyano, M. Hirose and F. Masuoka, IEEE Trans. on Electron Devices, Vol.39-8 p. 1876, (1992)

39   H.I. Hanafi, S. Tiwari, S. Burns, W. Kocon, A. Thomas, N. Garg, and K. Matsushita, Technical Digest of IEDM, p. 657, (1995)

40   S. Watanabe, K. Tsuchida, D. Takashima, Y. Oowaki, A. Ntayama, K. Hieda, H. Takato, K. Sunouchi, F. Horigushi, K. Ohuchi, F. Masuoka and H. Hara, IEEE Journal of Solid-State Circuits, Vol. 30-9, p. 960, (1995)

41   Y. Omura, S. Horiguchi, M. Tabe and K. Kishi, IEEE Electron Device Letters, Vol. 14-12, p. 569, 1993

42   Y. Tosaka, K. Suzuki and T. Sugii, IEEE Electron Device Letters, Vol. EDL-15, p. 466, (1994)

43   K. Suzuki, T. Tanaka, Y. Tosaka, H. Horie and Y. Arimoto, IEEE Trans. on Electron Devices, Vol. 40-12, p. 2326, (1993)

44   R.H. Yan, E. Ourmazd and K.F. Lee, IEEE Trans. on Electron Devices, Vol. 39, p. 1704, (1992)

45   A.O. Adan, S. Ono, H. Shibayama and R. Miyake, Technical Digest of IEDM, p. 399, (1990)

46   F. Balestra, S. Cristoloveanu, M. Benachir and T. Elewa, IEEE Electron Device Letters, Vol. EDL-8, p. 410, 1987

47   S. Venkatesan, G. W. Neudeck and R.F. Pierret, IEEE Electron Device Letters, Vol. EDL-813, p. 44, (1991)

48   P. Francis, A. Terao, D. Flandre, and F. Van de Wiele, Microelectronic Engineering, Vol. 19, p. 815, (1992)

49   P. Francis, A. Terao, D. Flandre, and F. Van de Wiele, IEEE Trans. on Electron Devices, Vol. 41-5, p. 715, (1994)

50   P. Francis, A. Terao, D. Flandre, and F. Van de Wiele, Proceedings of the 23rd ESSDERC, Editions Frontières, p. 621, (1993)

51   P. Francis, A. Terao, D. Flandre, and F. Van de Wiele, Solid-State Electronics, Vol. 38-1, p. 171, (1995)

52   A. Terao, D. Flandre, E. Lora-Tamayo, and F. Van de Wiele, IEEE Electron Device Letters, Vol. 12, p. 682, 1991

53   T. Tanaka, H. Horie, S. Ando and S. Hijiya, Technical Digest of IEDM, p. 683, (1991)

54   K. Suzuki and T. Sugii, IEEE Trans. on Electron Devices, Vol. 42-11, p. 1940, (1995)

55   D. Hisamoto, T. Kaga, Y. Kawamoto and E. Takeda, Technical Digest of IEDM, p. 833, (1989)

56   T. Tanaka, K. Suzuki, H. Horie and T. Sugii, IEEE Electron Device Letters, Vol. EDL-15, p. 386, (1994)

57   G. Roos and B. Höfflinger, Electronics Letters, Vol. 29-9, p.2103, (1993)

58   J.P. Denton and G.W. Neudeck, Proceedings IEEE International SOI Conference, p. 135, (1995)

59   S. Venkatesan, G.W. Veudeck and R.F. Pierret,IEEE Electron Device Letters, Vol. 13, p. 44, (1992)

60   J.C. Chang, J.P. Denton, and G.W. Neudeck, Proceedings of the IEEE International SOI Conference, p. 88, 1996

61   J.P. Denton and G.W. Neudeck, IEEE Electron Device Letters, Vol. 1, No. 11, p. 509, 1996

62   J.P. Colinge, M.H. Gao, A. Romano, H. Maes and C. Claeys, Technical Digest of IEDM, p. 595, 1990

63   P. Francis, A. Terao, B. Gentinne, D. Flandre, JP Colinge, Technical Digest of IEDM, p. 353, (1992)

64   P. Francis, D. Flandre, J.P. Colinge and F. Van de Wiele, Proceedings of the 25th ESSDERC, Editions Frontières, p. 225, (1995)

65   J.P. Colinge and A. Terao, IEEE Transactions on Nuclear Science, Vol. 40, p. 78, (1993)

66   J.P. Colinge, in "Silicon-on-Insulator Technology and Devices", ed. by. P.L.F. Hemment, S. Cristoloveanu, K. Izumi, T. Houston, and S. Wilson, The electrochemical Society Proceedings, Vol. 96-3, p. 271, 1996

67   P. Francis, J.P. Colinge, and G. Berger, IEEE Transactions on Nuclear Science, Vol. 42-6, p. 2127, 1995

68   J.P. Colinge, X. Baie and V. Bayot, IEEE Electron Device Letters, Vol. 15, p. 193, (1994)

69   P. Francis, X.Baie and J.P. Colinge, Extended Abstracts of the International Conference on Solid-State Devices and materials (SSDM), p. 277-279, Yokohama Japan, (Aug. 1994)

70   U. Magnusson, H. Norström, W. Kaplan, S. Zhang, M. Jargelius, and D. Sigurd Proceedings of the 23rd ESSDERC, Ed. by J. Borel, P. Gentil, J.P. Noblanc, A. Nouailhat, and M. Verdone, Editions Frontières, p 683, 1993

71   S. Feindt, J.J.J. Hajjar, M. Smrtic, and J. Lapham, in "Semiconductor Wafer Bonding: Science, Technology, and Applications, Ed. by. M.A. Schmidt, C.E. Hunt, T. Abe, and H. Baumgart, Electrochemical Society Proceedings Vol. 93-29, p. 189, 1993

72   S.J. Gaul, J.A. Delgado, G.V. Rouse, C.J. McLachlan, and W.A. Krull, Proceedings IEEE SOS/SOI Technology Conference, p. 101, 1989

73   M.D. Church, Proceedings IEEE SOS/SOI Technology Conference, p. 175, 1989

74   J.P. Blanc, J. Bonaime, E. Delevoye, J. Gauthier, J. de Pontcharra, R. Truche, E. Dupont-Nivet, J.L. Martin, and J. Montaron, Proceedings IEEE SOS/SOI Technology Conference, p. 85, 1990

75   K. watanabe, T. Hashimoto, M. Yoshida, M. Usami, Y. Sakai, and T. Ikeda, in "Semiconductor Wafer Bonding: Science, Technology, and Applications", Ed. by U. Gösele, T. Abe, J. Haisma, and M.A.A Schmidt, Proceedings of the Electrochemical Society, Vol. 92-7, p. 443, 1992

76   JP Colinge, Electronics Letters, Vol. 22, p. 886, 1986

77   W.M. Huang, K. Klein, M. Grimaldi, M. Racanelli, S. Ramaswami, J. Tsao, J. Foerstner, and B.Y. Hwang, Technical Digest of IEDM, p. 449, 1993

78   G.G. Shahidi, D.D. Tang, B. Davari, Y. Taur, P. McFarland, K. Jenkins, D. Danner, M. Rodriguez, A. Megdanis, E. Petrillo, M. Polcari, and T.H. Ning, Technical Digest of IEDM, p. 663, 1991

79   C.J. Patel, N.D. Jankovic and J.P. Colinge, in Physical and Technical Problems of SOI Structures and Devices edited by J.P. Colinge, V. S. Lysenko and A. N. Nazarov, NATO ASI Prtnership Sub-Series: 3. Vol. 4, Kluwer Academic Publishers, Dordrecht, p. 211, 1995

80   M.Chan, S.K.H. Fung, C. Hu,and P.K. Ko, Proceedings of the IEEE International SOI Conference, p. 90, 1995

81   B. Edholm, J. Olsson, and A. Söderbärg, IEEE Transactions on Electron Devices, Vol. 40, No. 12, p. 2359, 1993

82   B. Edholm, J. Olsson, and A. Söderbärg, Microelectronic Engineering, Vol. 22, p. 379, 1993

83   R. Dekker, W.T.A. v.d. Einden, and H.G.R. Maas, Technical Digest of IEDM, p. 75, 1993

84   S.A. Parke, C. Hu, and P.K. Ko, IEEE Electron Device Letters14, p. 234, 1993

85   *ibidem*

86   R. Dekker, W.T.A. v.d. Einden, and H.G.R. Maas, Technical Digest of IEDM, p. 75, 1993

87   W.M. Huang, K. Klein, M. Grimaldi, M. Racanelli, S. Ramaswami, J. Tsao, J. Foerstner, and B.Y. Hwang, Technical Digest of IEDM, p. 449, 1993

88   G.G. Shahidi, D.D. Tang, B. Davari, Y. Taur, P. McFarland, K. Jenkins, D. Danner, M. Rodriguez, A. Megdanis, E. Petrillo, M. Polcari, and T.H. Ning, Technical Digest of IEDM, p. 663, 1991

89   U. Magnusson, H. Norström, W. Kaplan, S. Zhang, M. Jargelius, and D. Sigurd Proceedings of the 23rd ESSDERC, Ed. by J. Borel, P. Gentil, J.P. Noblanc, A. Nouailhat, and M. Verdone, Editions Frontières, p 683, 1993

90   S. Feindt, .J.J. Hajjar, M. Smrtic, and J. Lapham, in "Semiconductor Wafer Bonding: Science, Technology, and Applications, Ed. by. M.A. Schmidt, C.E. Hunt, T. Abe, and H. Baumgart, Electrochemical Society Proceedings Vol. 93-29, p. 189, 1993

91   J.C. Sturm and J.F. Gibbons, in "Semiconductor-On-Insulator and Thin Film Transistor Technology", Chiang, Geis and Pfeiffer Eds., (North-Holland), MRS Symposium Proceedings, Vol. 53, p. 395, 1986

92   T. Arnborg and A. Litwin, IEEE Transactions on Electron Devices, Vol. 42, No. 1, p. 172, 1995

93   K. Yallup, S. Edwards and O. Creighton, Proceedings of the 24th ESSDERC, Ed. by C. Hill and P. Ashburn, Editions Frontières, p. 565, 1994

94   X. Xiao, J.C. Sturm, P.V. Schwartz, and K.K. Goel, Proceedings IEEE SOS/SOI Technology Conference, p. 171, 1990

95   Y. Omura, IEEE Transactions on Electron Devices, Vol. 43, No. 3, p. 436, 1996

96   Y. Nakajima, Y. Takahashi, S. Horiguchi, K. Iwadate, H. Namatsu, K. Kurihara, and M. Tabe, Extended Abstracts of the International Conference on Solid-State Devices and Materials, p. 538, 1994

97   Y. Nakajima, Y. Takahashi, S. Horiguchi, K. Iwadate, H. Namatsu, K. Kurihara, and M. Tabe, Japanese Journal of Applied Physics, Vol. 34, p. 1309, 1995

98   Y. Nakajima, Y. Takahashi, S. Horiguchi, K. Iwadate, H. Namatsu, K. Kurihara, and M. Tabe, Applied Physics Letters, Vol. 65, No. 22, p. 2833, 1994

99   M. Büttiker, Y. Imry, R. Landauer, and S. Pinhas, Phys. Rev. B, Vol. 31, p. 6207, 1985

100   J.P. Colinge, X. Baie, V. Bayot, E. Grivei, Solid-State Electronics, Vol. 39, pp. 49-51, 1996

101   J.P. Colinge, Microelectronic Engineering, Vol. 28, pp. 423-430, 1995

102   X. Baie, J.P. Colinge, V. Bayot, and E. Grivei, Proceedings of the IEEE International SOI Conference, p. 66, 1995

103   Y. Takahashi, H. Namatsu, K. Kurihara, K. Iwadate, M. Nagase, and K. Murase, IEEE Transactions on Electron Devices, Vol. 43, No. 8, p. 1213, 1996

104   K. Murase, Y. Takahashi, Y. Nakajima, H. Namatsu, M. Nagase, K. Kurihara, K. Iwadate, S. Horiguchi, M. Tabe, and K. Izumi, Microelectronic Engineering, Vol. 28, No. 1-4, p. 399, 1995

105   H. Ishikuro, T. Fujii, T. Sayara, G. Hashiguchi, T. Iramoto, and T. Ikoma, Applied Physics Letters, Vol. 66, No. 24, p. 3585, 1966

106   T. Hiramoto, H. Ishikuro, T. Fujii, T. Sayara, G. Hashiguchi, and T. Ikoma, Physica B: Condensed Matter, Vol. 227, p. 95, 1996

107   A. Nakajima, T. Futatsugi, K. Kosemura, T. Fukano, and N Yokoyama, Technical Digest of IEDM, p. 952, 1996

108   L. Guo, E. Leobandung, and S.Y. Chou, Technical Digest of IEDM, p. 955, 1996

109   J.J. Welser, S. Tiwari, K.Y. Lee, and Y. Lee, IEEE Electron Device Letters, Vol. 18, No. 6, p. 278, 1997

CHAPTER 7 - The SOI MOSFET Operating in a Harsh Environment

SOI MOSFETs present several properties which allow them to operate in harsh environments where bulk devices would typically fail from operating satisfactorily. These interesting properties of the SOI MOSFETs are due to the small volume of silicon in which the devices are made, to the small area of the source-body and drain-body junctions, and to the presence of a back gate. In this Chapter, we will describe the behavior of the SOI MOSFET operating in two cases of extreme environments: the exposure to radiations and high-temperature operation.

7.1. Radiation environment

One of the major niche markets where SOI circuits and devices are currently employed is the aerospace/military market, because of the high hardness of SOI technology against transient radiation effects. The effect of radiation on an electronic device depends on the type of radiation (neutrons, heavy particles, electromagnetic radiations,...) to which the device is submitted. Unlike bipolar devices, MOSFETs are relatively insensitive to neutron irradiation (neutrons basically kill the carrier lifetime in silicon through inducing displacement of atoms within the crystal lattice). MOS devices are much more sensitive to the exposure to single-event upset (SEU), single-event latchup (SEL) single-event burnout (SEB), gamma-dot upset, and total-dose exposure than to neutrons. The effects created in bulk silicon MOSFETs by such radiation exposures are well documented and can be found in Reference [1], for instance. Table 7.1 presents the effects produced by radiations in semiconductor devices. The following Sections will compare the hardness of SOI and bulk devices to single-event phenomena, gamma-dot and total-dose irradiation conditions.

Particle	Physical effect	Result	Environment
Photon	Creation of pairs Photocurrent (high dose rate) Creation of interface states	Oxide charges Devices turned on Interface states	Space, nuclear Nuclear explosion Space, nuclear
Heavy ion	Creation of pairs	Single-event effects	Space
Neutron	Displacement of atoms Recoil atoms	Lifetime reduction SEU	Nuclear Avionics
Proton	Recoil atoms Nuclear interaction Creation of pairs Displacement of atoms	SEU SEU Oxide charges Lifetime reduction	Space (solar flares) (earth's radiation belts)
Electron	Creation of pairs Displacement of atoms	Oxide charges Lifetime reduction	Space (earth's radiation belts)

Table 7.1: Radiation types and their effect on semiconductor devices.[2,3]

7.1.1. Single-event phenomena

Single-event upset (SEU) is caused by the penetration of an energetic particle, such as an alpha particle or a heavy ion (cosmic ray) within a device. Indeed, when such a particle penetrates a reverse-biased junction, its depletion layer and the bulk silicon underneath it, a plasma track is produced along the particle path, and electron-hole pairs are generated.[4] The presence of this track temporarily collapses the depletion layer and distorts it in the vicinity of the track. The distortion of the depletion layer is called a "funnel" (Figure 7.1.1).

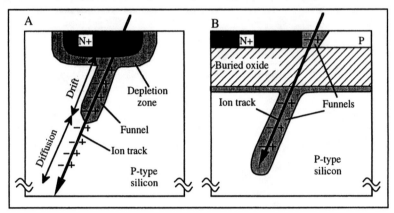

Figure 7.1.1: Penetration of a bulk (A) and an SOI (B) device by an energetic ion.

The funnel extends the depletion zone along the particle track such that the electrons created in the funnel drift towards the junction (Figure 7.1.1.A). The holes move downwards and create a substrate current. The collected electrons give rise to a current transient which can upset the logical state of a node. The duration of the collection of the electrons by the node is on the order of a fraction of a nanosecond. The *drift* current created in this process is called "prompt current". The length of the track in the silicon is typically on the order of ten micrometers. Afterwards, free electrons generated along the particle track underneath the funnel can diffuse towards the depletion region, where they create a second current (*diffusion* current), called "delayed current". This current is smaller in magnitude compared to the prompt current, but it lasts much longer (up to hundreds of nanoseconds or microseconds).[5] Multiple bit upsets (MBU) are sometimes observed, as a result of the penetration of a heavy ion into an integrated circuit. [6,7] In an SOI device, the impinging particle ionizes the silicon along its track as well. However, because of the presence of a buried insulator layer between the active silicon film and the substrate, none of the charges generated in the substrate can be collected by the junctions of the SOI devices. The only electrons which can be collected are those produced within the thin silicon film, the thickness of which is typically 150-300 nm in rad-hard applications. The ratio of the lengths of the tracks along which the electrons are collected gives a first-order approximation of the advantages of SOI over bulk in terms of SEU hardness (*e.g.*: 10μm/200nm = 50 in the case of a 200 nm-thick SOI device).

The energy deposited by a particle along the track is expressed in linear energy transfer (LET) units. It is defined by the following relationship: $\text{LET} = \dfrac{1}{m_v} \dfrac{dW}{dx}$, where x is the linear distance along the particle track, dW is the energy lost by the particle and absorbed by the silicon, and m_v is the volumic mass of silicon. The LET is usually expressed in MeV·cm²/mg. The number of electrons or holes created by a single-event upset is given by:

$$\frac{dN}{dx} = \frac{dP}{dx} = \frac{m_v}{w} \, \text{LET},$$ where w is the energy needed to create an electron-hole pair [8]. As an illustration, a single carbon ion with an energy of 1 GeV (LET $\cong$ 0.24 MeV·cm²/mg) produces 1.5×10^4 pairs/μm in silicon. If it hits a bulk silicon device, it produces 1.5×10^5 electron-hole pairs ($\cong 0.3$ pC) along 10 μm of track. If the electrons migrate towards the node within a time scale of 10-100 ps, an SEU current spike of 1-10 mA is then created. Because of the reduced length of the track along which the electrons are collected in SOI, the SEU current spike will be a factor ...50... smaller in an SOI device than in a bulk device. The magnitude of the impact of SEU on a circuit is measured by the upset cross section (units: cm²/bit). The upset cross section represents the area per bit which is sensitive to SEU. For example, in a memory chip, it represents the area of the junctions which can be upset by an SEU mechanism within a single memory cell. The smaller this cross-section, the less the devices are sensitive to particle irradiation. Light particles such as protons, electrons and neutrons have a LET which is usually too low to ionize the silicon. Indeed, a LET<1 MeV.cm²mg⁻¹ generates a charge which is too small to have significant effect on a device (approximately 0.01 pC/μm), except maybe in CCDs and DRAMs. But these particles cause silicon atom recoil through a "direct hit" mechanism or induce nuclear reactions which produce recoil nuclear fragments. These recoil atoms or nuclei fragments can, in turn, act as heavy ions and cause SEU. Thus particles such as protons usually cause SEU not through direct ionization (their LET is too low), but rather through nuclear reactions resulting in recoils which can deposit enough energy in the sensitive volume of a device to generate an upset.[9]

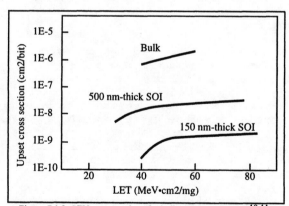

Figure 7.1.2: SEU cross sections for bulk and SOI circuits.[10,11]

The single-event upset cross-sections of several bulk and SOI circuits are presented in Figure 7.1.2. One can see that 150 nm-thick SOI devices are approximately 10 times less sensitive than thicker (500 nm) devices which, in turn, are about 50 times harder than bulk CMOS devices.

It is worth noting that the generation of a funnel in the substrate underneath the buried oxide of an SOI structure can also influence the characteristics of a device submitted to SEU. Indeed, if the substrate (back-gate) bias is such that the substrate under the buried oxide is depleted, the electrons created along the track will move upwards towards the buried oxide (Figure 7.1.1.B). These electrons immediately induce a positive mirror charge in the upper silicon layer. As a result, electrons are injected into the device by the external circuit to restore equilibrium, and the occurrence of a transient current is measured. This effect is not observed if the surface of the substrate underneath the buried oxide is either inverted or accumulated.[12]

The photocurrent - or "ionocurrent" - generated by the impact of a particle in an SOI MOSFET can be amplified by |the parasitic lateral bipolar transistor present in the device. Indeed, the hole current I_B created within the body of an SOI MOSFET acts as a base current for the parasitic lateral NPN bipolar transistor (n-channel device case). In response to the base current pulse induced by the particle, a collector current $I_C=\beta\ I_B$ is produced. This current adds to the current pulse caused by the SEU-induced electrons collected by the drain, such that the drain current pulse actually becomes equal to $(\beta+1)\ I_{SEU}$, where I_{SEU} is the electron current initially generated by the particle (in an n-channel device). Therefore, the presence of any bipolar transistor action, even with low gain ($\beta<1$) contributes to enhance the transient current produced by SEU [13], but the problem is, of course, more pronounced in short-channel devices where β is large. In that case the bipolar gain can be so high that SOI transistors can actually become more sensitive to SEU than bulk devices.[14] The solutions proposed to remedy this problem are the use of a body tie through which part of the base current can be evacuated, and the use of SOI material with poor carrier lifetime. Circuit solutions (in contrast to device or technology solutions) include the increase of RC time constant of the different circuit nodes, but this technique presents the drawback of degrading the circuit speed performances.[15]

Single-event latchup (SEL) can be caused by a heavy-ion strike in bulk CMOS integrated circuits where a parasitic n-p-n-p path exists. SEL is triggered by excess current in the base of either a parasitic npn or pnp transistor following a heavy-ion strike. Because of the regenerative feedback loop which exists between those two transistors latchup can occur. Latchup is triggered on the order of nanoseconds, and can cause destructive burnout within hundreds of microseconds. The holding voltage for latchup is typically on the order of 1 volt. Thus, unless the power supply voltage is interrupted, the low-resistance conduction path from the power supply to ground will be maintained.[16,17] SEL can cause permanent damage (hard error) in an integrated circuit. In SOI CMOS devices there is no p-n-p-n structure that can latchup. Therefore, SEL does not exist in SOI circuits.

Single-event burnout (SEB) can occur in power bipolar or power MOS devices. Power MOS devices contain a "parasitic" bipolar transistor structure.[18] If a heavy ion strikes the bipolar transistor the charge generated will cause current to flow in the base and raise the potential of the emitter-base junction. If the current flow is high enough it can forward bias the emitter-base junction, and the bipolar transistor turns on. After the parasitic bipolar transistor is turned on, second breakdown of the bipolar transistor can occur. This second breakdown has been referred to as current-induced avalanche.[19,20,21] Depending on the current density during current-induced avalanche, the current induced in the parasitic transistor by the heavy ion will either dissipate without device degradation or will regeneratively increase until (in absence of current limiting elements) the device is destroyed. Protons have been shown to cause SEB as well.[22] There is no basic difference between bulk and thick SOI power devices, from an SEB point of view.

Single-event gate rupture (SEGR) can occur as a heavy ion passes through a gate oxide. [23,24,25] It occurs only at high oxide electric fields, such as those present during a write or clear operation in a nonvolatile E²PROM cell, or in a power MOSFET. It is caused by the combination of the applied electric field and the energy deposited by the particle. As an ion passes through a gate oxide it forms a highly conducting plasma path between the silicon and the gate. If the energy is high enough, it can cause localized heating of the dielectric and potentially a thermal runaway condition. If this occurs the temperature along the plasma track will be high enough to cause the dielectric to locally melt or evaporate. Most likely, there is no basic difference between bulk and thick SOI power devices, from an SEGR point of view.

Single-event snapback (SES) can occur in partially depleted SOI transistors, and is enhanced by the bipolar effect. In an n-channel transistor which is turned on, the energy deposited by a particle can create enough free holes to increase the forward bias of the source-body junction, and enough electrons to increase the drain current. These conditions increase impact ionization mechanisms, which can lead to snapback (sometimes called "single-transistor latchup"). This effect has been predicted for fully depleted devices, but has not yet been observed.[26,27,28]

7.1.2. Total dose

Total-dose effects are caused by the cumulative exposure to ionizing radiation such as X-rays and gamma rays. The dose unit is the rad(Si) which is defined by the deposition of 100 erg of radiation energy per gram of silicon. The rad(Si) is a CGS unit, but there exists a MKS unit for dose called "Gray" (Gy). One Gy is defined as the deposition of 1 joule of radiation energy per kilogram of matter. The equivalence between the two units is straightforward: 1 Gy = 100 rads. The number of electron-hole pairs which are generated is related to the energy dW absorbed by volume unit dv of the material: $\dfrac{dN}{dv} = \dfrac{dP}{dv} = \dfrac{1}{w}\dfrac{dW}{dv}$ where w is the effective energy needed to produce a pair in silicon (w=3.6 eV). The relationship between the dose, D, and the number of pairs generated is given by: $\dfrac{dN}{dv} = \dfrac{dP}{dv} = \dfrac{m_v}{w} D$.[29] It is generally admitted that 1 rad(Si) generates 4×10^{13} pairs cm^{-3} in silicon and 7.6×10^{12} pairs cm^{-3} in SiO$_2$. To get an idea of the magnitude of the levels of irradiation to which devices can be submitted, one can mention the following numbers: a medical or dental X-ray corresponds to less than 0.1 rad(Si). The human being becomes sick after being exposed to 100 rad(Si) and falls in an instant coma if submitted to 10 krad(Si). During their lifetime, satellites orbiting around the earth receive total doses ranging between 10 krad(Si) and 1 Mrad(Si), depending on the orbit parameters. Interplanetary spacecrafts and some electronics in nuclear reactors can be exposed to doses in excess of 10 Mrad(Si).[30] Some SOI circuits have been tested at doses up to 500 Mrad(Si).[31,32]

The main effect caused by total dose in MOS devices is the generation of charges in the oxides and the generation of interface states at the Si/SiO$_2$ interfaces. If the rate of energy deposition is high, a sufficient amount of electron-hole pairs can be created in the silicon to produce photocurrents. This case, where dD/dt is high, will be considered in the next section (gamma-dot effects). Ionizing electromagnetic radiations such as X-rays and gamma rays (produced *e.g.* by a ^{60}Co source) create electron-hole pairs in silicon dioxide. Electrons are fairly mobile in SiO_2, even at room temperature, and can move rapidly outside the oxide (towards a positively biased gate electrode, in the case of a gate oxide, for instance). Holes, on the other hand, remain trapped within the oxide and contribute to the creation of a positive oxide charge, Q_{ox}. If we take the example of a gate oxide, the charge Q_{ox} is proportional to the thickness of the oxide, t_{ox}, and the resulting threshold voltage shift is, therefore, proportional to t_{ox}^2 since $\Delta V_{th} = -\dfrac{Q_{ox}\, t_{ox}}{\varepsilon_{ox}}$. The relationship between the threshold voltage shift and the dose can be written:

$$\Delta V_{th} = -\alpha\, \frac{q\, m_v}{w\, \varepsilon_{ox}}\, t_{ox}^2\, D \tag{7.1.1}$$

where w is the effective energy needed to produce a pair in oxide (18 eV), m_v is the volumic mass of oxide and where the factor α accounts for the fact that only a fraction of the charges become trapped within the oxide. α is a technology-dependent parameter. Typical values of α are 0.15 for normal, unhardened oxides and 0.05 or less for rad-hard oxides. SIMOX buried oxides have an α of approximately 0.05. The physics of the irradiation of devices is, unfortunately, much more complex than what is described in the few equations above. Dose irradiation also creates surface states at the Si-SiO_2 interfaces. Contrarily to oxide charges, which are always positive (which decreases the threshold voltage in n-channel devices), interface states are amphoteric and trap electrons in an n-channel device, which increases the threshold voltage. Another type of traps, called "border traps", located in the oxide very close to the silicon/oxide interface, can trap electrons as well.[33] Oxide charges anneal out with time and even contribute to the creation of additional interface traps. This creates an effect which is called "rebound" and it is illustrated in Figure 7.1.3.[34]

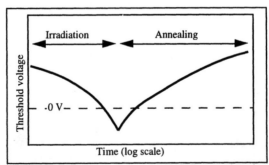

Figure 7.1.3: "Rebound" effect in an n-channel MOSFET exposed to dose irradiation. Annealing occurs after irradiation has stopped.

This effect is, of course, temperature dependent. A rebound effect can also be seen if the dose is increased without interruption above a given value (typically around 1 Mrad(Si)). In that case the generation of oxide charges saturates, while the creation of interface states does not. As a result, a rebound is seen in n-channel devices when the dose is increased.[35] The dose rate at which the device is irradiated is important as well. At low dose rates oxide charges have time to migrate to the silicon/oxide interface and to transform into interface states. As a result, the rebound effect occurs at lower doses than when higher dose rates are used. This makes a difference in the radiation response between devices used in the field (*e.g.* in a spacecraft, low dose rates) and in lab conditions (higher dose rates).

Furthermore, the dose-induced effects are highly dependent on the front- and back-gate bias of the device. Worst-case conditions correspond to positive gate biases (which push the holes in the oxide towards the Si-SiO_2 interfaces). In p-channel transistors, the generation of both positive charges and interface states cause an increase of the absolute value of the threshold voltage (*i.e.*: it becomes more negative). Bias gate conditions are also different (front- and back-gate biases are usually zero or negative) than in n-channel devices.

The SOI MOSFET, with its many Si-SiO_2 interfaces (gate oxide, buried oxide and edge (field) oxide), is quite sensitive to total-dose exposure, unless special hardening techniques are used. While SOI gate oxide hardening techniques are similar to those employed in bulk (*e.g.*: low-temperature oxide growth), the techniques for prevention of back and edge leakage are specific to SOI. The classical solution for avoiding the formation of an inversion layer at the bottom of the silicon film in n-channel devices is to increase the threshold voltage of the back transistor by means of a back boron impurity implant. In some instances, use of a back-gate bias can be made to compensate for the radiation-induced generation of positive charges in the buried oxide. The creation of a peak of boron doping at the back interface implies the use of partially depleted devices. Indeed, fully depleted are usually too thin to allow one to realize such an implant (see Section 4.3). In addition, the presence of charges in the buried oxide and interface states at the silicon - buried oxide interface causes front threshold voltage shifts and degrades the performance of the devices if the front and back interfaces are electrically coupled, which is the case in thin-film, fully depleted, devices. SOI circuits (non-fully depleted) made in a silicon film thickness as low as 150 nm and capable of withstanding doses up to 300 Mrad(Si) have, however, been reported.[36]

The control of edge leakage currents in n-channel devices poses an even more severe problem. Indeed, the parasitic MOS transistors can be quite sensitive to dose irradiation because of the relatively thick oxides (such as LOCOS or oxidized mesa oxides) which are in contact with the edges of the silicon active areas (see Section 4.2). An example of radiation-induced edge-leakage current is presented in Figure 7.1.4.[37,38]

Edge leakage problems can be eliminated through optimization of the sidewall doping and field oxide growth processes [39] or through the use of edgeless devices or transistors where a P^+ diffusion interrupts edge leakage paths between the N^+ source and drain diffusions (see Section 4.6). There is usually no edge leakage problem in p-channel devices.

229

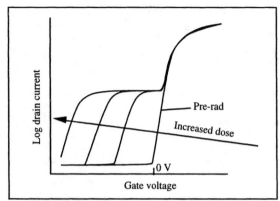

Figure 7.1.4: Edge leakage current caused by dose irradiation.

7.1.3. Dose-rate

Dose-rate (D') effects take place when a large dose of electromagnetic energy (X-rays, gamma rays) is deposited within a short time interval by events such as nuclear explosions. Dose-rate effects are also often referred to as gamma-dot effects. The dose-rate unit is the $rad(Si).s^{-1}$. As mentioned in the previous Section, 1 rad(Si) generates approximately $4x10^{13}$ pairs cm^{-3} in silicon. If the energy is absorbed by the silicon in a short time, a significant amount of photocurrent can be collected within the depletion zones of the devices (prompt current). In bulk devices a delayed diffusion current component is observed as well. The generated photocurrent is expressed by: $I_{ph}=q \cdot V_{depl} \cdot g \cdot D'(t)$ where q is the electron charge, V_{depl} is the volume of the depletion zone, and g is the carrier generation constant in silicon, which is equal to $4.2x10^{13}$ hole-electron pairs cm^{-3} $rad(Si)^{-1}$.[40] In the most dramatic cases, all transistors are flooded with free carriers and become so conducting that the supply voltage rail is virtually shorted to ground, and supply voltage collapse occurs. The major difference between bulk and SOI devices resides in the much smaller depletion volume, V_{depl}, found in SOI devices (Figure 7.1.5). As a result, the dose-rate-induced photocurrent is significantly smaller in SOI transistors than in bulk devices.

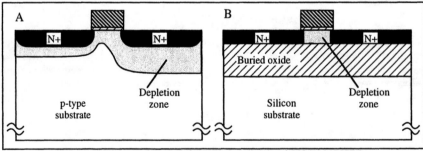

Figure 7.1.5: Volume of the depletion zones in bulk (A) and SOI (B) MOSFETs.

As in the case of SEU, the photocurrent generated in an SOI MOSFET can be amplified by the parasitic lateral bipolar transistor present in the device. Indeed, the hole current created within the body of an SOI MOSFET acts as a base current for the parasitic lateral NPN bipolar transistor (n-channel device case). In response to the base current pulse I_B induced by the particle, a collector current $I_C=\beta\ I_B$ is produced. This current adds to the photocurrent caused by the gamma-dot event. Therefore, the presence of any bipolar transistor action, even with low gain ($\beta<1$) contributes to enhance the transient current produced by a gamma-dot event but the problem is, of course, more pronounced in short-channel devices where β is large. The solutions proposed to remedy this problem are the use of a body contact through which part of the base photocurrent can be evacuated, and that of SOI material with poor carrier lifetime.

7.2. High-temperature operation

Some applications (well logging, avionics, automotive,...) require electronic circuits capable of operating at temperatures up to 300°C. SOI devices and circuits present three advantages in this field over their bulk counterparts: the absence of thermally-activated latch-up, reduced leakage currents and, in thin-film, fully depleted devices, a smaller variation of threshold voltage with temperature. The absence of latch-up in SOI is quite obvious and will not be discussed here, although it is an important argument for the use of SOI circuits for high-temperature applications.

7.2.1. Leakage currents

The increase of junction leakage current is one of the main causes of failure in circuits operating at high temperature. The junction leakage current is, of course, proportional to the area of the junctions. The strongest leakage component found in bulk CMOS operating at high temperatures originates from the "huge" well junctions.[41] These cause loss of functionality in CMOS circuits at temperatures above ...180°C... while the same SOI CMOS circuits were still operating at 300°C. Operation of SOI CMOS ring oscillators at temperatures even up to 500°C has been demonstrated.[42] The leakage current of a reverse-biased N^+-P junction is expressed by [43]:

$$I_{leak} = q\ A \left(\frac{D_n}{\tau_n}\right)^{1/2} \frac{n_i^2}{N_a} + q\ A\ \frac{n_i\ W}{\tau_e}$$

where q is the electron charge, A is the junction area, D_n is the electron diffusion coefficient, τ_n is the electron lifetime in p-type neutral silicon, n_i is the intrinsic carrier concentration, N_a is the doping concentration in the p-type material, W is the depletion width, and $\tau_e=(\tau_n+\tau_p)/2$ is the effective lifetime related to the thermal generation process in the depletion region. The temperature dependence of n_i is given by equation 7.2.3. The first term in the above equation (diffusion component) is proportional to n_i^2, and the second term (generation component) is proportional to n_i. It is experimentally observed [44] that the leakage of SOI MOSFETs and reverse-biased diodes varies as n_i below a temperature of ...100-150°C... and as n_i^2 above that temperature. It is worth noting that A, the area of the junction, and $A \cdot W$, the volume of the space-charge region associated with the diode, are both much smaller in SOI transistors than in bulk devices. Therefore, extremely low leakage current can be obtained in SOI devices. The leakage current ($V_{DS} = 5$ V) of bulk and SOI n-channel transistors with same geometries are compared in Figure 7.2.1, where the slightly higher leakage in the SOI device at room temperature is attributed to a relatively poor value of τ_e in the measured device. At high temperatures, however, where diffusion current dominates, [45] the leakage current is three orders of magnitude lower in the SOI devices than in the bulk device. In thin-film SOI devices it has been observed that the junction leakage current is proportional to $n_i(T)$ if the body of the device is fully depleted when the transistor is turned off. This is always the case in accumulation-mode SOI MOSFETs (Figure 7.2.2).

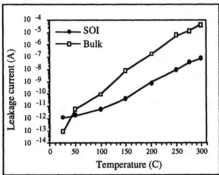

Figure 7.2.1: Leakage current in bulk and SOI n-channel transistors, of same geometries, as a function of temperature.

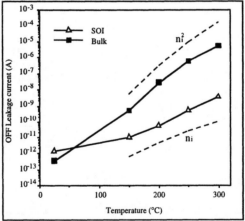

Figure 7.2.2: Leakage current dependence on temperature of an SOI accumulation-mode pMOS transistors for zero back gate and -3 V drain bias, and of a pMOS bulk transistor with -3 V drain and zero substrate and well bias. Evolutions proportional to n_i and n_i^2 are indicated (dashed lines).

The device OFF current, which is equal to the reverse-biased drain junction in enhancement-mode SOI MOSFETs, is markedly smaller in SOI than in bulk transistors, owing, on one hand, to the reduced junction area and, on the other hand, to the $n_i(T)$ current increase with temperature.

This observation is also valid for accumulation-mode devices. Figure 7.2.3 presents the drain current as a function of gate voltage in an accumulation-mode p-channel SOI device, where I_{ON}/I_{OFF} ratios in excess of 100,000 and 300 are obtained at 200 and 300°C, respectively. Owing to these high I_{ON}/I_{OFF} ratios SOI logic circuits are still functional at high temperatures. For comparison, the I_{ON}/I_{OFF} ratio in a bulk transistor is only on the order of 40 at 250°C, while it is on the order of 5,000 in an SOI device.

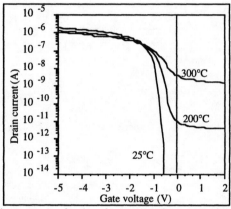

Figure 7.2.3: Drain current as a function of gate voltage in an accumulation-mode p-channel transistor at different temperatures.[46]

In addition to the fact that the area of source and drain junctions is much smaller in SOI than in bulk (by a factor of 15 to 100, depending on the design rules), it is also worthwhile noting that the largest of all junctions, the well junctions, are totally absent from SOI CMOS. This contributes to a drastic reduction in the overall standby current consumption of SOI circuits, compared to bulk CMOS. Figure 7.2.4 presents the junction leakage paths in bulk and SOI CMOS inverters. In the bulk device leakage currents flow to the substrate from the reverse-biased well junction and from the reverse-biased drain of the n-channel MOSFET (the case where the output is "high" is considered here). In the SOI inverter the only leakage current to be considered is the n-channel drain junction leakage. Consequently, the high-temperature standby power consumption is much smaller in the SOI inverter than in the bulk device.

233

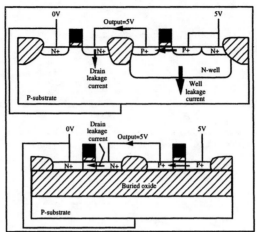

<u>Figure 7.2.4</u>: Leakage current paths in bulk (top) and SOI (bottom) CMOS inverters. No current flow to the substrate is allowed in SOI.

<u>7.2.2. Threshold voltage</u>

The threshold voltage of an n-channel MOSFET (bulk or SOI) is given by:

$$V_{th} = \Phi_{MS} + 2\Phi_F - Q_{ox}/C_{ox} - Q_{depl}/C_{ox} \qquad (7.2.1)$$

where Φ_{MS}, Φ_F, Q_{ox}, Q_{depl} and C_{ox} are the metal-semiconductor work function difference, the Fermi potential, the charge density in the gate oxide, the depletion charge controlled by the gate, and the gate oxide capacitance, respectively. Equation (7.2.1) is valid for partially depleted devices where Q_{depl} is replaced by $q\,N_a\,x_{dmax}$

x_{dmax} is the maximum depletion depth allowed for the doping concentration, N_a, under

consideration, and is equal to $\sqrt{\dfrac{4\,\varepsilon_{si}\,\Phi_F}{q\,N_a}}$. In thin, fully depleted, films Q_{depl} depends on

back-gate bias, and has a value which can range between $q\,N_a\,t_{si}$ and $q\,N_a\,t_{si}/2$, q being the electron charge, k the Boltzmann constant, N_a the doping concentration and t_{si} the silicon film thickness. If an N^+ polysilicon gate is assumed, the value of the work function difference is given by:

$$\Phi_{MS} = -\frac{E_g}{2} - \Phi_F \,, \text{ with } \Phi_F = \frac{kT}{q}\,\ln(N_a/n_i) \qquad (7.2.2)$$

with E_g, T and n_i being the silicon bandgap, the temperature and the silicon intrinsic carrier concentration, respectively. The temperature dependence of the intrinsic carrier concentration, n_i, is given by [47]:

$$n_i = 3.9 \times 10^{16}\ T^{3/2}\ e^{-(E_g/2kT)} \qquad (7.2.3)$$

The temperature dependence of the threshold voltage can be obtained from equations (7.2.1) and (7.2.2). For simplification, we will assume that Q_{ox} shows no temperature dependence over the temperature range under consideration and we will not take into account the presence of surface states at the Si-SiO$_2$ interfaces. E$_g$ is assumed to be independent of temperature over the temperature range under consideration (20 - 250°C). Its actual variation is 0.3%. In the case of a bulk or a thick-film SOI device, one obtains:

$$\frac{dV_{th}}{dT} = \frac{d\Phi_F}{dT} \left[1 + \frac{q}{C_{ox}} \sqrt{\frac{\varepsilon_{si} N_a}{k\, T\, \ln(N_a/ni)}}\,\right] \tag{7.2.4}$$

with
$$\frac{d\Phi_F}{dT} = 8.63 \times 10^{-5} \left[\ln(Na) - 38.2 - \frac{3}{2}\{1 + \ln(T)\}\right] \tag{7.2.5}$$

In the case of a thin-film, fully depleted, device, the depletion charge, Q$_{depl}$, is equal to q N$_a$ t$_{si}$/n, where t$_{si}$ is the silicon film thickness, which is independent of temperature, and the value of n is ranging between 1 and 2, depending on oxide charge and back-gate bias conditions. If one assumes that n is independent of temperature, the following dependence for the thin-film device is obtained: $\dfrac{dV_{th}}{dT} = \dfrac{d\Phi_F}{dT}$ $\tag{7.2.6}$

It can be seen by mere comparison of (7.2.4) and (7.2.6), that dV$_{th}$/dT is smaller in thin SOI devices than in bulk or thick SOI devices. The ratio of threshold voltage variation in bulk devices to that in thin SOI devices is given by the bracketed term of (7.2.4), and is typically ranging from 2 to 3, depending on the gate oxide thickness and the channel doping concentration. The right-hand term of the bracketed expression of (7.2.4) is indeed caused by the temperature-induced variation of the depletion zone depth, variation which is obviously nonexistent in fully depleted devices. Typical values for dV$_{th}$/dT range between -0.7 and -0.8 mV/K in thin SOI MOSFETs. Much larger values of dV$_{th}$/dT are observed in bulk and thick-film SOI devices, such as presented in table 7.2.1.

	SOI, T=25°C	Bulk, T=25°C	SOI, T=200°C	Bulk, T=200°C
t$_{ox}$=19nm N$_a$=1.6 10^{17} cm^{-3}	-0.74	-1.87	-0.8	-2.3
t$_{ox}$=32.5 nm N$_a$=1.2 10^{17} cm^{-3}	-0.76	-2.42	-0.82	-3.05

Table 7.2.1: Calculated dependence of threshold voltage on temperature, dV$_{th}$/dT, (in mV/K), in bulk and thin (t$_{si}$=100 nm) SOI MOSFETs. The data are presented for two temperatures (25 and 200°C) and for the two sets of technological parameters (t$_{ox}$ and N$_a$).[48]

When the temperature increases in thin-film devices the intrinsic carrier concentration increases, and Φ_F decreases. This gives rise to a decrease in the maximum depletion width, in such a manner that the devices are no longer fully depleted above a given critical temperature, T$_k$, where x$_{dmax}$ becomes smaller than t$_{si,eff}$, where t$_{si,eff}$ is the effective electrical film thickness "seen" from the top gate, *i.e.* the silicon film thickness minus the backside depletion zone generated by both the interface oxide charges and the influence of the back gate. Above T$_k$, the transistor behaves as a thick-film SOI MOSFET, since the film is no longer fully depleted, and the temperature dependence of the threshold voltage becomes similar to that observed in bulk or PD SOI devices. dV$_{th}$/dT is then governed by equation (7.2.4) (Figure 7.2.5).

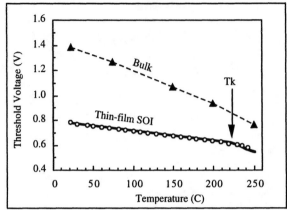

Figure 7.2.5: Variation of the threshold voltage of bulk and thin-film SOI n-channel MOSFETs with temperature.

7.2.3. Output conductance

The output conductance of transistors is a parameter playing an important role in the performances of analog CMOS ICs. Because the body of the devices is electrically floating, impact ionization-related effects (kink effect, parasitic bipolar action,...) tend to degrade the output conductance of SOI MOSFETs. This degradation is minimized, but nevertheless present, if fully depleted devices are used. It is observed that the output conductance of SOI MOSFETs actually improves when the temperature is increased [49], as can be seen in Figure 7.2.6. This is explained by several mechanisms: high temperature reduces impact ionization near the drain, excess minority carrier concentration in the device body is reduced through increased recombination, and the body potential variations are reduced owing to an increase of the saturation current of the source junction.

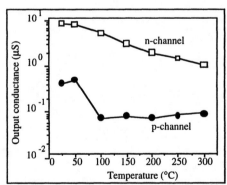

Figure 7.2.6: Variation of the output conductance in thin-film SOI MOSFETs (n- and p-channel) with temperature. $W/L = 20\mu m/10\mu m$ and $V_{GS} = \pm 1.5$ V.

References

1   G.C. Messenger and M.S. Ash, <u>The Effects of Radiation on Electronic Systems</u>, Van Nostrand Reinhold Company, New York, 1986

2   J. R. Schwank, Short course on Silicon-on-Insulator circuits, IEEE International SOI Conference, p. 5.1, 1996

3   J. Olsen, P.E. Becher, P.B. Fynbo, P. Raaby, and J. Schultz, IEEE Transactions on Nuclear Science, Vol. 40, No. 2, p. 74, 1993

4   G.C. Messenger and M.S. Ash, <u>The Effects of Radiation on Electronic Systems</u>, Van Nostrand Reinhold Company, New York, 1986, p. 307

5   J. R. Schwank, Short course on Silicon-on-Insulator circuits, IEEE International SOI Conference, p. 5.1, 1996

6   Y. Song, K.N. Vu, J.S. Cable, A.A. Witteles, W.A. Kolasinski, R. Koga, J.H. Elder, J. VO. Osborn, R.C. Martin, and N.M. Ghoniem, IEEE Transactions on Nuclear Science, Vol. 35, No. 6, p. 1673, 1988

7   O. Musseau, F. Gardic, P. Roche, T. Corbière, R.A. Reed, S. Buchner, P. McDonald, J. Melinger, and A.B. Campbell, IEEE Transactions on Nuclear Science, Vol. 43, No. 6, p. 2879, 1996

8   J.L. Leray, Microelectronics Engineering, Vol. 8, p. 187, 1988

9   E. Normand, IEEE Nuclear and Space Radiation Effects Conference Short Course on Radiation Effects in Commercial Electronics, p. V-1, 1994

10   G.E.Davis, L.R. Hite, T.G.W. Blake, C.E. Chen, H.W. Lam, R. DeMoyer, IEEE Trans. on Nuclear Science, Vol. 32, p. 4432, 1985

11   J.L. Leray, E. Dupont-Nivet, O. Musseau, Y.M. Coïc, A. Umbert, P. Lalande, J.F. Péré, A.J. Auberton-Hervé, M. Bruel, C. Jaussaud, J. Margail, B. Giffard, R. Truche, and F. Martin, IEEE Trans. on Nuclear Science, Vol. 35, p. 1355, 1988

12   *ibidem*

13   L.W. Massengill, D.V. Kerns, Jr., S.E. Kerns, and M.L. Alles, IEEE Electron Device Letters, Vol. 11, p. 98, 1990

14   Y. Tosaka, K. Suzuki, and T. Sugii, Technical Paper Digest of the Symposium on VLSI Technology, p. 29, 1995

15   G.E. Davis, in "Silicon-On-Insulator and Buried Metals in Semiconductors", Sturm, Chen, Pfeiffer and Hemment Eds., (North-Holland), MRS Symposium Proceedings, Vol. 107, p. 317, 1988

16   J. R. Schwank, Short course on Silicon-on-Insulator circuits, IEEE International SOI Conference, p. 5.1, 1996

17   P.V. Dressendorfer and A. Ochoa, IEEE Transactions on Nuclear Science, Vol. 28, p. 4288, 1981

18   M. Allenspach, C. Dachs, G.H. Johnson, R.D. Schrimpf, E. Lorfèvre, J.M. Palau, J.R. Brews, K.F. Galloway, J.L. Titus, and C.F. Wheatley, IEEE Transactions on Nuclear Science, Vol. 43, No. 6, p. 2927, 1996

19   J. R. Schwank, Short course on Silicon-on-Insulator circuits, IEEE International SOI Conference, p. 5.1, 1996

20   T.F. Wrobel, FG.N. Coppage, G.L. Hash, and A.J. Smith, IEEE Transactions on Nuclear Science, Vol. 32, p. 3991, 1985

21   G.H. Johnson, R.D. Schrimpf, and K.F. Galloway, IEEE Transactions on Nuclear Science, Vol. 39, p. 1605, 1992

22   J.W. Adolphsen, J.L. Barth, and G.B. Gee, IEEE Transactions on Nuclear Science, Vol. 43, No. 6, p. 2921, 1996

23   C.F. Wheatley, J.L. Titus, D.I. Burton, and D.R. Carley, IEEE Transactions on Nuclear Science, Vol. 43, No. 6, p. 294, 1996

24   T.F. Wrobel, IEEE Transactions on Nuclear Science, Vol. 34, p. 1262, 1987

25   J. R. Schwank, Short course on Silicon-on-Insulator circuits, IEEE International SOI Conference, p. 5.1, 1996

26   *ibidem*

27   J. Gautier and A.J. Auberton-Hervé, IEEE Electron Device Letters, Vol. 12, p. 372, 1991

28   O. Musseau, IEEE Transactions on Nuclear Science, Vol. 43, p. 603, 1996

29   J.L. Leray, Microelectronics Engineering, Vol. 8, p. 187, 1988

30   F. Wulf, D. Braunig, and A. Boden, ECFA STUDY WEEK on Instrumentation Technology for High-Luminosity Hadron Colliders, Ed. by E. Fernands and G. Jarlskog, Proc. Vol. 1, p. 109, 1989

31   J.L. Leray, E. Dupont-Nivet, J.F. Péré, Y.M. Coïc, M. Raffaelli, A.J. Auberton-Hervé, M. Bruel, B. Giffard, and J. Margail, IEEE Transactions on Nuclear Science, Vol. 37, No. 6, p. 2013, 1990

32   J.L. Leray, E. Dupont-Nivet, J.F. Péré, O. Musseau, P. Lalande, and A. Umbert, Proceedings SOS/SOI Technology Workshop, p. 114, 1989

33   D.M. Fletwood, IEEE Transactions on Nuclear Science, Vol. 39, p. 269, 1992

34   D.M. Fletwood, S.S. Tsao, and P.S. Winokur, IEEE Trans. on Nuclear Science, Vol. 35, p. 1361, 1988

35   G.C. Messenger and M.S. Ash, <u>The Effects of Radiation on Electronic Systems</u>, Van Nostrand Reinhold Company, New York, 1986, p. 243

36   J.L. Leray, E. Dupont-Nivet, O. Musseau, Y.M. Coïc, A. Umbert, P. Lalande, J.F. Péré, A.J. Auberton-Hervé, M. Bruel, C. Jaussaud, J. Margail, B. Giffard, R. Truche, and F. Martin, IEEE Trans. on Nuclear Science, Vol. 35, p. 1355, 1988

37   *ibidem*

38   V. Ferlet-Cavrois, O. Musseau, J.L. Leray, J.L. Pelloie, and C. Raynaud, IEEE Transactions on Electron Devices, Vol. 44, No. 6, p. 965, 1997

39   G.E. Davis, in "Silicon-On-Insulator and Buried Metals in Semiconductors", Sturm, Chen, Pfeiffer and Hemment Eds., (North-Holland), MRS Symposium Proceedings, Vol. 107, p. 317, 1988

40   G.C. Messenger and M.S. Ash, <u>The Effects of Radiation on Electronic Systems</u>, Van Nostrand Reinhold Company, New York, 1986, p. 267

41   W.A. Krull and J.C. Lee, Proceedings IEEE SOS/SOI Technology Workshop, p. 69, 1988

42   W.P. Maszara, Proc. of the 4th International Symposium on Silicon-on-Insulator Technology and Devices, Ed. by D. Schmidt, the Electrochemical Society, Vol. 90-6, p. 199, 1990

43   S.M. Sze, Physics of Semiconductor Devices, 2nd Edition, J. Wiley & Sons Eds, p. 91, 1981

44   D.P. Vu, M.J. Boden, W.R. Henderson, N.K. Cheong, P.M. Zavaracky, D.A. Adams, and M.M. Austin, Proceedings IEEE SOS/SOI Technology Conference, p. 165, 1989

45   T.E. Rudenko, V.I. Kilchitskaya, and A.N. Rudenko, Microelectronic Engineering, Vol. 36, No. 1-4, p. 367, 1997

46   D. Flandre, in High-Temperature Electronics, Ed. by. K. Fricke and V. Krozer, Materials Science and Engineering, Elsevier, Vol. B29, p. 7, 1995

47   R.S. Muller and T.I. Kamins, Device Electronics for Integrated Circuits, 2nd Edition, J. Wiley & Sons Eds, p. 56, 1986

48   G. Groeseneken, J.P. Colinge, H.E. Maes, J.C. Alderman and S. Holt, IEEE Electron Device Letters, Vol. 11, p. 329, 1990

49   P. Francis, A. Terao, B. Gentinne, D. Flandre, JP Colinge, Technical Digest of IEDM, p. 353, 1992

CHAPTER 8 - SOI Circuits

Thin-film, fully depleted SOI MOSFETs and circuits present the following advantages over bulk devices:

◊ absence of latchup
◊ higher soft-error immunity
◊ higher transconductance
◊ less processing yield hazards
◊ reduced electric fields
◊ reduced parasitic capacitances
◊ reduced short-channel effects
◊ sharper subthreshold slope
◊ shorter and easier CMOS processing

This Chapter will demonstrate how these properties can be taken advantage of to produce high-performance integrated circuits.

8.1. General considerations

8.1.1. Fabrication yield

An issue has to be addressed before SOI can be considered as a serious competitor to bulk silicon: the issue of fabrication yield. Indeed, the defect density in SOI films is larger than in bulk silicon. High-quality bulk silicon wafers present defect densities on the order of 10 dislocations/cm^2. A quick look at Figure 2.7.10 shows that state-of-the-art SIMOX wafers have a 300-5000 cm^{-2} dislocation density. It is also worthwhile giving some thoughts to the effects produced by defects in MOS devices. These can be of different natures, such as degradation of the gate oxide quality or junction leakage. In a bulk device, a defect can produce a yield hazard if it is located in the active area, *i.e.*, if it is located in the channel region, the source or the drain. If λ is the minimum design rule for all levels, the area where the presence of a defect can be felt is given by: A = device length x device width = [2(field isolation to contact hole spacing + contact hole length + contact hole to gate spacing) + gate length)] x [field isolation to contact hole spacing + contact hole width + field isolation to contact hole spacing] = $[2(3\lambda)+\lambda][3\lambda]=21\lambda^2$. In the case of a thin-film SOI device with reach-through junctions, the area where the presence of a defect can be felt is given by the area of the channel region only, because of the full dielectric isolation of the source and drain junctions. This area is equal to $3\lambda^2$ in our example, and is, therefore, significantly smaller than in a bulk device. According to such considerations, the use of SOI wafers permits one to obtain higher fabrication yield than bulk silicon if the defect densities in both materials are comparable. The cost of SOI wafers is higher than that of bulk silicon wafers, but the increased demand stimulates competition between suppliers and the increased throughput of fabrication techniques will force the wafer price to drop significantly. For example, the price of a UNIBOND® wafer is predicted to be equal to that of a bulk epi wafer in a near future. The cost of front-end CMOS processing, on the other hand, increases steadily as smaller and smaller devices are fabricated. This cost increase is significantly larger in bulk CMOS processing than in thin-film SOI processing. Indeed, deep-submicron CMOS processing necessitates heavy investments in terms of R&D efforts in order to produce adequate device isolation, shallow junctions, anti-latchup and anti-punchthrough techniques. SOI devices, on the other hand, are scalable in their current form to deep-submicron dimensions without the need for any major new technique development, which reduces the need for costly R&D investments. Front-end thin-film SOI processing is already currently cheaper than front-end bulk processing, and it will become even more so in the future. In addition, higher packing density (and, therefore, higher yield) as well as increased performance can be obtained when SOI is used. When the considerations of wafer and processing cost, potential higher yield and higher performance are added one can reach the conclusion that higher benefits can be made by switching from bulk to SOI.[1,2,3,4] Examples of high-density or high-performance SOI circuits are given in Tables 2.7.1 and 8.3.1.

8.1.2. Performance

A definite advantage of SOI technology is the increase of circuit speed due to the reduction of source and drain capacitances and, in the case of fully depleted SOI devices, to the increase of transconductance. Figure 8.1.1 compares the propagation delay in some state-of-the art CMOS ring oscillators from the literature as a function of the gate length × gate oxide thickness product. The SOI oscillators show a steady 2× improvement of speed over the bulk devices.[5,6] Larger circuits, such as SRAMs and microprocessors, show 15% to 30% improvement of access time or maximum operating frequency, even if the design is directly transferred from bulk to SOI without optimization.[7]

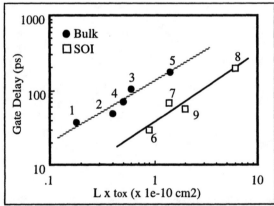

Figure 8.1.1: Delay per gate in advanced bulk and SOI CMOS ring oscillators as a function of the product of gate length and gate oxide thickness. The references are: 1-IBM, 1988; 2-AT&T, 1988, 3: TI, 1988; 4-AT&T, 1989; 5-TI, 1988, 6-Hughes, 1988; 7-HP, 1988; 8-Mitsubishi, 1988; 9-Northern Telecom 1989.[8]

In addition to the increase in speed, SOI technology permits one to reduce both the standby power consumption and the operating power consumption of the circuits. The standby power consumption is lower than in bulk devices because of the reduced area of the reach-through SOI source and drain junctions. Standby currents of 10 nA and even 6.5 nA have been obtained in 64k and 256k SRAMs, respectively (V_{DD}=5V).[9,10] The operating power consumption is reduced mainly by the lower value of the source and drain capacitances. Indeed, a 50% reduction of the current consumption is observed in SOI 16k SRAMs, compared to the equivalent bulk parts (for a constant access time). Similarly, a 30% decrease of access time is obtained if the current consumption is maintained constant in the comparison between bulk and SOI devices.[11] From another standpoint, it can be said that a generation of CMOS SOI delivers the same performances as the next generation of bulk circuits.

8.1.3. ESD protection

The input and output pads of MOS circuits have to be protected against electrostatic discharge (ESD). Specifications are that the circuit should, for instance, survive a 2,000 volt pulse delivered by a RC network corresponding to the capacitance and the resistance of a human body (*human body model, HBM*). Under these conditions significant power is dissipated in the input/output stages of the circuit, and device destruction (or: "zapping") can occur. Since SOI devices have a reduced junction area and are thermally insulated from the substrate, there is a concern about the efficiency and the reliability of SOI ESD protection structures.

It has been suggested to use the snapback characteristics of the $I_D(V_D)$ curve of a MOSFET to realize ESD protection structures (Figure 8.1.2).[12] Indeed, the SOI MOSFET presents a double snapback characteristics. If the gate is grounded and the drain voltage is increased up to a certain value, called snapback voltage (V_{sb}), drain current suddenly flows in the device and the drain voltage drops to a holding value (V_h). If one attempts to further increase the drain voltage, a second snapback mechanism can be observed for $V_D=V_{sb2}$ (second snapback voltage). The first snapback is due to classical avalanche multiplication and impact ionization mechanisms, while the second snapback is caused by current amplification in the parasitic bipolar transistor of the device.[13] Modelling of the drain breakdown characteristics of SOI MOSFETs can be found in References [14,15,16].

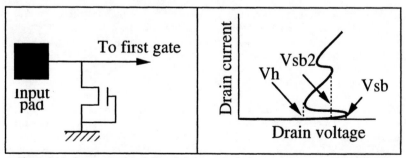

Figure 8.1.2: Input protection (left) and drain snapback characteristics of an SOI MOSFET (right).

These snapback characteristics can be used to clamp ESD pulses and can, therefore, be used as protection devices (Figure 8.1.2). The holding voltage, snapback voltages and the level of protection reached by SOI snapback n-channel MOSFETs are presented in Figure 8.1.3. The protection levels are expressed in ESD pulse volts per micron of protection device width. The protection level obtained from these devices ranges from 50 to 80% of that provided by similar protections made in bulk silicon. Higher protection levels can be reached by using more elaborate structures combining snapback MOSFETs and Zener diodes.[17] Another alternative consists in opening a window in the buried oxide and to fabricate the ESD protection devices in the bulk silicon substrate.[18] 4000-V ESD protection (HBM) of fully depleted SOI CMOS circuits (L = 0.3 μm) has been demonstrated.[19]

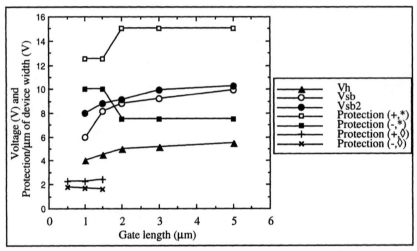

Figure 8.1.3: V_{sb}, V_{sb2}, V_h, and protection level as a function of gate length. *: 19 nm-thick gate oxide[20]; ◊: 8 nm-thick gate oxide[21]; +: positive ESD pulses; -: negative ESD pulses.

8.2. High-temperature circuits

The excellent behavior of thin-film SOI MOSFETs at high temperature makes SOI technology highly suitable for high-temperature IC applications. Indeed, the major causes of failure in bulk CMOS logic at high temperature, *i.e.* excess power consumption and degradation of logic levels and noise margin, are observed to be much reduced in SOI circuits. Latch-up is of course totally suppressed when SOI technology is used. SOI CMOS inverters exhibit full functionality and very little change in static characteristics at temperatures up to 320°C.[22] The switching voltage remains stable, owing to the remarkably weak and symmetrical variation of the n- and p-channel threshold voltages. The output voltage swing is reduced from only a few millivolts at 320°C, due to the slightly increased leakage current of the OFF devices and the reduced carrier mobility of the ON devices. This degradation is, however, totally negligible when compared to what is observed in bulk devices. In logic gates with series transistors, such NAND gates, the increase of standby supply current with temperature remains even more limited than what can expected on basis of the leakage current of all constituent devices (Figure 8.2.1). This is because in SOI circuits, the drain leakage current of each individual transistor flows towards its source, and thus into the following transistor, unlike in bulk circuits, where all drain leakage currents are collected by the substrate.

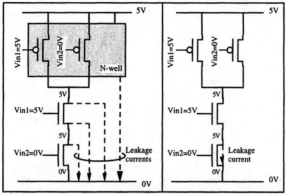

Figure 8.2.1: Leakage current paths in a bulk (left) and an SOI (right) CMOS NAND gate. Both gates have a high and a low input bias. All bulk junction leakage currents flow in parallel towards the substrate. The leakage of the SOI gate is limited to that of a single transistor.[23]

The best way to assess the influence of temperature on the performances SOI CMOS circuits is to study the behavior of simple basic circuits. For example, circuit speed has been tested on toggle-chain frequency dividers (divide-by-32). Bulk dividers implemented in static CMOS logic start to behave erratically around 180°C and fail to function at about 225°C. A similar implementation of the dividers in SOI technology is still functional at 320°C (Figure 8.2.2). The maximum frequency of operation at 320°C is half that achieved at room temperature due to the reduction of carrier mobility.

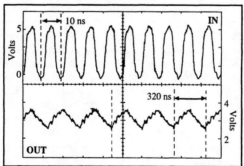

Figure 8.2.2: Measured input (100 MHz) and output (3.125 MHz) signals of a SOI CMOS divide-by-32 circuit operating at 300°C.

In a similar way, the holding time characteristics of dynamic gates (clocked NOR gates) have been showed to degrade by a factor 10 only at 320°C, compared to room temperature. This excellent result is, of course, due to the limited increase of the leakage current in SOI devices. As mentioned earlier, functionality of larger circuits, such as 16k and 256k SRAMs at high temperatures (up to 300 °C) has been demonstrated as well.[24,25]

The performances of analog circuits depend on other device parameters than digital circuits. For instance, standby current consumption is fixed by the operating point of the circuit rather than by junction leakage. One important parameter is the output conductance of the transistors. Indeed, the dc voltage gain of an operational amplifier (Operational Transconductance Amplifier, OTA) is proportional to the product of the transconductance of the input transistors by the output impedance of the output transistors. As the temperature is increased, the transconductance of the input transistor decreases because of the reduction of carrier mobility with temperature. The output impedance (the inverse of the output conductance), on the other hand, increases as temperature is increased (Figure 7.2.6). Owing to this kind of compensation effect in the input transconductance-output impedance product, the resulting dc gain of the overall amplifier is expected to be relatively independent of temperature. The measured characteristics of the device are consistent with what can be expected from theory: as the gate transconductance decreases and the output impedance increases, the dc gain of the SOI CMOS amplifier remains quite stable.[26] It is also worth noting that the offset voltage of the amplifier remains between 0 and 2 millivolts for all temperatures between 20 to 300°C. Such amplifiers are suitable for A-to-D converters and switched capacitor circuits having to operate in a high-temperature environment. The operation of SOI A-to-D converters at a temperature of 350°C has been demonstrated.[27]

The excellent behavior of both analog and digital SOI CMOS circuits at high temperatures suggests the use of this technology for different applications. Classical industrial applications such as well logging can take advantage of SOI CMOS for *in-situ* signal amplification and data processing in the high-temperature environment encountered deep inside earth's crust. The temperature ranges within which electronic devices may have to work in oil wells, gas wells, steam injection processes and geothermal energy plant applications are listed in Table 8.2.1.

Automotive and aerospace industries are also big potential consumers of SOI CMOS. Indeed, car electronics (engine and bake control systems, as well as underhood electronics) has a growing need for both analog and digital circuits which are able to withstand temperatures of 200°C or above. Underhood electronics must be able to withstand relatively high temperatures when a car is parked in the hot sun (up to 200°C). Much higher temperature tolerance is asked from engine monitoring electronics, such electronic injection systems, where precision control of the injector position is controlled by electronic components located close or within the engine block. Modern anti lock breaking systems (ABS) also require electronic control systems placed close to the brakes and, therefore, submitted to high temperatures when the brakes are activated. The ability of SOI CMOS to operate at high-temperatures, combined with the possibility of integrating power devices with low-power logic on a single, dielectrically isolated substrate, renders SOI the ideal technology for such purposes.[28,29] High-temperature airplane applications include on-board electronics and, of course, engine control and surface control (wing temperature, ...). Space applications are also concerned. Indeed, thermal shielding for satellites is quite heavy, and the weight of a satellite has a direct impact on how much it costs to place it on orbit. In some special applications (Venus or Mercury probes), shielding is no longer possible and high-temperature electronics is absolutely required. Commercial nuclear applications (power plants,...) also require high-temperature electronics for core control.[30]

Application	Temperatures
Well logging	75-600°C
Oil wells	75-175°C
Gas wells	150-225°C
Steam injection	200-300°C
Geothermal energy	200-600°C
Automotive	150-600°C
Underhood	-50-200°C
Engine sensors	up to 600°C
Combustion & exhaust sensors	up to 600°C
ABS	up to 300°C
Aircraft	150-600°C
Internal equipment	150-250°C
Engine monitoring	300-600°C
Surface controls	300-600°C
Satellites (Venus probe)	150-600°C
Commercial nuclear	30-550°C

Table 8.2.1: Temperature requirements for electronics components used in several consumer or industrial applications.

In the case of analog circuits such as operational amplifiers, the increase of leakage current with temperature may result in the loss of the operating point, in bulk circuits. To illustrate this effect on can refer to Figures 8.3.9 and 8.3.10, where the amplifier functions properly when a bias current of 1 µA flows in each of the three branches of the circuit. In a bulk implementation, some (or most) of this bias current will flow to the substrate if the junction leakage increases. As a result the devices are no longer properly biased and the circuit no longer operates correctly (unless maybe the bias current is increased to compensate for the leakage losses). If SOI technology is used the leakage current can only flow through the branches of the circuit, and the circuit operates without loss of the bias operating point (Figure 8.3.10).

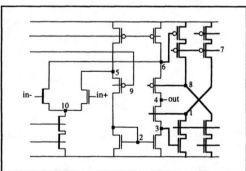

Figure 8.2.3: Schematic diagram of a CMOS OTA designed with a gain-boosting stage. The numbers correspond to several critical nodes of the circuit.

Better performances can be obtained using circuit design techniques, such as the use of a cascode gain-boosting stage, which decreases the output conductance (g_D) of the output stage and, therefore, increases the overall gain of the amplifier, and the zero-temperature coefficient (ZTC) concept, according to which there exists a bias point where the current is independent of temperature (the reduction of mobility is compensated by the reduction of threshold voltage). Figure 8.2.3 presents such an amplifier. It is not a low-power device (25 mW), but it produces a high gain (115 dB - which corresponds to an amplification factor of 560,000 - and a large bandwidth (100 MHz)). The dc gain is still 50 dB at a temperature of 400°C (Figure 8.2.4).

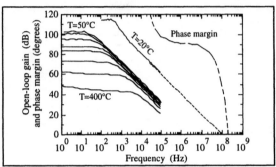

Figure 8.2.4: Measured Bode diagram of the gain-boosting OTA presented in the previous Figure. Measurement was carried out with a 10 pF load at 20°C and a 250 pF load at all other temperatures. The phase margin was measured at 20°C.[31]

8.3. Low-voltage, low-power (LVLP) circuits

As more and more systems become portable there is a general trend towards the reduction of the supply voltage and the power consumption of integrated circuits. SOI CMOS offers there unique opportunities. The assets of SOI MOSFETs in this area are their reduced drain capacitance, low body effect, sharp subthreshold slope and high drive current. Both digital and analog circuits made using SOI CMOS technology offer excellent LVLP performances.

8.3.1. Digital circuits

It is well known that fully depleted (FD) SOI devices have a near-optimal subthreshold slope. This allows one to reduce the threshold voltage (and to increase the current drive) without increasing the OFF current (compared to bulk devices). Low leakage current and low threshold voltage can thus be obtained simultaneously (Figure 8.3.1).

A simple analysis can be used to demonstrate the potential of FD SOI technology for LVLP CMOS applications. Simulation will be performed next using the EKV model.[32] Using this model, the characteristics of transistors and simple circuit elements can readily be modeled. Figure 8.3.2 presents the $I_D(V_G)$ characteristics of bulk and fully depleted SOI n-channel MOSFETs plotted using the EKV model. The varying parameter between the different curves is the (uniform) doping concentration in the channel region, ranging from 10^{16} to 1.2×10^{17} cm^{-3}. The gate oxide thickness is 15 nm.[33]

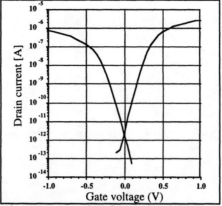

<u>Figure 8.3.1</u>: Measured current vs. gate voltage for an n-channel and a p-channel SOI MOSFET with W=L=3 μm at V_{DS} = ±50 mV. Because of the sharp subthreshold slope, OFF currents less than 1 pA/μm and low threshold voltage (0.4 V) can be obtained simultaneously.

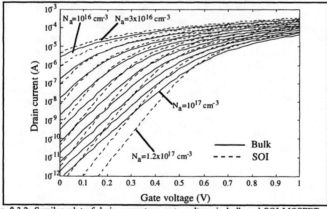

<u>Figure 8.3.2</u>: Semilog plot of drain current vs. gate voltage in bulk and SOI MOSFETs for ten values of channel doping concentration (10^{16} to 10^{17} cm^{-3} for the bulk device and 3×10^{16} to 1.2×10^{17} cm^{-3} for the SOI device). The gate oxide thickness is 15 nm. V_{DS} = 100 mV.

It can be seen that, for any given threshold voltage, the SOI device presents a lower OFF current (at $V_G = 0$ V) and a higher ON current (at $V_G = 1$ V) than the bulk device. The I_{ON}/I_{OFF} ratio in bulk and SOI saturated MOSFETs is presented in Figure 8.3.3 as a function of threshold voltage. Again, it can be seen that the SOI transistor presents better switching characteristics than the bulk device, regardless of the value of the threshold voltage. The I_{ON}/I_{OFF} ratio advantage of the SOI device becomes smaller as the threshold voltage is reduced, but it still is ten times larger than in the bulk transistor for a threshold voltage of 250 mV.

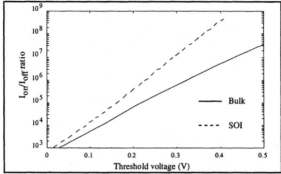

<u>Figure 8.3.3</u>: $I_{ON}(V_G=1V)/I_{OFF}(V_G=0V)$ ratio in bulk and SOI MOSFETs as a function of threshold voltage.

In order to estimate the "ultimate" theoretical reduction of supply voltage and power consumption the EKV model has been used to calculate the power needed to discharge a constant load capacitor by a MOSFET in a realistic circuit environment. A constant discharge time constant of 100 picoseconds has been chosen, which is consistent with a typical value of 250 MHz for the clock frequency. The load capacitor of the transistor being a combination of the next gate input capacitance (which is almost the same in SOI and in bulk) and of drain/source diffusion, polysilicon and metal line capacitances, the value of the load capacitance was chosen to be 25% higher in bulk than in SOI. This value is quite conservative, since the improvement of speed performance of SOI over bulk observed in the literature is in the 30 to 100% range, which tends to indicate capacitance reduction much over 25%.[34] Similarly, a previous study indicates that compared to bulk, a reduction of total node capacitances by a factor of 2 is statistically observed in SOI, when the layout is optimized taking full advantage of the absence of wells and substrate/well bias contacts to bring n- and p-MOSFETs closer and minimize the number of interconnections and their length.[35]

The average power dissipated by the device is given by the sum of the static and dynamic power dissipation:

$$P_{total} = P_{stat} + P_{dyn} = I_{OFF} \, V_{DD} + \alpha \, f \, C_L \, V_{DD}^2 \qquad (8.3.1)$$

where V_{DD} is the supply voltage, f the input signal frequency, C_L the load capacitance and α the degree of activity of the gate. An activity of 1% is assumed here, and both SOI and bulk devices operate at the same clock frequency. The algorithm used for calculating the power consumption is the following. In a first step, the current I_{ON} required to discharge the load capacitor with a time constant of 100 psec is estimated. This current being obtained for a gate voltage equal to V_{DD}, I_{OFF} is obtained from Figure 8.3.2 by setting the origin of the x-axis at the value $V_G(ON) - V_{DD}$. Knowing the position of x-axis origin, the threshold voltage needed to achieve constant switching speed under varying V_{DD} conditions can be extracted (Figure 8.3.4). A threshold voltage of $V_{DD}/4$ (for example) is shown as well. Finally the static and dynamic power consumption can readily be obtained from Equation (8.3.1).

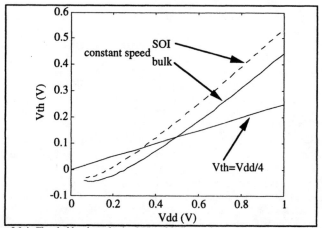

<u>Figure 8.3.4</u>: Threshold voltage in the SOI and bulk transistors vs. supply voltage, in order to maintain the switching speed constant. A threshold voltage equal to $V_{DD}/4$ is presented as well.

The results of this computation are presented on Figure 8.3.5. The following observations can be made. Firstly, the dynamic power consumption of the SOI device is 25% lower than that of the bulk transistor, which is quite logical considering our previous assumption on the load capacitance values. Secondly, the static power consumption is lower in SOI than in bulk, which results from the better I_{ON}/I_{OFF} ratio of SOI devices.

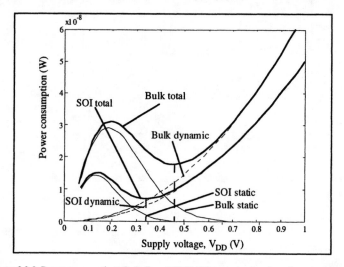

<u>Figure 8.3.5</u>: Power consumption of a bulk or SOI CMOS gate as a function of supply voltage.

Finally, the point of minimum power dissipation occurs for a lower supply voltage in SOI than in bulk, at which point the consumption is twice as small in SOI than in bulk, even though the load capacitance was chosen to be only 25% lower in SOI than in bulk. These

251

results depend of course on the values chosen for the clock frequency and the gate activity factor. For higher values of these parameters, the dynamic power consumption will completely dominate the overall consumption, but, in any case, the consumption of the SOI device remains lower than that of the bulk device.

This brief analysis of gate switching outlines the benefits which low-voltage low-power CMOS digital circuits can draw from the low body-effect coefficient of FD SOI MOSFETs and reduced load capacitance in SOI CMOS designs. [36]

If one considers the point of view of maximum speed performance, the allowed reduction of threshold voltage in SOI devices has a positive, but modest effect on the speed if the circuits (through an increase of I_{Dsat}, given by Equation (5.9.1)) when V_{DD}=5V. This effect, however, becomes more pronounced as the supply voltage is reduced. Figure 8.3.6 presents the ratio of saturation currents between an SOI MOSFET with a body factor n=1.05 and a threshold voltage V_{th}=0.5V and a bulk transistor with n=1.15 and V_{th}=0.7V. The I_{Dsat} enhancement due to the V_{th} reduction is quite remarkable at low supply voltage.

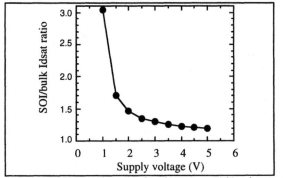

Figure 8.3.6: SOI/bulk I_{dsat} ratio vs. supply voltage. n_{SOI} =1.05 and n_{bulk} =1.15

In parallel to the reduction of V_{DD}, a reduction of power consumption is desirable. The dissipated dynamic power of a circuit is equal to $f \cdot C \cdot V^2$ where f is the frequency, V is the supply voltage, and C is the sum of all the capacitances in the circuit. Since all capacitances but the gate oxide capacitance are smaller in SOI than in bulk,[37] less power is dissipated in SOI circuits.

It has been reported, for instance, that a frequency divider implemented on SOI is twice as fast and consumes half of the power of the equivalent bulk circuit. and that ASICs and 16k SRAMs deliver the same speed performance at 2V than the equivalent bulk circuits at 3.3V, at 25% of the power of the bulk circuit.[38] Several other examples are available.[39,40] Table 3 presents the performances of some recent low-voltage, low-power (LVLP), high-speed SOI CMOS circuits. The choice between fully depleted (FD) and partially depleted (PD) devices is not fully resolved yet. Some favor the better performances of the FD devices, and others prefer the better uniformity of threshold voltage control observed in PD devices (see Figure 4.3.3 for comments on this matter).

Circuit	L (μm)	V_{DD}	Speed	Power	Company	Ref.
Frequ. Divider	0.15	1V	1.2 GHz	50 μW	NTT	41
Frequ. Divider	0.15	2V	2.5 GHz	130 μW	NTT	idem
Frequ. Divider	0.1	1V	1.2 GHz	60 μW	NTT	idem
Frequ. Divider	0.1	2V	2.6 GHz	350 μW	NTT	idem
Prescaler	0.4	1V	1 GHz	0.9 mW	NTT	42
Prescaler	0.4	2V	2 GHz	7.2 mW	NTT	idem
PLL	0.4	1.2V	1 GHz	1.4 mW	NTT	idem
PLL	0.4	2V	2 GHz	8.4 mW	NTT	idem
PLL	0.24	1.5V	2.2 GHz	4.5 mW	NTT	43
512k SRAM	0.2	1V	3.5 ns access time		IBM	44
64k DRAM	0.6	1.5V	-	-	Mitsubishi	45
μcontroller CPU	0.5	0.9V	5.7 MHz	-	Motorola	46
8k-gate 16b ALU*	-	0.5V	40 MHz	5 nW to 350 μW	NTT	47
300k gate array	0.25	1.2V	38 MHz	30 mW	NTT	48
16x16b multiplier*	-	0.5V	18 ns	4 mW	Toshiba	49
PLL	0.35	1.5V	1.1 GHz	-	Sharp	50
32b ALU*	-	0.5V	260 MHz	2.5 mW	Toshiba	51
8x8 ATM switch	0.25	2V	40 Gb/s	8.4 W	NTT	52
560k master array	0.35	1V	50 MHz	50 nW/MHz/gate	Mitsubishi	-

Table 8.3.1: Low-voltage, low-power, and high-speed SOI CMOS circuits. "*" means that hybrid bipolar-MOS devices were used.

8.3.2. Analog circuits

The dc gain(A_0) and the transition unit-gain frequency (f_T) of a MOSFET are given by $A_0 = -\dfrac{g_m}{I_D} V_A$ and $f_T = \dfrac{g_m}{2\pi C_L}$ where V_A is the Early voltage of the device, and C_L is its load capacitance. The g_m/I_D ratio is a measure of the efficiency to translate current (hence power) into transconductance. The larger the g_m/I_D ratio, the larger the transconductance for a given current value. The high g_m/I_D value of fully depleted SOI devices should, therefore, allow one to realize near-optimal micropower analog designs (g_m/I_D values of 35 V^{-1} are obtained, while g_m/I_D reaches typical values of 25 V^{-1} in bulk MOSFETs - Figures 5.4.1 and 8.3.6).[53]

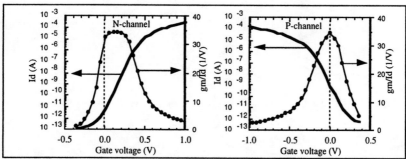

Figure 8.3.6: g_m/I_D and I_D in 600μm/20μm SOI transistors. V_{back}= 0V (fully depleted n-channel) and V_{back} = -1.5V (accumulation-mode p-channel). Silicon and gate oxide thickness are 80 and 30 nm, respectively. g_m/I_D reaches values of 35 V^{-1} in both devices right below threshold.

The CMOS analog switch (pass gate) provides a rough estimation of the lowest acceptable analog supply voltage, V_{dd}, that can be used in a circuit (Figure 8.3.7). In order to transmit a signal without alteration through a switch with the gate of the nMOS transistor held at V_{dd} and that of the pMOS transistor at 0 V, V_{dd} must be larger than a minimum value which can be expressed as a function of the threshold voltage V_T and body factor n of the n- and p-MOSFETs. The following relationship must be satisfied:[54]

$$V_{dd} \geq \frac{V_{Tp}.n_n + V_{Tn}.n_p}{n_p + n_n - n_p.n_n} \tag{8.3.2}$$

In bulk CMOS with V_T being close to 0.7 V and n close to 1.5, V_{dd} is classically limited to a minimum supply voltage of 3 V. In fully depleted SOI CMOS with n equal to 1.1 and the possibility to lower V_T to 0.5 V without degrading the leakage current performance, V_{dd} can be as low as 1.2 V. This confirms the potential of FD SOI CMOS for low-voltage analog applications, for example when using switched-capacitor circuits. Circuits based on SOI pass gates and operating at a frequency of 260 MHz under a 0.5 V supply (hybrid bipolar-MOS transistors, $V_{TH} = 0.15$ V in the off state and -0.05 V in the on state) have been demonstrated.[55]

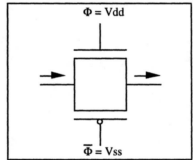

Figure 8.3.7: CMOS analog switch

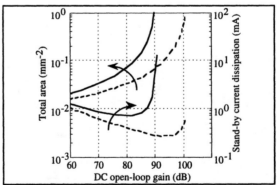

Figure 8.3.8: Synthesis of bulk (—) and SOI (- -) Miller amplifier for $C_L = 10$ pF and $f_T = 10$ MHz.

CMOS operational amplifiers (op amps) can be designed on basis of the knowledge of g_m/I_D and the Early voltage, V_A, as a function of the scaled drain current, $I_D/(W/L)$.[56] This design technique was used to synthesize a simple 1-stage opamp with specified active device size (W/L) and output capacitance (C_L). Figure 8.3.8 clearly shows that the A_{V0} of the SOI amplifier is 25 to 35% larger than in the bulk counterpart, while the static bias current I_{dd} is reduced in the same proportion. Figure 8.3.9 shows the results of such a design, carried out under the assumptions that n is close to 1.1 and 1.5 for fully depleted SOI and bulk MOSFETs, respectively, and that V_A is similar for both device types. Taking into account the reduction of parasitic source/drain and intrinsic gate capacitances[57] found in SOI in addition to the lower n value, it can be further demonstrated that a 2-stage Miller opamp can be synthesized to achieve an A_{V0} 15 dB larger and a I_{dd} 3 times smaller in SOI than in bulk for similar area and identical bandwidth considerations (Figure 8.3.8). Conversely, total area and I_{dd} can be divided by a factor up to 2-3 when all other specifications are kept constant. These theoretical predictions have been confirmed by the realization of a 100 dB-A_{V0}, 3 μA-I_{dd} Miller opamp using a fully depleted SOI CMOS process (Figure 8.3.10). These results establish the potential of FD SOI CMOS to boost the speed (f_T), gain (A_{V0}) and power ($I_{dd} \times V_{dd}$) performances of opamps well over bulk implementations.

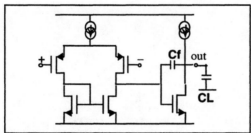

Figure 8.3.9: Layout of a simple low-power Miller operational amplifier. V_{dd}=3V and I_{dd}=3μA (1μA in each branch of the circuit).

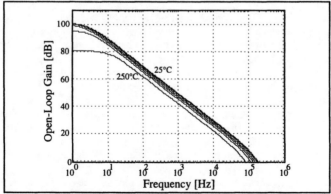

Figure 8.3.10: Measured Bode diagram of the Miller operational amplifier presented in the previous Figure. V_{dd}=3V and I_{dd}=3μA (1μA in each branch of the circuit).

255

Measurement was carried out a 6 different temperatures (25, 50, 100, 150, 200 and 200°C). C_{load}=60 pF.

Table 8.3.2 presents the performances of several operational amplifiers realized using FD SOI CMOS technology. It is most unlikely that high bandwidth, high dc gain and low power consumption can all be achieved at the same time, but the three of them have been demonstrated separately. Indeed, a dc gain of 115 dB, a bandwidth of 1.1 GHz and a power consumption of 3.6 μW have been demonstrated.

Type	dc gain (dB)	f_T (MHz)	V_{DD} (V)	Power (mW)	Ref.
OTA	**115**	100 (*)	5	25	58
OTA	100	0.3 (+)	3	**0.009**	59
OTA	21	**1100** (◊)	4	280	60
OTA	44	0.35 (•)	1.2	**0.0036**	61

Table 8.3.2: Performance of SOI CMOS operational transconductance amplifiers (OTA)
(*,•) 10 pF load; (+) 60 pF load; (◊) 2 pF load

Finally, it is worth mentioning that the use of an SOI instead of bulk allows for significant reduction of crosstalk in the chips.[62,63,64] Crosstalk immunity is enhanced if high-resistivity substrates are used.[65] This feature is particularly attractive for mixed-mode circuits.

8.4. Rad-hard circuits

One of the major niche markets where SOI circuits and devices are currently used is the aerospace/military market, where radiation hardness is a key issue. High-level radiation hardness is also a prerequisite for circuits used in core instrumentation for nuclear power reactors (a containment accident may lead to more than 100 Mrad(Si) of dose exposure), nuclear well logging, control electronics in spacecraft nuclear reactors, reaction chamber instrumentation for particle colliders, and fusion reactor instrumentation. In some cases, these high-level irradiation environments may be associated with high temperature conditions [66]. Under these conditions, SOI has an advantage over bulk CMOS devices, since the reduced active silicon volume and the full dielectric isolation gives rise to lower leakage currents in SOI than in bulk. As we have seen in Section 7.1, SOI MOSFETs are inherently harder against SEU and gamma-dot than bulk devices. SOI circuits are also free of latchup problems, which can be triggered by an energetic heavy ion (SEL) or gamma-dot photocurrents in bulk CMOS circuits. Upon SEU or gamma-dot exposure SOI circuits such as 4k x 1 SIMOX SRAMs show bit error rates comparable to SOS but better than those obtained in bulk CMOS [67]. The main reason for not obtaining performances superior to those of SOS (a smaller silicon film thickness can indeed be utilized in SIMOX than in SOS) is the amplification of the SEU-induced photocurrent by the parasitic lateral bipolar transistor present in the SIMOX devices, where the minority carrier lifetime is high, compared to that of SOS material. Modern SIMOX memories (64k SRAMs) offer SEU error rates of 10^{-9} errors/bit-day (worst-case geosynchronous orbit error rate) and retain functionality at dose-rate exposures up to 10^{11} rad(Si)/sec [68]. These numbers represent an hardness improvement by a factor $\cong$100 over bulk silicon.

Total-dose hardening, on the other hand, does not come for free but can be obtained using an optimized fabrication process and special transistor design [69,70]. The degradation phenomena generated by dose exposure on CMOS circuits are a degradation of the response time (*e.g.* an increase of the access time in SRAMs) and the increase of power consumption. These are mostly due to the threshold voltage shift and the increase of leakage current in MOSFETs upon irradiation. Most rad-hard 16k and 64k CMOS SOI SRAMs are designed to remain functional after exposures to doses up to 1 Mrad(Si) [71,72,73], but more recent parts show 10 Mrad(SiO_2) hardness levels [74] (1 rad(SiO_2) = 0. 56 rad(Si)). The highest total-dose hardness results reported to date for large circuits mention the operation of 29101 16-bit microprocessors at doses up to 100 Mrad(SiO_2). These circuits exhibit a 30% and 100% increase of worst-case propagation delay after 10 and 30 Mrad(SiO_2) exposure, respectively, and some parts are still functional after 100 Mrad(SiO_2) irradiation. Smaller CMOS/SIMOX circuits (frequency dividers) can show no propagation delay degradation at doses up to 1 Mrad(Si) and can even survive doses of 0.5 Grad(Si) [75,76]. Figure 8.1.1 presents the static current consumption and the gate propagation delay of divide-by-16 CMOS/SIMOX circuits as a function of ^{60}Co dose exposure (dose-rate=2.8 Mrad(Si)/hour).[77]

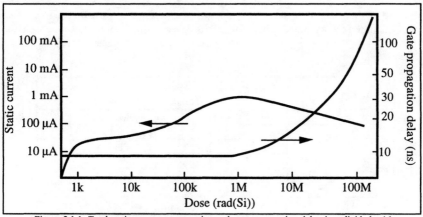

Figure 8.1.1: Total static current consumption and gate propagation delay in a divide-by-16 circuit realized using a CMOS/SIMOX hardened process.

The typical performances of an SOI SRAM are given below (Honeywell HLX6228, 128K × 8 SRAM):

Total dose hardness .... 1 Mrad(SiO_2)
Neutron hardness .... 10^{14} cm^{-2}
Dose rate .... 10^{11} rad(Si)/s
SEU rate .... <10^{-10} upsets/bit-day
SEL.. latchup free
Read/write cycle time............ 14 ns (typical)
Power supply........................ 3.3 V ±0.3 V, <12mW/MHz
Gate length .... 0.6 μm (L_{eff} = 0.45 μm)

As another example, the specifications of the DMILL circuit family is listed below:[78,79]

Available devices: CMOS, NPN bipolar, and P-channel JFET
Input/Output analog & Digital structures available in the design kit
Hardened 4000V ESD protection included
Radiation assurance up to 10Mrad(Si) and 10^{14} neutrons cm^{-2}
Noise Monitoring
Military temperature range -55 to +125°C

8.5. Smart power circuits

The full dielectric isolation provided by the SOI substrate allows one to fabricate high-voltage devices without the burden of having to use a complicated junction isolation process. Compared to bulk, the advantages of the SOI full dielectric isolation are a higher integration density, no latch-up, less leakage current, high-temperature operation, improved noise immunity, easier circuit design, and the possibility to integrate vertical DMOS transistors. Furthermore, power devices can be integrated on the same substrate as CMOS logic to produce "smart power" circuits. Relatively thick films of SIMOX or BESOI with thicknesses ranging from 5 to 20 μm are usually employed (Table 8.5.1), although some devices have been fabricated in thin SOI films as well. It is also possible, using a mask during oxygen ion implantation, to produce SIMOX material with bulk regions (i.e. unimplanted areas). Using such a technique it is possible to combine on a same chip SOI CMOS devices and bulk power devices.

Power device	t_{si} (μm)	V_{max} (V)	I_{max} (A)	R_{on} ($\Omega.mm^2$)	Company	Ref.
VDMOS	6	50	-	0.15	FhG IMS	80
VDMOS	15	150	2	2	IMS	81
LDMOS	5	400	-	15	Philips	82
LIGBT	5	400	-	5	Philips	idem
VDMOS	"bulk"	500	10	18	FhG IMS	83
LDMOS	20	640	-	-	Daimler	84
n-MOSFET	thin	44	0.5	19 $\Omega.mm$	AlliedSignal	85
VDMOS	"bulk"	1200	-	-	FhG IMS	86

Table 8.5.1: Characteristics of power devices used in some smart-power SOI circuits

The flexibility of dielectric isolation allows for the fabrication of mixed chips featuring *e.g.:* CMOS + NPN bipolar + VDMOS devices (Figure 8.5.1) using a process which requires less masking steps that in the equivalent bulk process.

The main drawback of dielectric isolation lies probably in the fact that it is accompanied by an increased thermal resistance between the device and the heat sink, because the thermal conductivity of SiO_2 is lower than that of silicon. As a result, the temperature raise in SOI power devices is higher than that observed in bulk power devices. The differences in self heating between SOI and bulk power devices is largest for short power transients, because the initial temperature raise in SOI devices is more rapid. As the pulse length increases, the difference in temperature rise between SOI and bulk devices converges to a constant value, which is proportional to the thickness of the buried oxide. For oxide thicknesses under 2 μm, the steady-state temperature is essentially determined by the thermal properties of the substrate and the device package, and not by the buried oxide.[87]

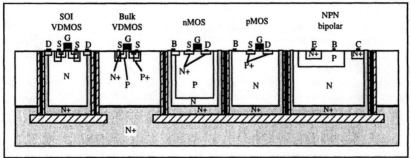

Figure 8.5.1: Process featuring SOI quasi-vertical DMOS, bulk vertical DMOS, CMOS and NPN devices.[88]

8.6. SRAMs and DRAMs

SOI memory chips have evolved quite rapidly during the last few years. Until 1992 only SRAMs chips had been demonstrated, but later on, when the quality of SIMOX material reached a high enough quality to ensure low leakage current levels in pass transistors, DRAM chips were fabricated as well. Table 8.6.1 presents the evolution of SOI SRAMs and DRAMs over the years.

Year	Company	RAM	Ref.
1982	NTT	1kb SRAM	89
1988	Harris	High-temperature CMOS 4kb SRAM	90
1989	TI / Harris	64kb SRAM	91
1989	LETI	Thin-film 16kb SRAM	92
1990	TI	Commercial 64kb SRAM	-
1991	TI	256kb SRAM	93
1991	IBM	256kb fully depleted SRAM	94
1993	TI	1Mb SRAM, 20 ns access time @ 5V	95
1993	IBM	512 kb SRAM, 3.5 ns acc. time @ 1V	96
1993	Mitsubishi	64 kb DRAM, 1.5V operation	97
1995	Samsung	16 Mb DRAM	98
1996	Mitsubishi	16 Mb DRAM, 0.9 V operation	99, 100
1997	Westinghouse	QML* qualification of 256kb SRAM	101
1997	Hyundai	1 Gb DRAM , 2 V operation	102

Table 8.6.1: Milestones in the evolution of SOI memory chips. (*)QML= Qualified Manufacturers Listing.

The advantages of SOI for memory chips are multiple. In the case of SRAMs, the use of SOI reduces the rate of SEU (single-event upset) and rules out SEL (single-event latchup). Hence the use of SOI for memory chips used in space applications. In addition, the reduced parasitic capacitances and the reduced body effect allow for faster and lower-voltage operation. In the case of DRAMs, the advantages of SOI are a reduction of the bit line capacitance (by 25% compared to bulk), a reduction of the access transistor leakage current, and the reduction of the soft-error sensitivity. As a result, storage capacitance of 12 fF is sufficient for a 256 Mb, 1.5 V DRAM realized in SOI, while 34 fF are required for the equivalent bulk device. The advantage of using SOI, in terms of reducing the memory cell area, is then obvious. It is also possible to realize DRAMs which can operate with a sub-1 V power supply (which has not yet been demonstrated in bulk). In order to achieve this level of performance, a reduced C_b/C_s ratio must be achieved (C_b and C_s are the bit line and the storage capacitances, respectively). The reduction of the bit line capacitance comes "for free" with the use of SOI, while the increase of C_s is achieved through the cell design. Considering the cell retention time, it has be shown that thin-film SOI MOSFETs have a much lower leakage current than bulk or thicker SOI devices. Off-state leakage currents smaller than 1 fA/μm are indeed observed in fully depleted, thin-film devices.[103]

DRAM architecture can also take advantage of the low body effect found in SOI devices. Indeed, bulk pass transistors suffer from body effect during the storage capacitor charging state such that the word line voltage has to be boosted to $V_{WL} = V_{th} + V_{ds} + \Delta V$, where ΔV compensates for the V_{th} increase in the bulk pass transistor design. Thus the body effect becomes a major issue in bulk pass transistor design.[104] As we already know from Equation 8.3.2, the use of fully depleted SOI significantly improves the performance of pass gates because of the low body effect found in SOI devices.

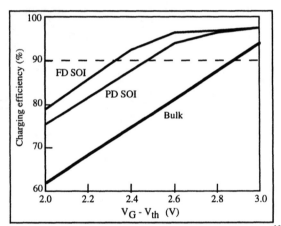

Figure 8.6.1: Charging efficiency as a function of gate voltage overdrive.[105]

In contrast to what happens in a bulk device, the body potential of a partially depleted transistor follows the source node potential during the charging, so that the body-to-source potential difference is kept at a constant value. For fully depleted SOI transistors there is no body potential change. Therefore, SOI transistors have a higher charging efficiency compared to bulk transistors. If the charging current is designed to be 1 μA/μm, then the charging efficiency, defined as the ratio of source (the storage node) voltage at 1 μA/μm to the drain (the bit line) voltage, is 80% for bulk transistors, and 88% and 98% for partially depleted and fully depleted transistors, respectively (Figure 8.6.1). The higher charging efficiency of SOI transistors results in higher programming speed for the DRAM cell. At time t=1ns after being activated by the word line a fully depleted transistor can transfer 10 times more charge than a bulk transistor. For a bulk pass transistor to the same charging efficiency as an SOI transistor, the word line voltage has to be increased during the charging. As shown in Figure 8.6.1, to reach a 90% charging efficiency, the gate voltage of a bulk transistor has to be 0.6 V higher than that of a fully depleted SOI MOSFET. Such an increased word line voltage can degrade gate oxide integrity and is also not favorable for low-power operation.

References

1   T. Stanley, Proceedings of the IEEE International SOI Conference, p. 166, 1992

2   T. Stanley, in "Silicon-on-Insulator Technology and devices", Ed. by. S. Cristoloveanu, The Electrochemical Society, Proceedings Vol. 94-11, p. 441, 1994

3   T.D. Stanley, Proceedings of the Second International Symposium on "Semiconductor Wafer Bonding: Science, Technology, and Applications", Ed. by M.A. Schmidt, C.E. Hunt, T. Abe, and H. Baumgart, The Electrochemical Society, Proceedings Vol. 93-29, p. 303, 1993

4   Y. Kado, H. Inokawa, Y. Okazaki, T. Tsuchiya, Y. Kawai, M. Sato, Y. Sakakibara, S. Nakayama, H. Yamada, M. Kitamura, S. Nakashima, K. Nishimura, S. Date, M. Ino, K. Takeya, and T. Sakai, Technical Digest of IEDM, p. 635, 1995

5   B.Y. Hwang, M. Racanelli, M. Huang, J. Foerstner, S. Wilson, T. Wetteroth, S. Wald, J. Rugg, and S. Cheng, Extended Abstracts of the International Conference on Solid-State Devices and Materials, Yokohama, p. 268, 1994

6   G.G. Shahidi, T.H. Ning, R. Dennard, and B. Davari, Extended Abstracts of the International Conference on Solid-State Devices and Materials, Yokohama, p. 265, 1994

7   M. Guerra, A. Wittkower, J. Stahman, and J. Schrankler, Semiconductor international, May 1990

8   J.P. Colinge, Technical Digest of IEDM, p. 817, 1989

9   T.W. Houston, H. Lu, P. Mei, T.G.W. Blake, L.R. Hite, R. Sundaresan, M. Matloubian, W.E. Bailey, J. Liu, A. Peterson, and G. Pollack, Proceedings IEEE SOS/SOI Technology Conference, p. 137, 1989

10   H. Gotou, A. Sekiyama, T. Seki, S. Nagai, N. Suzuki, M. Hayasaka, Y. Matsukawa, M. Miyazima, Y. Kobayashi, S. Enomoto, and. K. Imaoka, Technical Digest of IEDM, p. 912, 1989

11   A.J. Auberton-Hervé, Proceedings IEEE SOS/SOI Technology Conference, p. 149, 1990

12   K. Verhaege, G. Groeseneken, J.P. Colinge, and H.E. Maes, IEEE Electron Device Letters, Vol. 14, No. 7, p. 326, 1993

13   J.S.T. Huang, Proceedings of the IEEE International SOI Conference, p. 122, 1993

14   J.S.T. Huang, H.J. Chen, and S.J. Kueng, IEEE Transactions on Electron Devices, Vol. 39, p. 1170, 1992

15   H.K. Yu, J.S. Lyu, S.W. Kang, and C.K. Kim, IEEE Transactions on Electron Devices, Vol. 41, No. 5, p. 726, 1994

16   N. Kistler and J. Woo, IEEE Transactions on Electron Devices, Vol. 41, No. 7, p. 1217, 1994

17   J.C. Smith, M. Lien, and S. Veeraraghavan, Proceedings of the IEEE International SOI Conference, p. 170, 1996

18   M. Chan, J.C. King, P.K. Ko, and C. Hu, IEEE Electron Device Letters, Vol. 16, No. 1, p. 11, 1995

19   A.O. Adan, T. Naka, S. Kaneko, D. Urabe, K. Higashi, and A. Kasigawa, Proceedings IEEE International SOI Conference, p. 116, 1996

20   K. Verhaege, G. Groeseneken, J.P. Colinge, and H.E. Maes, Microelectronics and Reliability, Special Issue 'Reliability Physics of Advanced Electron Devices', Vol. 35, No. 256, Issue 3, p. 555, 1994

21   M. Chan, S.S. Yuen, Z.J. Ma, Y. Hui, P.K. Ko, and C. Hu, IEEE Transactions on Electron Devices, Vol. 42, No. 10, p. 1816, 1995

22   P. Francis, A. Terao, B. Gentinne, D. Flandre, and J.P. Colinge, Technical Digest of IEDM, p. 353, 1992

23   A.J. Auberton-Hervé, J.P. Colinge and D. Flandre, Japanese Solid State Technology, pp. 12-17, Dec. 1993 (In Japanese)

24   W.A. Krull and J.C. Lee, Proceedings IEEE SOS/SOI Technology Workshop, p. 69, 1988

25   H. Gotou, A. Sekiyama, T. Seki, S. Nagai, N. Suzuki, M. Hayasaka, Y. Matsukawa, M. Miyazima, Y. Kobayashi, S. Enomoto, and K. Imaoka, Technical digest of IEDM, p. 912, 1989

26   P. Francis, A. Terao, B. Gentinne, D. Flandre, and J.P. Colinge, Technical Digest of IEDM, p. 353, 1992

27   A. Viviani, D. Flandre, and P. Jespers, Proceedings of the IEEE SOI Conference, p. 110, 1996

28   J. Weyers and H. Vogt, Technical Digest of IEDM, p. 225, 1992

29   A. Nakagawa, N. Yasuhara, I. Omura, Y. Yamaguchi, T. Ogura, and T. Matsudai, Technical Digest of IEDM, p. 229, 1992

30   D.B. King, "A summary of high-temperature electronics needs, research and development in the United States", presented at the EUROFORM seminar, May 1992, Darmstadt, Germany

31   B. Gentinne, J.P. Eggermont, and J.P. Colinge, Electronics Letters, vol. 31-24, p. 2092, 1995

32   C.C. Enz, in Low-power HF microelectronics: a unified approach, edited by G.A.S. Machado, IEE circuits and systems series 8, the Institution of Electrical Engineers, p. 247, 1996

33   J.P. Colinge, in Low-power HF microelectronics: a unified approach, edited by G.A.S. Machado, IEE circuits and systems series 8, the Institution of Electrical Engineers, p. 139, 1996

34   J.P. Colinge, Technical Digest of IEDM, p. 817, 1989

35   D. Flandre, C. Jacquemin and J.P. Colinge, Proceedings of the IEEE International SOI Conference, 1992, pp. 164-165

36   J.P. Colinge, in Low-power HF microelectronics: a unified approach, edited by G.A.S. Machado, IEE circuits and systems series 8, the Institution of Electrical Engineers, p. 139, 1996

37   A.J. Auberton-Hervé, Proc. Vol. 90-6, The Electrochemical Society, p. 455, 1990

38   E.D. Nowak, L. Ding, Y.T. Loh and C. Hu, Proceedings of the IEEE International Conference, p. 41, 1994

39   G.G. Shahidi, T.H. Ning, R.H. Dennard, and B. Davari, Extended abstracts of the International Conference on Solid-State Devices and Materials, Yokohama, Japan, p. 265, 1994

40   B.Y. Hwang , M. Racanelli, M. Huang, J. Foerstner, S. Wilson, T. Wetteroth, S. Wald, J. Rugg, and S. Cheng, Extended abstracts of the International Conference on Solid-State Devices and Materials, Yokohama, Japan, p. 268, 1994

41   M. Fujishima, K. Asada, Y. Omura, and K. Izumi, IEEE Journal Solid-State Circuits, Vol. 28, No. 4, p. 510, 1993

42   Y. Kado, M. Suzuki, K. Koike, Y. Omuran, and K. Izumi, IEEE Journal Solid-State Circuits, Vol. 28, No. 4, p. 510, 1993

43   Y. Kado, T. Ohno, M. Harada, K. Deguchi, and T. Tsuchiya, Techn. Digest of IEDM, p. 243, 1993

44   GG. Shahidi, T.H. Ning, T.I. Chappell, J.H. Comfort, B.A. Chappell, R. Franch, C.J. Anderson, P.W. Cook, S.E. Schuster, M.G. Rosenfield, M.R. Polcari, R.H. Dennard, and B. Davari, Techn. Digest of IEDM, p. 813, 1993

45   T. Eimori, T. Oashi, H. Kimura, Y. Yamaguchi, T. Iwamatsu, T. Tsuruda, K. Suma, H. Hidaka, Y. Inoue, S. Satoh, and H. Miyoshi, Techn. Digest of IEDM, p. 45, 1993

46   W.M. Huang, K. Papworth, M. Racanelli, J.P. John, J. Foerstner, H.C. Shin, H. Park, B.Y. Hwang, T. Wetteroth, S. Hong, S. Shin, S. Wilson, and S. Cheng, Techn. Digest of IEDM, p. 95, 1995

47   T. Douseki, S. Shigematsu, Y. Tanabe, M. Harada, H. Inokawa, and T. Tsuchiya, Digest of Technical Papers, IEEE International Solid-State Circuits Conference, p. 84, 1996

48   M. Ino, H. Sawada, K. Nishimura, M. Urano, H. Suto, S. Date, T. Ishiara, T. Takeda, Y. Kado, H. Inokawa, T. Tsuchiya, Y. Sakakibara, Y. Arita, K. Izumi, K. Takeya, and T. Sakai, Digest of Technical Papers, IEEE International Solid-State Circuits Conference, p. 86, 1996

49   T. Fuse, Y. Oowaki, M. Terauchi, S. Watanabe, M. Yoshimi, K. Ohuchi, and J. Matsunaga, Digest of Technical Papers, IEEE International Solid-State Circuits Conference, p. 88, 1996

50   A.O. Adan, T. Naka, S. Kaneko, D. Urabe, K. Higashi, and A. Kasigawa, Proceedings IEEE International SOI Conference, p. 116, 1996

51   T. Fuse, Y. Oowaki, Y. Yamada, M. Kamoshida, M. Ohta, T. Shino, S. Kawanaka, M. Terauchi, T. Yoshida, G. Matsubara, S. Oshioka, S. Watanabe, M. Yoshimi, K. Ohuchi, and S. Manabe, Digest of Technical papers, International Solid-State Circuits Conference, p. 286, 1997

52   Y. Ohtomo, S. Yasuda, M. Nogawa, J. Inoue, K. Yamakoshi, H. Sawada, M. Ino, S. Hino, Y. Sato, Y. Takei, T. Watanabe, and K. Takeya, Digest of Technical papers, International Solid-State Circuits Conference, p. 154, 1997

53   F. Silveira, D. Flandre, and P.G.A. Jespers, IEEE Journal of Solid-State Circuits, Vol. 31, No. 9, p. 1314, 1996

54   E.A. Vittoz, Proceedings of 23rd ESSDERC, Ed. by J. Borel, P. Gentil, J.P. Noblanc, A. Nouhaillat, and M. Verdone, Editions Frontières, p. 927, 1993

55   T. Fuse, Y. Oowaki, Y. Yamada, M. Kamoshida, M. Ohta, T. Shino, S. Kawanaka, M. Terauchi, T. Yoshida, G. Matsubara, S. Oshioka, S. Watanabe, M. Yoshimi, K. Ohuchi, and S. Manabe, Digest of Technical papers, International Solid-State Circuits Conference, p. 286, 1997

56 D. Flandre, B. Gentinne, J.P. Eggermont, and P.G.A. Jespers, Proceedings of the IEEE International SOI Conference, p.99, 1994

57 Y. Omura and K. Izumi, IEEE Electron Device Letters, Vol. EDL-12, p. 655, 1991

58 B. Gentinne, J.P. Eggermont and J.P. Colinge, Electronics Letters, vol. 31-24, p. 2092, 1995

59 J.P. Colinge, J.P. Eggermont, D. Flandre, P. Francis and P.G.A. Jespers, Technical Digest of ISSCC, p. 194, 1995

60 J.P. Eggermont, D. Flandre, R. Gillon and J.P. Colinge, Proceedings of the International SOI Conference, p. 127, 1995

61 J.P. Eggermont, D. De Ceuster, D. Flandre, B. Gentinne, P.G.A. Jespers and J.P. Colinge, IEEE Journal of Solid-State Circuits, Vol. 31, No. 2, 1996

62 I. Rahim, I. Lim, J. Foerstner, and B.Y. Hwang, Proceedings of the IEEE International SOI Conference, p. 170, 1992

63 K. Joardar, Solid-State Electronics, Vol. 39, No. 4, p. 511, 1996

64 I. Rahim, B.Y. Hwang, and J. Foerstner, Proceedings of the IEEE International SOI Conference, p. 170, 1992

65 A. Viviani, J.P. Raskin, D. Flandre, J.P. Colinge, and D. Vanoenacker, Technical Digest of IEDM, p. 713, 1995

66 D.M. Fletwood, F.V. Thome, S.S. Tsao, P.V. Dressendorfer, V.J. Dandini, and J.R. Schwank, IEEE Trans. Nucl. Sci, Vol. 35, p. 1099, 1988

67 G.E.Davis, L.R. Hite, T.G.W. Blake, C.E. Chen, H.W. Lam, R. DeMoyer, IEEE Trans. on Nuclear Science, Vol. 32, p. 4432, 1985

68 W.F. Kraus and J.C. Lee, Proceedings SOS/SOI Technology Conference, p. 173, 1989

69 J.L. Leray, E. Dupont-Nivet, M Raffaelli, Y.M. Coïc, O. Musseau, J.F. Péré, P. Lalande, J. Brédy, A.J. Auberton-Hervé, M. Bruel, and B. Giffard, Annales de Physique, Colloque n°2, Vol. 14, suppl. to n°6, p. 565, 1989

70 J.L. Leray, E. Dupont-Nivet, J.F. Péré, Y.M. Coïc, M. Rafaelli, A.J. Auberton-Hervé, M. Bruel, B. Giffard and J. Margail, IEEE Trans. Nucl. Sci., Vol. 37, no. 6, p. 2013, 1990

71 G.T. Goeloe, G.A. Hanidu, and K.H. Lee, Tech. Prog. of IEEE Nuclear and Space Radiation Effects Conference (NSREC), paper PG-5, p. 35, 1990

72 M.A. Guerra, Proc. of the 4th International Symposium on Silicon-on-Insulator Technology and Devices, Ed. by D. Schmidt, the Electrochemical Society, Vol. 90-6, p. 21, 1990

73 W.F. Kraus, J.C. Lee, W.H. Newman, and J.E. Clark, Tech. Prog. of IEEE Nuclear and Space Radiation Effects Conference (NSREC), paper PG-6, p. 35, 1990

74 L.J. Palkuti, J.J. LePage, IEEE Transactions on Nuclear Science, Vol. 29, p. 1832, 1982

75 J.L. Leray, E. Dupont-Nivet, J.F. Péré, Y.M. Coïc, M. Rafaelli, A.J. Auberton-Hervé, M. Bruel, B. Giffard and J. Margail, IEEE Trans. Nucl. Sci., Vol. 37, no. 6, p. 2013, 1990

76 J.L. Leray, E. Dupont-Nivet, J.F. Péré, O. Musseau, P. Lalande, and A. Umbert, Proceedings SOS/SOI Technology Conference, p. 114, 1989

77 *ibidem*

78   M. Dentan, E. Delagnes, N. Fourches, M. Rouger, M.C. Habrard, L. Blanquart, P. Delpierre, R. Potheau, R. Truche, J.P. Blanc, E. Delevoye, J. Gautier, J.L. Pelloie, J. de Pontcharra, O. Flament, J.L. Leray, J.L. Martin, J. Montaron, and O. Musseau, IEEE Transactions on Nuclear Science, Vol. 40, No. 6, p. 1555, 1993

79   O. Flament, J.L. Leray, and O. Musseau, in <u>Low-power HF microelectronics: a unified approach</u>, edited by G.A.S. Machado, IEE circuits and systems series 8, the Institution of Electrical Engineers, p. 185, 1996

80   J. Weyers, H. Vogt, M. Berger, W. Mach, B. Mütterlein, M. Raab, F. Richter, and F. Vogt, Proceedings of the 22nd ESSDERC, Microelectronic Engineering, Vol. 19, p 733, 1992

81   C. Harendt, U. Apel, T. Ifström, H.G. Graf, and B. Höfflinger, Proceedings of the Second International Symposium on "Semiconductor Wafer Bonding: Science, Technology, and Applications", Ed. by M.A. Schmidt, C.E. Hunt, T. Abe, and H. Baumgart, The Electrochemical Society, Proceedings Vol. 93-29, p. 129, 1993

82   H. Pein, E. Arnold, H. Baumgart, R. Egloff, T. Letavic, S. Merchant, and S. Mukherjee, Proceedings ot the IEEE International SOI Conference, p. 146, 1992

83   F. Vogt, B. Mütterlein, and H. Vogt, Proceedings of the fifth International Symposium on Silicon-on-Insulator Technology and devices, Ed. by: K. Izumi, S. Cristoloveanu, P.L.F. Hemment, and G.W. Cullen, The Electrochemical Society Proceedings volume 92-13, p. 77, 1992

84   W. Wondrak, E. Stein, and R. Held, in "Semiconductor Wafer Bonding: Science, Technology, and Applications", Ed. by U. Gösele, T. Abe, J. Haisma, and M.A.A Schmidt, Proceedings of the Electrochemical Society, Vol. 92-7, p. 427, 1992

85   W.P. Maszara, D. Boyko, A. Caviglia, G. Goetz, J.B. McKitterick, and J. O'Connor, Proceedings of the IEEE International SOI Conference, p. 131, 1995

86   M. Radecker, H.L. Fiedler, F.P. Vogt, and H. Vogt, Digest of Technical papers, International Solid-State Circuits Conference, p. 378, 1997

87   E. Arnold, H. Pein, and S.P. Herko, Technical Digest of IEDM, p. 813, 1884

88   J. Weyers, H. Vogt, M. Berger, W. Mach, B. Mütterlein, M. Raab, F. Richter, and F. Vogt, Proceedings of the 22nd ESSDERC, Microelectronic Engineering, Vol. 19, p 733, 1992

89   K. Izumi, Y. Omura, M. Ishikawa, and E. Sano, Techn. Digest of the Symposium on VLSI Technology, p. 10, 1982

90   W.A. Krull and J.C. Lee, Proc. IEEE SOS/SOI Technology Workshop, p. 69, 1988

91   T. W. Houston, H. Lu, P. Mei, T.G.W. Blake, L.R. Hite, R. Sundaresan, M. Matloubian, W.E. Bailey, J. Liu, A. Peterson, and G. Pollack, Proc. IEEE SOS/SOI Technology Workshop, p. 137, 1989

92   A.J. Auberton-Hervé, B. Giffard, and M. Bruel, Proceedings IEEE SOS/SOI Technology Workshop, p. 169, 1989

93   W.E. Bayley, H. Lu, T.G.W. Blake, L.R. Hite, P. Mei, D. Hurta, T.W. Houston, G.P. Pollack, Proceedings of the IEEE International SOI Conference, p. 134, 1991

94   L.K. Wang, J. Seliskar, T. Bucelot, A. Edenfeld, N. Haddad, Techn. Digest of IEEE Internat. Electron Device Meeting (IEDM), p. 679, 1991

95   H. Lu, E. Yee, L. Hite, T. Houston, Y.D. Sheu, R. Rajgopal, C.C. CHen, J.M. Hwang, G. Pollack, Proceedings of IEEE International SOI Conference, p. 182, 1993

96   G.G. Shahidi, T.H. Ning, T.I. Chappell, J.H. Comfort, B.A. Chappell, R. Franch, C.J. Anderson, P.W. Cook, S.E. Schuster, M.G. Rosenfield, M.R. Polcari, R.H. Dennard, and B. Davari, Techn. Digest of IEDM, p. 813, 1993

97   T. Eimori, T. Oashi, H. Kimura, Y. Yamaguchi, T. Iwamatsu, T. Tsuruda, K. Suma, H. Hidaka, Y. Inoue, T. Nishimura, S. Satoh, and M. Miyoshi, Technical Digest of IEDM, p. 45, 1993

98   H.-S. Kim, S.-B. Lee, D.-U. Choi, J.-H. Shim, K.-C. Lee, K.-P. Lee, K.-N. Kim, and J.-W. Park, Digest of Technical Papers of the Symposium on VLSI Technology, p. 143, 1995

99   T. Oashi, T. Eimori, F. Morishita, T. Iwamatsu, Y. Yamaguchi, F. Okuda, K. Shimomura, H. Shimano, S. Sakashita, K. Arimoto, Y. Inoue, S. Komori, M. Inuishi, T. Nishimura, and H. Miyoshi, Technical Digest of IEDM, p. 609, 1996

100 K. Shimomura, H. Shimano, F. Okuda, N. Sakashita, T. Oashi, Y. Yamaguchi, T. Eimori, M. Inuishi, K. Arimoto, S. Maegawa, Y. Inoue, T. Nishimura, S. Komori, K. Kyuma, A. Yasuoka, and H. Abe, Digest of Technical papers, International Solid-State Circuits Conference, p. 68, 1997

101 http://www.hiten.com/hiten/news/Hwell.QML.html (January 1997)

102 C.S. Kim (Semiconductors Asia/Pacific), Dataquest - Korea, May 1997

103 Y. Hu, C. Teng, T.W. Houston, K. Joyner, and T.J. Aton, Digest of Technical papers, Symposium on VLSI Technology, p. 128, 1996

104 A. Chatterjee, J. Liu, S. Aur, P.K. Mozumder, M. Rodder, and I.C. Chen, Technical Digest of IEDM, p. 87, 1994

105 Y. Hu, C. Teng, T.W. Houston, K. Joyner, and T.J. Aton, Digest of Technical papers, Symposium on VLSI Technology, p. 128, 1996

INDEX